AF842545

Farming in a Global Economy

International Studies in Sociology and Social Anthropology

Editors

Tukumbi Lumumba-Kasongo, Rubin Patterson, Masamichi Sasaki

VOLUME 101

Farming in a Global Economy

A Case Study of Dutch Immigrant Farmers in Canada

by

Frans Schryer

BRILL
LEIDEN • BOSTON
2006

This book is printed on acid-free paper.

Cover photo was provided by Mr. Andy Veldman, sales representative of Re/Max a-b Realty Ltd. in Stratford, Ontario

Library of Congress Cataloging-in-Publication Data

Schryer, Frans.
Farming in a global economy : a case study of Dutch immigrant farmers in Canada / by Frans Schryer.
p. cm. — (International studies in sociology and social anthropology, ISSN 0074-8684 ; v.101)
Includes bibliographical references (p.).
ISBN-13: 978-90-04-15170-3
ISBN-10: 90-04-15170-2 (pbk. : alk. paper)
1. Farmers—Ontario. 2. Dutch—Ontario—Social conditions. 3. Agriculture—Ontario. I. Title. II. Series.

HD8039.F32C275 2006
338.1086'91209713—dc22

2006047568

ISSN 0074-8684
ISBN-10: 90 04 15170 2
ISBN-13: 978 90 04 15170 3

PRINTED IN THE NETHERLANDS

Contents

Preface

This book would have been impossible without the help of numerous people. I am indebted to the farmers who opened their doors and shared their stories. Not only did I learn from them about life in the Dutch countryside, but they dispelled any preconceptions I had about farming in Canada as a way of life and a means of making a living. I also want to thank my research assistants: Michael Fallon interviewed farmers of Dutch background in the Niagara peninsula and in Brant County, while working on his doctorate in history. Lisa Dent-Couturier, then an MA student in rural extension, conducted interviews in Essex County, and Margaret VanderSchot, a farmer, helped me in the southern half of Perth and the northern tip of Oxford counties. Their initials will appear in footnotes, when I refer to those interviews. Michael Johnston's contribution included mapping out the locations of Dutch farmers in Grey County. His MA thesis on the perception of recent Dutch immigrant farmers is cited in this book. Fiep de Biep did one interview and conducted observations in Waterloo and Wellington counties as part of her training in fieldwork methods while enrolled in a BA program at the University of Guelph. At various times undergraduate students worked for me as part of a student-work program. Andrew Westert helped me for several weeks and Wilma Hovius was my part-time research assistant for close to a year. She conducted interviews, sent letters to various farm organizations, and obtained detailed information on the geographical distribution of farmers in the Listowel area. A year later Elizabeth Allingham did fieldwork in Eastern Ontario and listened to taped interviews housed in the Multicultural Historical Society of Ontario archives.

This book has benefited greatly from Kate George's meticulous copy-editing work, Marie Puddister's skill in map making, and my daughter Emily's assistance with final proofreading. I am also grateful for the encouragement and feedback I received from the editorial board of Brill and their external reviewers.

List of Figures and Tables

List of Figures

List of Tables

Abbreviations

1) Institutions, labels

AMRO	Amsterdam-Rotterdam Bank
ARDA	Agricultural Rehabilitation and Development Act
CBTB	*Christelijke Jonge Boeren en Tuinders Bond* (Union of Christian Farmers and Gardeners)
CIBC	Canadian Imperial Bank of Commerce
CFA	Christian Farmers Association
CFFO	Christian Farmers Federation of Ontario
CNIC	Canadian Netherlands Immigration Council
CNR	Canadian National Railroad
CPR	Canadian Pacific Railroad
CRLC	Catholic Rural Life Conference
DP	Displaced Person
GFO	General Farmers Organization
HVA	*Handels Vereniging Amsterdam* (Amsterdam Trade Organization)
MHSO	Multicultural Historical Society of Ontario
NAFTA	North American Free Trade Agreement
OAC	Ontario Agricultural College (part of the University of Guelph since 1964)
OFA	Ontario Federation of Agriculture
OFVGA	Ontario Fruit and Vegetable Growers' Association
OFU	Ontario Farmers Union
OMAF(RA)	Ontario Ministry of Agricultural and Food (and Rural Affairs)
UCO	United Cooperative of Ontario

2) People who conducted interviews with farmers

EA (Elizabeth Allingham)
FS (Frans Schryer)
FdB (Fiep de Biep)
MF (Michael Fallon)
WH (Wilma Hovius)

Introduction

The Netherlands is well known for its farming and agribusiness, and the Dutch have an international reputation as being successful agriculturalists. For at least 200 years, in English-speaking countries, the words Dutch and Holland have been associated with tulips, cheese, irrigation (wind mills), and market gardening. Even nurseries with no Dutch connections carry such names as Dutch Mill Nursery or Holland Gardens. In other words, the Dutch have acquired a reputation or a set of positive expectations, which I have labeled as a positive stereotype. This stereotype is likewise associated with Dutch farm immigrants. For instance, on a recent visit to the United States I was told that "rich" families of Dutch descent own almost all of the dairy farms in the vicinity of the city of Riverside.[1] A positive image of Dutch immigrant farmers is also apparent on websites that portray Dutch dairy farmers setting up large dairy farms in Ohio and Michigan.[2] Academic writers have also portrayed Dutch immigrant farmers in a positive light. In Canada, a scholar who examined the history of Dutch immigrants in the province of New Brunswick entitled his book *They Farmed Well*.[3] However, as the case study to be presented in this book will illustrate, this positive stereotype has complex implications for both immigrant Dutch farmers and any group that is overrepresented in a social field.

DUTCH FARMING IN A BROADER GLOBAL CONTEXT

While Dutch immigrant farmers illustrate the dynamics of positive stereotyping, they also exemplify the connections among international migration, globalization, and the transformation of social identities. The Dutch have

---

[1] In 1998, I was visiting a nearby university, on matters completely unrelated to research on Dutch immigrants. Afterwards I drove past several farms, and visited one of them.

[2] The presence of recent Dutch farm immigrants in the U.S. is reflected in news coverage and on websites. See "Ohio State University Fact Sheet" (ohioline.osu.edu/as-fact/0010.html), "Who is Dutch Valley Growers, Inc.?" (www.dutchvalleygrowers.com/history.htm), and a 1994 article by Ann Mullen in *Metrotimes*, a Detroit newspaper, also available on internet (www.metrotimes.com/editorial/story.asp.id=2004).

[3] This expression was used in the title of a book written about Dutch-Canadian immigrant farmers in Nova Scotia by Gerrit. H. Gerrits: *They Farmed Well: The Dutch-Canadian Agricultural Community in Nova Scotia, 1945-1995* (Kentville, N.S.: Vineland Press, 1996).

experienced the effects of these social processes for a long time. Since the birth of the Dutch republic, Dutch farmers have moved to other parts of the world, often taking with them their way of life and their farming techniques. In the seventeenth century, the Dutch East India and Dutch West India companies brought farm settlers to such diverse places as South Africa and the eastern coast of North America. Mennonite Dutch farmers also ended up in Prussia where they were invited to reclaim swampland that had formed part of marginal estates. In the eighteenth century, this same group of farmers of Dutch descent established farms in Southern Russia at the request of the empress Catherine the Great. Throughout the nineteenth century, Dutch Calvinist dissenters, unhappy with the political situation at home, opened up land in Michigan and other parts of the mid-western United States. Dutch Catholic farmers emigrated to parts of North America and other European countries in the first half of the twentieth century. Dutch farm emigrants thus ended up on every continent, including Australia. This trend of Dutch farmers moving to other parts of the world continues today, as shown by the presence of Dutch farmers and horticulturalists in Eastern Europe and in Russia and other parts of the former Soviet Union.[4] These immigrant farms today operate within an increasingly integrated agri-business economy that operates on a global scale.

Farm migrants are but one component of large-scale movements of Dutch people from diverse socio-economic backgrounds and occupations. Yet, historians who have examined these large-scale movements generally do not focus on farm migrants, even when those of farm backgrounds (including farm workers) constituted a large proportion of those who left their homeland. Nor is it easy to find information about what that proportion might be. A notable exception is the work of Robert Swierenga, who compiled statistics of Dutch emigrants to the United States, South America, South Africa, and Southeast Asia between 1835 and 1880.[5] His figures show that the proportion of farmers in three emigration periods varied from a low of 15 percent to a high of 44 percent, while that of farm laborers varied from a low of 8 to a high of 29 percent. He also did a breakdown of occupation by social class, showing

[4] A large proportion of immigrant farmers in Poland originating in the EU are Dutch (see www.nlchamber.pl), and Dutch farmers are growing seed potatoes in the Chernigiv region of the Ukraine (see website www.welcometoukraine.com/2projects.html). The Dutch government is currently promoting the settlement of Dutch farmers in Belarus (see www.ce-review.org/00/25/belarusnews25.html).

[5] Robert P. Swierenga, *Dutch Emigrants to the United States, South Africa, South America and Southeast Asia, 1835-1880: An Alphabetical Listing by Household Heads and Independent Persons* (Wilmington, DE: Scholarly Resources, 1983).

that both farmers and farm labourers were more likely to be 'middling' as opposed to "well-to-do" or "needy".[6]

However, even that valuable study does not focus on farm immigrants per se. His study of Dutch Calvinist immigration and settlement in the United States between 1820 and 1920 reveals how a combination of religious conflict and pressure on agricultural resources motivated Dutch emigration in this period.[7] He also refers to changes in occupational status, with half of thirty-one farm laborers and a third of 181 craftsmen becoming farmers.[8] However, available archival records do not indicate whether farm laborers who became farmers (or even former white-collar workers who bought land), were more successful as commercial farmers than those who had been independent farmers in the Netherlands. Nor do previous case studies or historical accounts shed much light on the question of how such diverse groups of farm immigrants contributed to the development of a positive stereotype.

People from the Netherlands were not the only immigrants who became commercial farmers. The prairies of North America, which became the breadbasket for the world in the second half of the nineteenth century, were settled by Ukrainians, Russians, Germans, and many other nationalities. Certainly in Western Canada, Ukrainian immigrant farmers far outnumbered their Dutch counterparts. Nor did groups of Dutch settlers always succeed. An attempt by farmers from the Dutch province of Groningen to establish small farms in Suriname (formerly Dutch Guyana) during the nineteenth century was a total failure.[9] Nor are Dutch agrarian immigrants the only ones who displaced other groups of farmers. For example, in the twentieth century, French Canadians bought land previously owned by English-speaking farmers in both the eastern counties and the northern part of Ontario.[10] When we look beyond family-owned farms, to include other forms of land tenure, such as those based on slavery and indentured labor, the Dutch are also not unique. People from other countries in Western Europe set up plantations in the tropics.[11] So why are Dutch immigrants so closely associated with being the experts in agriculture?

---

[6] See tables in Robert P. Swierenga, *Faith and Family: Dutch Immigration and Settlement in the United States, 1820-1920.* (New York: Holmes & Meier, 2000), 62-63.

[7] Swierenga, *Faith and Family*, chap. 1.

[8] Swierenga, *Faith and Family*, 259-65.

[9] See www.sr.net/srnetInfoSurinam/history.html.

[10] See G. Elmore Reaman, *A History of Agriculture in Ontario* (Toronto: Saunders, 1970): 125.

[11] Although the role played by Dutch plantation owners in bringing in new forms of irrigation and drainage for the island of Java did reflect their special expertise.

I suspect that the special reputation of Dutch agriculturalists developed as a result of a combination of factors: the long-standing prominence of commercial agriculture and agriculturally related activities in the Netherlands; and the fact that so many groups of Dutch farm immigrants moved to various parts of the world at different times. Today farm immigrants from various countries in Western Europe are again emigrating to other parts of the world to set up commercial agricultural enterprises. For example, in Canada they come from Germany and Switzerland as well as the Netherlands. However, current Dutch immigrant farmers seem to far outnumber their counterparts from other countries in both Canada and other parts of the world, thus reinforcing the impression held by many people that "the Dutch make good farmers."

GROUP REPUTATION AND THE PITFALLS OF MAKING COMPARISONS

A positive stereotype goes hand in hand with the act of comparing and evaluating groups. In a recent article in an Irish electronic farm journal, a top student in a management course is quoted as saying that he was impressed by "the positive attitude of Dutch farmers," in contrast to "the negative image of farming in Ireland."[12] Group comparisons, including those made in academic writing, also portray some ethnic groups as more prosperous than others. One scholar, who examined ethnic stereotypes of various immigrant groups in the United States, argues that it is wrong to compare ethnic categories without at the same time dispelling the myths surrounding group differences.[13] Group stereotypes distort social phenomena. Yet stereotypes, particularly positive ones, can also reflect observable patterns. In the case of Canada, this stereotyped image corresponds to trends revealed in official census data indicating the predominance of recent immigrant farmers from the Netherlands in agriculture. The author of an article published by Statistics Canada (1999) sums up these trends noting, "Farming is Going Dutch."[14]

Group comparisons and any corresponding generalizations, even when they correspond to observable trends supported by statistics, must be handled with care. The very act of singling out an ethnic or national group, whose members have done well, glosses over the heterogeneous

[12] Kay Kevlihan, "Choice of job offers for young farm managers," in June 19, 2004, issue of the *Irish Farmers Journal (Interactive)*. (www.farmersjournal.ie).

[13] See Stephen Steinberg, *The Ethnic Myth: Race, Ethnicity and Class in America*. (Boston: Beacon Press, 1989).

[14] Charlene Lonmo, Farming is Going Dutch, in *Canadian Agriculture at a Glance* (Ottawa: Statistics Canada, 1999), 297-301.

nature of that group. Moreover, in selecting any group of people for study, the relationship between group reputation and the success or failure of individuals from that group must be taken into account. We should be aware of perceptions: how others perceive such a group, how its members see themselves, and how such perceptions (as well as their actual place in the broader social order) change over time. Such perceptions, including those held by social scientists, can be misleading. Thus, when members of an ethnic group are overrepresented in the top social stratum of a particular "field of action", to use a term coined by the French sociologist Pierre Bourdieu,[15] it is easy to overlook those at the bottom. Likewise, when researchers draw attention to a stigmatized group whose members have little economic or cultural capital, individuals from that same group who have risen to the top might well be overlooked. A tendency to conflate the trajectories of individuals with those of identifiable broad-based groups can likewise distort the study of class stratification and social mobility. Thus, when someone belonging to a group underrepresented in the field of education becomes the president of a prestigious university, that individual might be seen as representing an oppressed group. Yet, when someone who belongs to a group that is overrepresented in the upper echelons of society occupies a low socio-economic position, he or she may still be seen as a member of a privileged minority, such as in the case of members of the elite Brahman caste in India who are poor farmers. Such group labeling even applies to cases when people from a particular group are located in the broad middle range of society in terms of income or occupation. Political activists who use dichotomous models of social hierarchy tend to treat people who belong to the middle social strata as exploitative, or else as downtrodden and disadvantaged, depending on the stereotypes associated with the groups to which they belong. Moreover, a group that is overrepresented and occupies a higher position in one occupation may be underrepresented in another. For example, the Dutch, as an ethnic or national group, have a stronger reputation in agriculture, and to some extent in the visual arts, than in other fields of endeavor such as politics or architecture.

A CASE STUDY OF DUTCH FARM IMMIGRANTS IN ONTARIO, CANADA

Dutch farm immigrants are an ideal group to study in order to understand the emergence of a positive group stereotype, that of the "successful Dutch

---

[15] See Pierre Bourdieu and Loïc I.D. Wacquant, *An Invitation to Reflexive Sociology* (Chicago: University of Chicago Press, 1992), 16-18.

farmer". They also illustrate the impact of global forces on commercial agriculture and how Dutch immigrant farmers responded to such forces. Canada was chosen as the setting for such an investigation for several reasons. While Dutch farmers as a group are today well known to most Canadians, a significant Dutch presence in Canadian agriculture is a relatively recent phenomenon. Although people originating in the Netherlands settled in central Canada as early as the nineteenth century, little evidence of that presence remained at the time of postwar immigration. Only in Western Canada, in the province of Alberta, did some postwar Dutch immigrants encounter Neerlandia, a town whose inhabitants still identified with their Dutch ancestors, who had settled there in 1902. The situation of being a small, almost forgotten component of Canadian society changed dramatically after the massive population movement of people from the Netherlands following the Second World War. While there were few people of Dutch descent in Canada prior to 1945, by 1961, they represented the fourth largest immigrant ethnic category in Canada.[16] These postwar Dutch immigrants were not all farmers, or even from a farm background, but people of Dutch descent were still overrepresented (+ 10.3) in agriculture, compared to Canadians as a whole. By 1970, over a third of all Canadians of Dutch origin lived in rural areas, making them the least urbanized of all ethnic groups, not counting the native peoples.[17] Moreover, Dutch Canadians are overrepresented among full-time commercial agriculturalists. The recent influx of hundreds of Dutch farmers looking to relocate in Canada has reinforced this trend.

This book focuses on the experiences of Dutch farm immigrants in the Canadian province of Ontario, as this area received the majority of Dutch immigrants in the period under study, i.e. many Dutch farm immigrants who initially landed in other provinces later moved to Ontario. In contrast to the western provinces, settled by immigrant farmers of diverse origins during the first half of the twentieth century, there had been no major influx of farm immigrants to Ontario for at least the six preceding decades. Unlike the pioneers of earlier eras, postwar Dutch immigrant farmers did not come to rural Ontario as pioneers to clear new land. Instead, they encountered a landscape already shaped by earlier immigrant groups, who had long established themselves as integral parts of a new society. They thus exemplify a more contemporary

[16] In 1961, 170,177 people in Canada reported having Dutch as their mother tongue (compared to 563,713 for Germans, 361,496 for Ukrainians, and 339,928 for Italians). See Dominion Bureau of Statistics, 1961, *Census Bulletin*, Population, Citizenship and Immigration, Cat. No. 92-548, vol. 1, part 1.

[17] See Frans Schryer, *The Netherlandic Presence in Ontario: Pillars, Class and Dutch Ethnicity* (Waterloo: Wilfrid Laurier Press, 1998), 295.

phenomenon of farm immigrants displacing previous migrants. A case study of post-war Dutch immigrants who started farming in Canada in the second half of the twentieth century may provide some insights into earlier waves of agrarian immigrants who came from the Netherlands. However, the main goal of this book is to better understand the relationship between group reputation and the ability to survive in farming in an increasingly competitive environment during a period of rapid economic change and ongoing globalization. While dealing mainly with farmers who live in Ontario, I am confident that the experience of post-war Dutch immigrant farmers in Ontario is typical of their counterparts in other parts of Canada.

Stereotypes and group reputations can last for a long time; however, they are also subject to transformation. When Dutch farm immigrants arrived in Ontario and other parts of Canada just after the end of the Second World War, they were viewed as poor people anxious to leave behind a desperate situation in their homeland. The reputation of the Dutch as successful farmers was not yet well established in most of rural Canada. Rather, they were an ideal source of cheap labor for farming – another stereotype. A mere four decades later, they were seen as successful agriculturalists, whether in terms of size of their operations or their ability to survive the turbulence of the rural economy. In such cases of a changing group reputation, yet another stereotype may emerge – that of generalized upward mobility. According to new collective consciousness, plus their portrayal in both popular and academic books, Dutch immigrant farmers were once penniless farm laborers living in chicken coops who overcame the odds through hard work and perseverance. This new image likewise oversimplifies a more complex social reality. Moreover, the reputation of the same group may vary across space as well as time. For example, Dutch farmers already had a strong reputation as successful farmers in some parts of North America, such as the state of Michigan,[18] even if this was not apparent to the Canadian farmers who sponsored Dutch farm immigrants between 1947 and 1955. Some Canadian farmers who sponsored Dutch immigrant farm workers in the forties and fifties, or who sold their farms to these same immigrants, told me they found the Dutch to be stubborn, and too strict with their children. Several thought that the Dutch tried to use agricultural practices from the old county that did not work in Canada. They said that the Dutch had to undo old habits and learn from their non-Dutch neighbors in order to survive in farming. Yet, other Canadian farmers expressed

[18] Lucas, Henry S. "The Beginnings of Dutch Immigration to Western Michigan, 1846." *Michigan Historical Magazine*, 6: 642-674.

respect and admiration for their Dutch neighbors' ingenuity and work ethic. Dutch perceptions of Canadian farmers were just as varied. The challenge is to disentangle such contradictory images and at the same time examine what actually happened, including any observable patterns and trends over time.

THE DUAL NATURE OF A SOCIAL SCIENCE RESEARCH

This book demonstrates that farmers of Dutch background are over-represented among full-time commercial farmers in Canada, while at the same time scrutinizing the positive stereotype of the successful Dutch farmer. To do so meant paying special attention to the interaction of perceptions and observable patterns of social relations. Like Bourdieu, I treat social classification (including group stereotypes) and social relations as part of a single reality, which is subjective and objective at the same time. Objectivity (the observable part, which includes statistical analysis) refers to the perspective of those, such as academic researchers, who do not have a stake in the field of action under study. The subjective part (known as emic analysis in social anthropology) provides insights into an equally important dimension of the same social phenomenon which is missed by an objective approach: the insider's perspective. These two ways of examining social life must be done in tandem, since each by itself is inadequate. In the case of Dutch immigrant farmers, it is impossible to separate how they are perceived by others, how they see themselves, and how they are examined through the lens of social science research. In human society, thoughts and actions or beliefs and observable patterns rarely coincide, yet they are mutually constitutive.

My work, however, departs from Bourdieu in one respect. He recognizes that "the order of words never exactly reproduces the order of things",[19] meaning that symbolic systems are slow to change, resulting in a gap between classificatory systems and social positions.[20] This statement implies that discrepancies between the structural (objective) and cultural (subjective) aspects of social phenomena are a temporary divergence.[21] In the case of Dutch farmers, it would mean that stereotypes are always catching up with their actual socio-economic positions in the field of agriculture.

---

[19] Pierre Bourdieu, *Distinctions: A Social Critique of the Judgement of Taste* (Cambridge: Harvard University Press, 1984), 481.

[20] He calls such a gap "hysteresis." See Pierre Bourdieu, *Outline of a Theory of Practice* (Cambridge: Cambridge University Press, 1977), 78, 85.

[21] In this respect I agree with Sally Falk Moore's position. See her "Epilogue: uncertainties in situations, indeterminacies in culture." In *Symbol and Politics in Communal Ideology* ed. S.F. Moore and B.G Myerhoff (Ithaca: Cornell University Press, 1975) 210-49.

In contrast, I see such discrepancies as a permanent, not a temporary, disjunction between these two inexorably intertwined dimensions of social reality, with no assumptions about direction of causation. In other words, labels and stereotypes will never correspond to the clusters of observable traits identified by disinterested social scientists, even though these two aspects of social life are mutually interdependent.[22] Likewise, the process of globalization has as much to do with belief and perception as it does with the creation of new social connections and power differentials among groups and individuals in diverse geographical settings.

My case study examines how Dutch farmers are perceived as a single group. But it also documents the diverse nature of immigrant farmers of Dutch background, a diversity that reflects both variation among rural families from different regions in the Netherlands and different opportunities in Canada. I will examine how economic background, previous education, Dutch region of origin and religious affiliation each shaped the way these farm immigrants adapted to life in rural Ontario. The reader will learn how long it took them to start farming on their own and how some eventually moved to the top in the world of farming. It will become apparent that Dutch immigrant farmers, on average, have indeed been relatively successful in Ontario. But not all Dutch farmers are well off or well known in the field of agriculture; a researcher not interested in the ethnic dimension of farming per se would have no trouble finding many examples of farm families of Dutch descent who succumbed to the debt crisis of the early eighties, or who never expanded beyond a barely viable farm operation. Yet, such researchers would also discover that even those Dutch-Canadian farmers who could not adapt to an increasingly competitive global agriculture, nevertheless tried to live up to the reputation of their group.

Between 1950 and 1970, farmers in the Netherlands and Dutch farm immigrants alike faced new challenges, with the increasing concentration of land ownership, rapid technological change, and the further integration of agriculture into a single world economy. By the seventies, the number of farms had declined even further, yet farmers of Dutch descent in Canada became better known as they represented an even greater proportion of those farmers who remained. Only two decades later, when another, albeit smaller, wave of Dutch farm immigrants arrived in Canada, the global village had changed even more, with new patterns of international migration and the widespread use of telecommunications and

[22] I earlier coined the term duplexity to refer to this permanent form of hysteresis when it comes to human groups (see Schryer, Multiple hierarchies and the duplex nature of groups, in *The Journal of the Royal Anthropological Institute*, 7 (4) (2001), 713.

biotechnology. An examination of Dutch immigrant farmers in Ontario in the second half of the twentieth century can help us to better understand the relationship between this ongoing process of globalization and new forms of international migration.

The sociologist Adrian Favell has noted that globalization and international migration are intimately interlinked. However, Favell also argues that both topics, particularly globalization, have been subject to too much theorizing. He points out that the literature is short on empirical case studies.[23] My book provides such an empirical case study of international migration in the context of an increasingly global agriculture. It will show how and why Dutch agrarian immigrants who came to Canada in the second half of the twentieth century, and those who are still coming today, are more likely to stay in the business of farming than their non-Dutch counterparts. How it is it possible that so many Dutch immigrant farmers have prospered in Canada, particularly in Southwestern Ontario, where the industrial and service service sector provide more lucrative opportunities for making money and getting ahead? The findings of such a study have implications for other Dutch farm immigrants who have developed viable commercial farm enterprises in other parts of the world or are currently trying to do so.

METHODOLOGY

The human geographer, Kevin McHugh, advocates the use of an interdisciplinary and multi-method approach in migration research.[24] He is particularly interested in combining ethnographic methods with other techniques for collecting and analyzing data. My study likewise incorporates an ethnographic approach in an interdisciplinary study that combines a diverse set of qualitative and quantitative techniques of data analysis. The methods needed to determine whether or not Dutch immigrant farmers in Ontario are overrepresented in the field of agriculture are quite different from those that enable a researcher to examine group identity and reputation.[25] I also wanted to tackle the spatial dimensions of Dutch farm immigration. However, before listing the specific techniques

---

[23] Adrian Favell, Migration, mobility and globaloney: metaphor, and rhetoric in the sociology of globalisation, in *Global Networks* 1 (4) (2001), 389-98.

[24] Kevin E. McHugh, Inside, outside, upside down, backward, forward, round and round: a case of ethnographic studies in migration, in *Progress in Human Geography*, 24 (1), (2000), 71-89.

[25] The use of multiple techniques not only allows research to provide a more complete picture through triangulation, but also to reveal discrepancies between the cultural system and social structure, between what people say and believe, versus observable patterns of social interaction and behaviour. See Schryer, Multiple hierarchies, 715.

used in my research, it will be useful to first look at the evolution of the research over the course of several years.

The order in which the chapters of this book were written does not correspond to the chronology of events described or the sequence in which I conducted the research. For example, the background covered in the first part is based on published sources and other written documents consulted prior to, during, and well into, the period of writing. Moreover, although I had a good idea of what I wanted to investigate, my plans changed over time. Initially, my research focused on family and career histories, how immigrants learned to farm in Canada, and recollections of Dutch methods of agriculture. Later, I started asking people about farm politics, education, and public involvement. By the second year of my project, I was still looking only at Dutch farmers who emigrated in the decade following the Second World War, when it became apparent that I could no longer exclude the "old-timers" (immigrants who came to Canada prior to the Second World War) in my research. At a subsequent stage, when I became aware that a large number of farms in Southern Ontario were being bought by a new, albeit smaller, wave of Dutch immigrants, my research expanded to include this new category of immigrant farmers. I also wanted to pay more attention to the broader process of globalization and how that process affected immigrant farmers. The task of doing research is never as systematic as portrayed in textbooks.

Much of the information in this book comes from interviews conducted with individuals and couples in English or Dutch (often a combination of both languages), with the use of occasional phrases from a dialect of Dutch or Frisian (a separate language spoken in the Netherlands). These interviews, which lasted at least three hours, were open-ended and exploratory. Follow-up interviews or phone calls with people already interviewed ensured that the basic information collected for each household was consistent. In order to include as broad a range of people as possible, I approached farmers involved in different forms of agriculture or agriculture-related activities, as well as farmers who operate feed mills or do their own processing or packing. I later added part-time farmers and farmers who combine a farm enterprise with a full-time, non-agricultural occupation such as teacher, real estate agent, or agricultural extension officer. I also ensured that different geographical regions, age groups, and cohorts were represented. In the technical language of the social sciences such a method of selecting informants is called a non-random, stratified sample.

In order to better interpret the findings of my ethnographic research, I turned to existing census data, which enabled me to tabulate what proportion of farmers in each county was of Dutch descent or still spoke the Dutch language. I then started doing archival research to complement the oral histories already collected. The main source of historical information on Dutch farm immigrants consisted of tapes and personal correspondence that are part of the Dutch Collection of the Multicultural Historical Society of Ontario (MHSO) as well as other documents housed in the archives of Ontario. Such data also provided me with new information and additional insights. Four years after starting my study, I tapped into yet another source of archival data: after hearing about the involvement of Dutch immigrant farmers in public life, I undertook the painstaking task of examining lists of elected committees (both local and provincial) of provincial marketing boards and producers' organizations. I did so in order to determine the level of involvement of people with Dutch or Frisian surnames, and to detect trends both over time and among different commodity groups in different regions. In order to collect this information, I spent time in the head and district offices of farm producer organizations in Mississauga, Guelph, and Chatham. All of the data, whether on microfiche, typed in lists or included in annual reports, was recorded on a laptop computer. While there were gaps, this exercise provided enough evidence to corroborate my hunches. During this part of my research, I spoke to representatives and officials in these organizations to gather their impressions of Dutch farmers. They also helped me to identify Dutch agricultural leaders and prominent producers of Dutch background. These additional interviews were used to set my study into a broader context. The same can be said for my interviews with a handful of non-Dutch farmers who gave their impressions of the Dutch families they sponsored. As a result of such conversations, and the names I encountered in documents, my already long list of names of people to be interviewed became more extensive.

Altogether, my research assistants and I spoke to over 200 farmers over a period of ten years, including 161 detailed, in-depth interviews. These interviews were reconstructed from written notes and, in some cases, transcribed tape recordings. Footnotes referring to such interviews include the initials of the person who conducted the interview (see list of abbreviations at the end of the book). About halfway through the research project, the written texts thus produced were distributed to most people interviewed for feedback and comments. Twenty-five people returned these "feed-back" files, sometimes with extensive written comments. In addition, twelve people provided biographies or family histories (cited in my bibliography). All of these stories are the source of quotations and excerpts, allowing people to speak for themselves.

The research process, from data collection to final write-up, involves many steps. Life histories and documents obtained from fieldwork and archival research were filed, classified, and analyzed. Over 1,000 pages of text from the interviews were coded using a computer program called

Non-Numerical Unstructured Data Indexing, Searching and Theorizing. That program allowed me to index segments of text on any topic that came up in the interviews. For example, all references to Dutch immigrant farmers who had worked in the reclamation of new polders in the Netherlands could be quickly retrieved and printed up. Apart from this method of analyzing qualitative data, I also coded my interview data to conduct a non-inferential, exploratory form of quantitative (statistical) analysis of my data using a computer program called dual scaling.[26] That method of data analysis, more commonly used by sociologists in Europe than North America, is similar to *analyse de correspondence*, a technique used by the French sociologist Pierre Bourdieu. Dual scaling is a useful tool for exploring the relationships among such variables as gender, class and occupation, kinship, and education from an objectivist perspective. I used it to generate visual displays showing clusters or sub-groups of farmers who share specific combinations of economic and social attributes. The data analyzed using dual scaling consisted of a set of standard questions asked in the 161 interviews, which I treated as a "survey." I did not conduct a random survey by means of a send-in questionnaire. Rather, my data was extracted from these open-ended interviews. The population selected for examination includes farmers of Dutch descent in Ontario who emigrated or whose parents emigrated prior to 1980. Although I cannot claim that findings from the data are valid for this population within a certain margin of error, the patterns revealed provide insights that could not have been obtained through strictly qualitative (non-statistical) means. The unit of analysis is the household, although some of the items selected are associated only with the person who has prime responsibility for managing the farm. In most cases, this was a man. Since both husband and wife are involved in many aspects of the farm enterprise, I took into account level of involvement in, and support of, the spouse (in most, but not all cases, the woman). The spouse's contribution thus became one of the variables. Although this book does not focus on women per se, I did not want to exclude them as an important component of the field of agriculture. I tried, as much as possible, to interview women by themselves in order to include a more accurate picture of their perspectives and experience. I also had access to the data collected by one of my research assistants, who carried out a small research project on Dutch immigrant farm women.[27] However, the field

[26] For readers not familiar with this form of statistical analysis, I have included a short explanation of the technique in the appendix.

[27] The results of her research were reported in a paper written for a course at the University of Guelph. See Elizabeth Allingham, "Different Voices of Postwar Immigrant Women in Rural Ontario: victims of agrarian ideology and 'invisible' contributors to agricultural production." (unpublished paper, 2000).

of commercial agriculture in Ontario and elsewhere is still dominated by men, which is in turn reflected in the way this research was conducted and reported.

ORGANIZATION OF BOOK

This book is divided into three parts. The first part, consisting of three chapters, provides an overview of farming, migration, and rural life in the Netherlands and in the Canadian province of Ontario, including any past connections between these two parts of the world. Chapter 3 includes a brief overview of Dutch migration to other parts of Canada. These chapters, drawing mainly on secondary sources, are meant to help those readers not well acquainted with the history or geography of either Canada or the Netherlands (and its past empire) to better understand both the country of origin and the destination of Dutch immigrant farmers. I pay special attention in each case to the broader political institutions and social norms that were part of the lives of people living in the countryside. These chapters also outline the "push" and "pull" factors that are generally used to explain migration in terms of broader structural factors and as the outcome of conscious decisions made by the people who migrate.

The second and longest part of the book (chapters 4 through 11) examines the trajectories of individuals and families from the Netherlands who arrived in Ontario as a result of an agreement between the governments of Canada and the Netherlands. This agreement, which triggered a large-scale population movement, resulted in thousands of farm immigrants coming to Canada between 1947 and 1960. Many of these people were still alive when I began my research. However, while focusing on people who emigrated during these two decades, I did not overlook the smaller number of farm immigrants who moved to Ontario in the sixties and seventies, the two decades when a positive stereotype of Dutch farmers became well entrenched. Chapter 4 describes the experiences of farm immigrants during their first year in Canada, while chapters 5 through 9 explore various aspects of their subsequent immigration experience in different regions in Ontario, with an emphasis on families who farmed on their own. The reader will discover how Dutch immigrants learned how to farm in Canada and the types of enterprises they established. Other topics covered include changing gender and family dynamics, the development of Dutch-Canadian communities, and the founding of new institutions. This part of the book most closely resembles an anthropological monograph, albeit with a strong historical bent. Chapter 10 is more sociological insofar as it presents the findings of the dual scaling exercise highlighting the commonalities and differences among

Dutch farm immigrants. A presentation of technical material is interspersed with examples of farm families to illustrate the various clusters uncovered through the statistical analysis. These examples also illustrate how diverse groups of postwar immigrants each contributed to the formation of a positive stereotype. A narrative approach comes again to the forefront in chapter 11, which presents stories of three generations of families to illustrate continuity and discontinuity.

Part three, with four chapters, starts off by examining the Dutch presence in rural Ontario from a macro perspective. Chapter 12, which is largely based on official census data, includes an examination of the spatial distribution of people of Dutch descent in rural Ontario and looks at where Dutch farms are located in Ontario. Chapter 13, based on information in several archives, includes figures indicating the Dutch presence in various branches of agriculture and agribusiness, including Dutch representation on the executive of the various producers' organizations. Unlike the largely qualitative and diachronic presentation of part two, these two chapters are more synchronic and quantitative, although they include short case studies and touch on trends over time. Chapter 14, which is again more historical, deals with the role of European Dutch capital in Ontario agriculture. The last chapter of part three deals with the new cohort of farm immigrants who have been coming to Ontario over the past fifteen years, including a comparison between them and the immediate postwar cohort.

The final concluding chapter provides an overall summary, and returns to the question posed in this introduction: Why have Dutch immigrant farmers had a competitive advantage in so many countries, in different historical periods? I argue that the prominence of Dutch immigrant farmers in different parts of the world, can be largely explained in terms of the social structure, the economic characteristics, and, equally importantly, the reputation of their country of origin. I will also revisit the topic of group reputation. The performance and trajectories of individual Dutch farm immigrants cannot be disengaged from how they viewed themselves and how others saw them. Stereotypes can have an influence, but not always in the way that we anticipate.

PART I – FARMING, CULTURE, AND MIGRATION: THE NETHERLANDS AND ONTARIO

I. The Netherlands: the Country of Origin

It is impossible to do a systematic comparison between the Netherlands and Canada in terms of either geography or culture. While both are sovereign nations, the latter is almost 300 times as big as the former in size. The Canadian province of Ontario alone is 27 times bigger that the Netherlands (see figure 1). Yet, despite its small size, the Netherlands is diverse in terms of geography, cultural differences, and systems of farming. Moreover, the Netherlands was, until recently, characterized by a high level of institutional segregation.[1] Social divisions were particularly salient prior to 1970, when the majority of Dutch immigrants came to Ontario and other parts of the world. It would not be possible to understand these immigrants without knowing something about their country of origin and why they left to seek a new life in other parts of the world, particularly in the second half of the twentieth century.

THE DUTCH LANDSCAPE

The Netherlands (also referred to as Holland) is part of a sub-region of Europe known as the Low Countries, which includes Belgium (see figure 2).[2] Most of this sub-region is a delta where the waters of the Rhine River, originating in the Alps, flow into the North Sea. Rising water levels following the annual melting of snow and ice in other parts of Europe, periodic fluctuation in the level of the North Sea, and the

[1] A more extensive treatment of the Dutch pillar system and how it shaped Dutch immigration in Ontario in general can be found in my book, *The Netherlandic Presence in Ontario*, (Waterloo: Wilfrid Laurier Press, 1988). Part of this chapter consists of modified sections from chapter 3 of that book.

[2] Geographical boundaries rarely correspond neatly to political and administrative divisions, including national borders, which have changed over time. For example, the southern part of the Netherlands and the northern half of Belgium, or Flanders, share a common landscape and cultural fabric, and they belonged to a single, politically united Netherlands in the first part of the nineteenth century. Nevertheless this book will not include Belgium. Although they share some geographical features, and historical roots, the experiences of people in the Netherlands and Belgium, and their migration patterns, have diverged over the past 150 years.

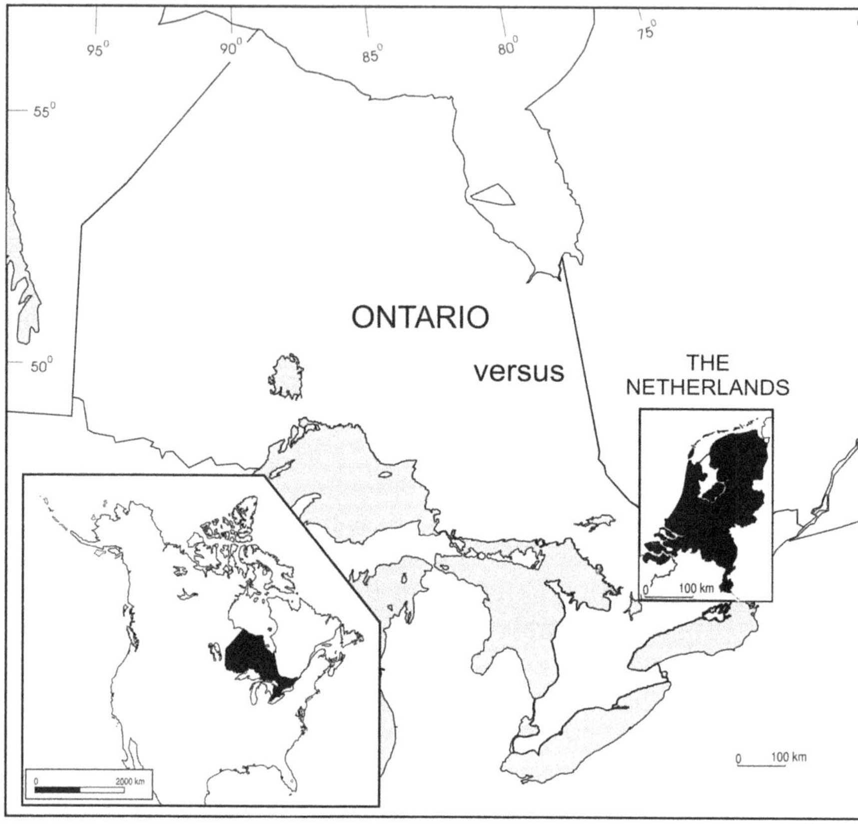

Figure 1. Ontario within North America.

onslaught of winter storms make much of the Low Countries susceptible to seasonal flooding. This is particularly true for the western half of the Netherlands, which lies below sea level. However, not all parts of the Netherlands are flat, nor do they all have the same kind of soil.

The low-lying areas of the Netherlands, including those below sea level, were once bogs and lagoons. New lagoons were created around AD 1000 when the forces of nature created newer and higher sand dunes which protected the coastal areas from seawater. These "young dunes" define the present coastline of Holland. The higher areas, which are one meter or more above sea level, are located in the eastern half of the country. They consist of sandy soils at one time covered with forests and swamps. These two geographical regions – clay ground and sandy soil – represent the most important difference when it comes to classifying Dutch agriculture. There are further topographical variations within both the clay and sand regions. In the case of the latter, hills and ridges give a distinct appearance to sections of the provinces of Overijssel, Gelderland, and Utrecht. Much of this hilly terrain consists of heaths (*heiden* in Dutch), poor soil covered with

Figure 2. The Low Countries within Europe.

coarse herbage. The higher sandy part of the Netherlands also includes flat, marshy areas that used to consist of peat moor (*hoogveen*) with a substratum of sand, such as in the Peel region between Noord Brabant and Limburg. In the clay region, we can make a further distinction between *laagveen*, or peat bog (on top of marine clay), and areas where any peat had long ago been washed away by seawater. However, past human action has shaped the Dutch landscape as much as have natural forces.

The reshaping of the Dutch landscape through human intervention goes back a long way. Around 300 BC, Frisian tribesmen who moved into the low-lying clay lands in the northern Netherlands (now Friesland and Groningen) started building refuge mounds (*terpen*) to protect their homes and cattle from flooding during high tides. Similarly, in the south, Batavians and other Germanic peoples built mounds for protection from the flooding of rivers, as did the invading Romans, who controlled the Netherlands between 12 BC and the fourth century AD. After the fall of the Roman Empire, rising sea levels and Viking attacks resulted in the abandonment of previous settlements along low-lying regions, while new waves of Germanic settlers (Franks and Saxons) continued to populate the river levees and higher regions. These settlers cut down forests to create pastures and cropland. However, marine regression did not permit the reclamation of land in the form of polders, protected by dikes, until after AD 900.[3]

The first dikes in the Netherlands, which served as roadways, were used to connect *terpen* and other kinds of mounds. By the fourteenth century, an extensive network of dikes, embankments, and river dams allowed for permanent cultivation and pasturage of areas previously under water or subject to seasonal flooding. At the same time, the cutting of peat bogs for turf (used for fuel and the construction of dwellings) caused the ground in other low-lying areas to sink, thus creating more inland lakes. These and other lakes were drained through the use of windmills pumps to create polders, especially during the seventeenth century.[4] By the middle of the nineteenth century, the use of steam power made possible the creation of more and bigger polders. The twentieth century witnessed the formation of even larger polders in the Zuiderzee, a body of salt water that once penetrated deep into the heart of the Netherlands. Its reclamation, done in stages, added over 100,000 hectares of fertile sea clay to the Netherlands.

---

[3] Much of the rest of this part of chapter one is based on Audrey M. Lambert's, *The Making of the Dutch Landscape: An Historical Geography of the Netherlands* (London: Seminar Press, 1971).

[4] The most ambitious large-scale draining of lakes in the seventeeth century created such areas as the Beemster, Purmer, Wormer, and Schermer (all in the province of Noord Holland). These lakes had been between two and four meters deep.

Polders, as well as low-lying clay regions bounded by rivers and protected by dikes, are kept dry through an extensive grid of drainage ditches and canals, whose surplus water is discharged into the sea, river, or other outer bodies of water. A typical feature of such polder landscapes is that farmer's fields are divided by ditches instead of fences. For the polders located below sea level, excess water is first pumped up to a higher level, into a *boezem*, which can be a canal, a river, or a lake that is part of a system of temporary reservoirs. Outer and inner (or "winter") dikes protect other low-lying regions, although the flood plains of major rivers were susceptible to inundation right up to the 1950s. For example, farmers with land located near the inner dikes close to Steenderen, on the left bank of the river Ijssel, always expected their farmland to be covered with a meter of water during the winter. Indeed, they welcomed the layer of slick deposited on their fields. Farmhouses in this area were located on higher land, although the village of Rha experienced regular flooding of at least one farm building. The local miller used to row his boat right into the barn attached to the house and step up to a platform where the farmer milked his cows. Each spring, the barn was cleaned and whitewashed once the water drained away.[5]

Soil Types and Forms of Land Use

Soil types are a good indication of the type of agriculture practised. Figure 3 shows the distribution of the major soil types, as well as the principal waterways (rivers and canals). Clay soils predominate in much of the western and northern provinces, but narrow clay strips follow the contours of the rivers in all provinces.[6] Well-drained marine clays in these low regions are more suitable for arable farming, while the wet *laagveen* of many polders has traditionally been used for dairy farming on grassland that is rarely ploughed.[7] Sandy soils are found in the eastern half of the country, as well as in the southern provinces of Noord Brabant and Limburg. However, the clay and sandy soil distinctions disregard further variations in soil types and topology. For example, the higher ground behind the dunes (*geestlanden*), where tulips are grown, is more fertile than other sandy regions. The various soil types and their corresponding forms of husbandry shape the prevailing forms of land tenure (who owns, controls, and/or uses the land) and technology. At the time of

---

[5] Interview (FS) with Herman Windmoller, 25 May, 1997 (who grew up in Rha, where his father was the miller).

[6] The Dutch term for the clay deposits near river levees is *komgrond*. The type of soil is referred to as *zavel*.

[7] This land is called *koeboerenland* in Dutch.

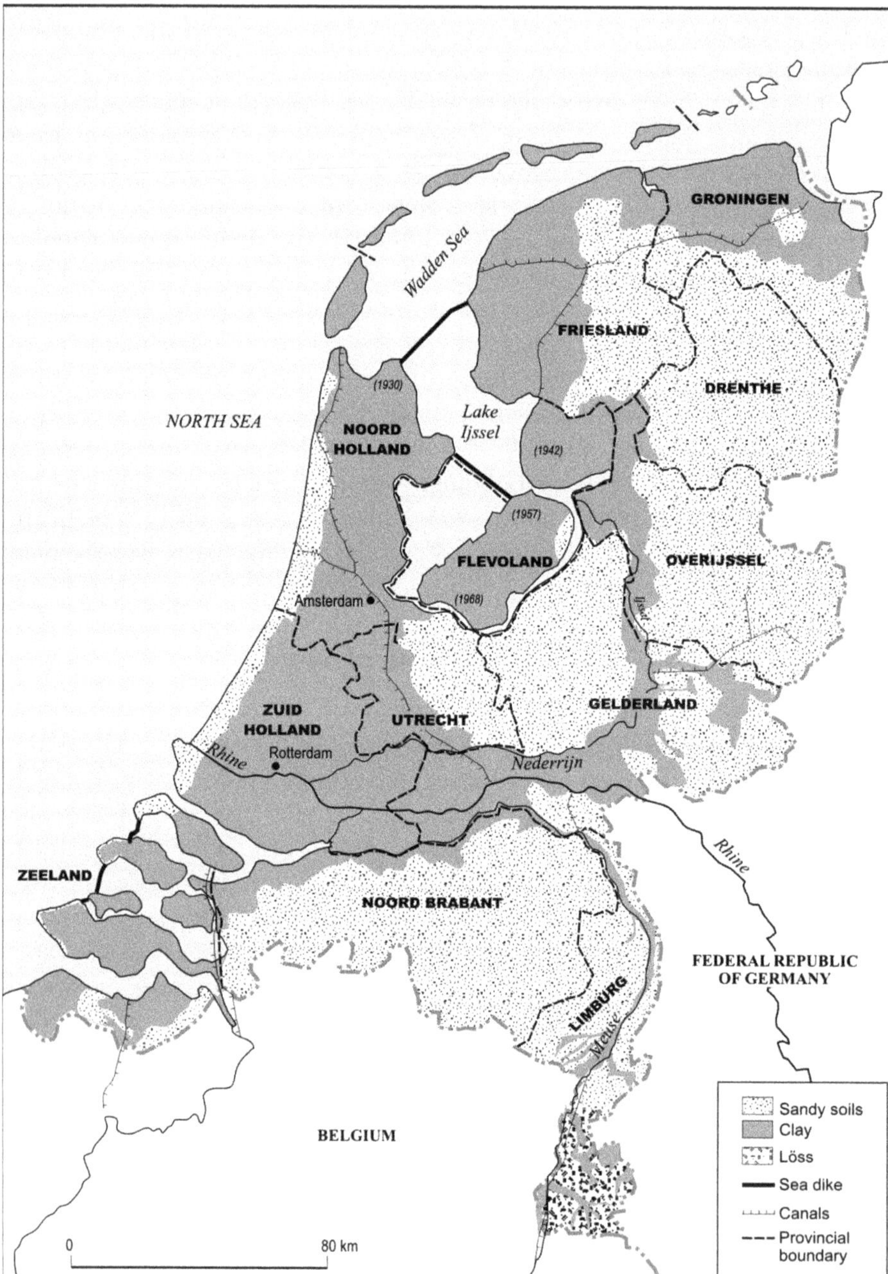

Figure 3. Dutch Provinces and Major Soil Types.

postwar immigration, farms in dairy regions were approximately of the same size (twenty to fifty hectares, depending on the area). In contrast, differences in farm size were, and still are, more pronounced in areas that specialize in cash cropping. Farmers in clay regions, regardless of type of farming, were usually more prosperous than their counterparts in sandy regions. Clay farmers (*kleiboeren*), especially those who grew cash crops, had more employees, and hired hands would rarely share a meal with the boss's family. In contrast, status distinctions among farmers and between farmers and workers were less sharply defined in the sandy regions. Such differences between clay and sand farming were still relevant around the time of Dutch postwar emigration to Canada.

The distinction between the rich clay farmers and poor sand farmers, while referred to most frequently in the literature, glosses over other forms of social inequality. In the predominantly sandy eastern and southern regions, the descendents of castle-dwelling, aristocratic landlords (*baronnen*) had considerable prestige and social influence well into the twentieth century. Such status distinctions were particularly noticeable in the Graafschap district of Gelderland, also known as the Achterhoek, where a hereditary class of estate managers, also involved in commerce, used to oversee the rental of small parcels of land.[8] In contrast, in the western clay regions, land was more of an investment than a source of status. The largest landowners were insurance companies or industrial corporations, whose properties were managed by *rentmeesters*. In still other regions, orphanages and religious institutions owned and rented out arable land.

THE HISTORY OF DUTCH AGRICULTURE

The long-standing distinction between two types of farmers (*kleiboeren* and *zandboeren*) dates back to the seventeenth century. At that time, the Netherlands entered a short-lived period of prosperity and world dominance known as "The Golden Age." With increasing commerce, the rural economy evolved into two distinct sectors: a cash crop and cattle economy oriented to the market and a largely self-sufficient one. This bifurcation was reinforced in the eighteenth century and beyond. The next major change in agriculture took place in the nineteenth century, when the opening of the American prairies to settlement made grain production in Northern Europe uneconomical. In the Netherlands, sugar beets replaced grains as the primary export crop. Yet a growing demand for dairy products

[8] This agrarian elite, who eventually became a form of gentry, survived up to the second half of the twentieth century. See Gerrit Wildenbeest, Recent Farmers' Protest in a Dutch Municipality: the Legacy of the Past, in *Dutch Dilemma*, Jeremy Boissevain and Jojada Verrips (Assen: Van Gorcum, 1989); 70-88.

and specialized crops in other parts of Europe stimulated a renewal of commercial agriculture in the Low Countries. The changes resulting from this process of expanding international interdependency culminated in forms of agriculture and patterns of rural life in the Netherlands that were still current in the decade following the Second World War. Land reclamation efforts were renewed with the use of steam-driven water pumps, especially in the Western part of the country. For example, the Haarlemmermeer polder in Noord Holland, created in the 1850s, rapidly became a showpiece of modern agriculture, with the use of the first stream ploughs and threshing machines on farms of up to 200 hectares.[9] In the sandy and peat regions, the introduction of artificial fertilizer facilitated more intensive forms of farming. Nevertheless, in both sandy and clay regions, agriculture continued to be labor-intensive, and larger farms typically had permanent, year-round, in addition to seasonal, workers.

The transformation of the Dutch countryside as a result of industrialization accompanied the formation of a labor market transcending both provincial and national borders. By the turn of the century, seasonal flax workers from Belgium regularly traveled to Zoetermeer, a clay polder in Zuid Holland dating from the seventeeth century.[10] Farmhands from Germany as well as from neighboring Groningen helped with haying in Friesland.[11] Likewise, in Pijnacker, in Zuid Holland, Belgians harvested flax, while workers from the Dutch region of Zeeuws Vlaanderen worked in potatoes and onions. These patterns of labor recruitment remained intact right up until the time of the Great Depression.

The Twentieth Century

The forms of agriculture that had emerged in the second part of the nineteenth century did not change much during the first four decades of the twentieth. The Netherlands remained neutral during the First World War, and so there was no war destruction. However, an old enemy, the North Sea, wreaked havoc in 1916, when a storm tide flooded the shores of the Zuiderzee. By the 1920s, the first phase of the reclamation project of this inland body of salt water was well underway, with the building of a dike between the island of Wieringer and the mainland of

[9] This polder also saw an influx of tenant farmers from the province of Noord Brabant, as well as Friesland. The Haarlemmermeer continued to attract farm workers from both of these regions right up until 1945. Interview (FS) with Martin Verkuyl, at his farm near Hickson, 23 November, 2000.

[10] Interview (FS) with Gus Sonneveld, 10 July, 1997 (originally from Pijnacker).

[11] Interview (FS) with Piet Bouwhuis, 30 April, 1998 (originally from Heerenveen, Friesland).

Noord Holland. Between 1927 and 1929, the first section was pumped dry to create a new polder with encircling dikes. In 1932, the closing of the sea dike which converted the remainder of the Zuiderzee into a freshwater lake (renamed the IJselmeer) was completed. This idea behind this government-run reclamation project was to turn over the land to farmers with school credentials and prior experience in modern farming. These farmers, who signed life-long leases, would never be able to own the land.

The preliminary work of desalinization and preparing the soil in the Wieringermeerpolder became a major make-work project during the Great Depression, and the many young farmers or farmers' sons who worked there hoped they would be selected as settlers. Many of them would later end up moving to Canada. With increasing pressure on land and rising unemployment, the Dutch government accelerated heath reclamation projects in the sandy regions. Those work programs provided some temporary relief to farm laborers and poor farmers, especially those in the sandy region. But, just as the economy started to recover and unemployment eased off, the Netherlands was invaded by Hitler's army.

The German occupation resulted in the closure of school buildings which were needed to billet soldiers in rural areas. Extension services, and courses in practical farming for the children of farmers, which had started prior to the war, almost disappeared. An order by the occupying forces to put as much land as possible into cultivation had an even greater impact on the parents of these children. Dairy farmers in *laagveen* polders with permanent pastures had to plough up sections of soil that had never been disturbed. Cattle restrictions triggered an underground economy. While much of the Dutch economy remained intact during the first few years of the war, the subsequent confiscation of horses and vehicles, and the destruction of dikes and roads had disastrous consequences. While some farmers cashed in on other peoples' misery during the last year of war, when food shortages forced city dwellers to exchange their prizes possessions for a few scraps of food, a significant part of the rural population saw their land flooded. At the time of the Allied liberation, the country was in ruins and a postwar coalition government had the challenging task of rebuilding. The school system for rural areas was expanded so that farm children would have the option to learn city trades. This renewed emphasis on education was part of a broader strategy to prepare the children of farmers to work in cities, given the lack of room in agriculture. Nevertheless, these children and their parents wanted to continue farming and many opted to emigrate rather than move to the city or face bankruptcy. Those who stayed in the Netherlands and continued to farm, adapted to new changes in agriculture. They later witnessed the transformation of many other aspects of Dutch social life, including sexual mores, religious practices, and politics.

DUTCH CULTURE AND THE PILLAR SYSTEM

Dutch society prior to 1960 had unique features; its major social groupings were segregated to a greater degree than any other Western democracy. Each religious group, as well those without religious affiliations, had its own school system and its own hospitals, labor unions, newspapers, and radio stations. Such institutional segregation, or *verzuiling*, has been labelled "pillarization" by English-speaking social scientists.[12] The three main *zuilen*, or pillars, were the Roman Catholics, orthodox Calvinists (*gereformeerden*), and "mainstream" Calvinists (*hervormden*). Their respective denominations were closely identified with several political parties. A neutral or "general" bloc encompassed a variety of smaller religious groups, including Dutch Mennonites and Jews, as well as socialists and liberals without any religious affiliation. People belonging to the major Dutch groups or pillars inhabited separate worlds and associated only with those connected with their own *zuil*. A young person was thus more likely to meet, and eventually marry, someone holding similar religious or non-religious views. There were also separate health care organizations.

Institutional segregation often dictated residential segregation. Catholics constituted the vast majority of people in the provinces of Noord Brabant and Limburg, where someone growing up in a small village might never encounter a Protestant. The various Calvinist denominations were predominant in the northern part of the country, although there were Catholic enclaves in provinces like Friesland and Noord Holland, such as the village of Volendam. Most of the larger towns in the rest of the north also had sizable Catholic minorities. The various Protestant groups, including Dutch Mennonites (*doopsgezinden*), were further segregated among each other in varying degrees. Many Protestant villages in the north had churches of more than one Calvinist denomination, and the level of co-operation or friction between *hervormden* and *gereformeerden* varied from place to place. Once in Canada, the degree of comfort of people with members of religious groups other than their own, and their level of prejudice, depended on the degree of regular interaction with members of other pillars in the Netherlands.

---

[12] See Arend Lijphart, *The Politics of Accommodation* (Berkeley: University of California Press, 1968), 17; William Z. Shetter, *The Pillars of Society: Six Centuries of Civilization in the Netherlands* (The Hague: Martinus Nijhoff, 1971), chap. 1.

Dutch Rural Institutions

In the Netherlands, the pillar system was an integral part of rural life. Farm organizations and financial co-operative institutions were either completely secular or based on religious conviction. The main organization for farmers and horticulturists who were Protestants was the *Christelijke Boeren en Tuinders Bond*, while Roman Catholic farmers joined their own *Boeren en Tuinders Bond*. There were also non-denominational associations for specialized branches of farming, such as the General Dutch Dairy Association (*Algemene Nederlandse Zuivelbond*). Each of these organizations had their own youth wings, such as the CJBTB (*Christelijke Jonge Boeren en Tuinders Bond*). Labor unions for agricultural workers were also set up along denominational lines.

This segregation also affected rural education to some extent. Agricultural education and home economics, taught at the village level (*dorpscholen*), dates back to the nineteenth century. By 1900, more elaborate, state-sponsored agricultural courses were given during the winter. With the introduction of the pillar system, at the turn of the century, agricultural schools at different levels were also set up along denominational lines. However, students moving beyond the elementary level (*lagere landbouw/tuinbouwschool*), were more likely to attend agricultural colleges not affiliated with their own pillars, since not every pillar offered agricultural training at the secondary (*middelbare*) level in each region. Yet, at the national level, each pillar set up institutions of higher learning (*hogere landbouwschool*) specializing in horticulture or agriculture.

Since the turn of the last century, two distinct forms of financial co-operatives (the Dutch equivalent of Ontario's credit unions), geared primarily to the rural sector, were likewise organized along denominational lines. The Raffeisenbank, named after the German founder of the co-operative movement, was predominantly Protestant in terms of both membership and leadership, while the *Coöperatieve Boerenleenbank* was exclusively set up for Catholics, with priests as spiritual advisors. Not until 1972, did their respective central (national) organizations merge to become the new Rabobank.[13] Prior to that date, many towns in the Netherlands had branches of two rival financial co-operative institutions. Other kinds of co-operatives, such as cooperative dairies (*zuivelfabrieken*) were likewise segregated.

To fully understand farmers of Dutch background in Canada and other parts of the world requires some familiarity with this Dutch pillar system. The model of religion-based organizations was particularly strong among postwar Dutch Calvinist farm immigrants, who founded the

[13] "De geschiedenis van driekwart eeuw coöperatieve bankieren," in *Rabokrant*, 16 December, 1982, 4-7.

Christian Farmers Federation of Ontario in 1954 (see chapter 8). Moreover, both Catholic and Calvinist farmers set up denominational credit unions in Canada, and the downtown core of Strathroy (near London) had branches of rival credit unions founded by these two groups. The amalgamation of the various agricultural organizations and co-operative banks in the Netherlands, giving farmers one voice, did not occur until two decades after these farmers had already gone to Canada.

Other Social Divisions

The pillar system, based on religious and political differences, was the most important, but not the only, basis of social division and group formation in the Netherlands prior to 1970. Dutch society displayed strong regional differentiation, and also had a set of class distinctions that were more pronounced than in England.

Dutch men and women depicted people from provinces other than their own in terms of stereotypes. The *Drentenaars* (people from Drenthe) were labelled as taciturn, introverted, and "hard." People from Gelderland were considered to be more enterprising than their counterparts from Noord-Brabant. Similar comparisons were made between people within the same province. In Brabant, farmers in the western and northern clay regions were deemed more enterprising and progressive compared to those in the sandy regions of the east. Similarly, in Limburg, people from the south were considered to be better off than those from the north. In contrast, people in each province or region portrayed themselves in a more positive light. Indeed, regions with their own names and identity, like the Peel, Twenthe, or the Achterhoek could be socially more significant than their respective provinces. Figure 4 shows some of these regions.

There were further distinctions at the local level. Up until three generations ago, most rural villages had unique customs, including the wearing of folk costumes, as well as dialects, which people spoke at home. People in many polders also had a strong sense of local identity, even if they were divided along religious lines, although some regions displayed more social cohesion than others. People born and raised in such regions were more likely to get together in the Canadian context (see chapter 8). Sub-national differences also come into play. Friesland has a separate language called Frisian, which resembles old English, and Frisians can often be identified by surnames ending in *-ma* or *-stra*.[14] While all Frisians

[14] There are people with Frisian names living in other parts of the Netherlands who would not identify with, or speak, Frisian. For an anthropological study of Frisian ethnic identity in rural Friesland, see Cynthia Keppley Mahmood's *Frisian and Free: Study of an Ethnic Minority in the Netherlands* (Prospect Heights, IL: Waveland Press, 1989).

Figure 4. Dutch Socio-Cultural Regions.

learn Dutch at school, Frisians are proud of their own history, going back long before the birth of the Netherlands as a nation state.[15]

Social stratification in the Netherlands is associated with marked status distinctions known as *standsverschil*.[16] In the 1950s distinctions based on education and occupation were particularly pronounced and there was little interaction between blue- and white-collar workers. Status distinction penetrated every facet of life, even in the smallest villages. In the clay regions of the countryside, there was a huge gap between the *herenboeren* (gentlemen farmers) and their *landarbeiders* (farm laborers). The latter did not eat at the same table as their employers, much less attempt to marry "up". There was further ranking among different categories of farm families, reflected in where one sat in church and the type of ornaments in women's headdresses: silver, gold, or diamonds depending on one's wealth and property.[17] Yet all farmers, including *pachters* (tenants), ranked higher than factory workers. In the other direction, farmers (*boeren*) were perceived as being "lower class", in comparison with "sophisticated city people". This system of social stratification is manifested in spoken language, since one's accent or inability to speak proper "high Dutch" was a conspicuous social marker. Family connections were equally important, especially among landlords and managers.

People who did not share the same class background were expected to "keep their distance," even if they belonged to the same region or province. For example, in Overijssel and Gelderland, which preserved remnants of the feudal system, tenant farmers and farmworkers were expected to act in a deferential manner, by removing their caps when speaking to the landowner or his representative.[18] A more subtle form of *standsverschil* was apparent among specialized dairy farmers in the *laagveen* regions, where the number of windows in a barn (an indication of number of

---

[15] Nevertheless, Dutch sociologists have argued that the most pronounced ethnic and regional cleavage is not between Frisians and non-Frisians, but between the southern provinces (Noord Brabant and Limburg) and the rest of the country. Indeed these two provinces continue to show signs of forming a separate regionally based subculture. See Johan Goudsblom's *Dutch Society* (New York: Random House, 1967).

[16] According to a study of nine countries conducted in the 1960s by UNESCO, both class feeling and inequality of income distribution were higher in the Netherlands than in Britain, France, or the U.S. See Lijphart, *The Politics of Accommodation*, 20-23; and Christopher Bagley, *The Dutch Plural Society: a comparative study in race relations* (London: Oxford University Press, 1973).

[17] Interviews (FS) with Katrina Bouwhuis, 30 April, 1998, and Peter Ellens, 30 November, 1993 (both originally from Friesland).

[18] Interviews (FS) with Bill Brunsveld, 5 December, 1990, and with Jeff Gerrits, 28 March, 1993 (originally from the Achterhoek in Gelderland and the land van Cuijk, in Noord Brabant, respectively).

cattle) would be used to gauge the suitability of their children for marriage.[19] Similarly, the number of greenhouse stovepipes served as an indicator of wealth and status among greenhouse growers in the Westland.[20] Such status distinctions were at odds with the informality and egalitarian ethos of Canadian society.

Common Elements and Other Variations

Despite variation along the lines of religion, region, and status, plus the distinctions between rural and urban, Dutch people shared common traits that could be described as a Dutch national culture. Various writers, including non-Dutch commentators, have attempted to generalize the ways of thinking, life style, and family life of this culture.

The ideal Dutch household is supposed to create a special atmosphere of cosiness and security. The Dutch word for that atmosphere (*gezellig*) is difficult to translate but easy to recognize when entering the house of someone born and raised in the Netherlands. One sees a lot of bric-a-brac, carpets, and pictures; an arrangement of chairs and tables allowing for maximum interaction; and many, many, indoor plants. The typical Dutch room features some Delft blue plates or tiles, usually hanging on the wall, decorative pieces of copper, table coverings resembling carpets, and lace curtains. The interior décor of the Dutch home is not only symbolic of national culture, but plays an important role in shaping the dispositions of Dutch people.[21] However, an emphasis on such common elements or a preoccupation solely with the pillar system, could lead a researcher astray, unless he or she also takes into account other cultural variations, including those in the countryside.

The topic of how the Dutch farm enterprise is passed on from generation to generation has not been systematically studied. Nevertheless, I was able to develop a rough sketch through discussion with a wide range of people of Dutch descent, as well as scattered references in the literature. I was told repeatedly that the north (Groningen and Friesland) was characterized by primogeniture (whereby the oldest son takes over ownership of the farm), in contrast to the south, where land was divided

---

[19] Interviews (FS) with Gerrit de Boer, 30 November, 1989; Jack and Teresa Van Nes, 17 June, 1998 (originally from Noord Holland and Utrecht respectively).

[20] Interview (FS) with Jerry Van Leeuwen, 14 July, 1995. (originally from the Westland region of Zuid Holland),

[21] For an examination of the historical origins of the Dutch household, see Simon Schama, *The Embarrassment of Riches: an Interpretation of Dutch Culture in the Golden Age* (New York: Knopf, 1988 edition), 399-400.

equally among all children. Another observation was that the women in the province of Zeeland had no legal rights whatsoever to inherit property.[22]

However, such generalizations about predominant trends and contrasting legal histories do not do justice to the complex mix of informal norms, practices, and legal rules that vary both within and across regions. Moreover, what is ideal or normative does not necessarily correspond to actual practice. We must further take into account whether farms were owned or rented, how long-term leases were regulated, and how farm size and type of soil and crops affected patterns of inheritance. In Groningen, Overijssel, and parts of Drenthe, long-term leases stipulated that only one person could assume the lease, with preference given to another family member.[23] It was expected this would fall to the oldest. Nevertheless, around the time of postwar emigration to Canada, a younger son still living on the farm when his parents were ready to retire was actually more likely to become the next tenant. Indeed, this custom of a younger child taking over (whether the farm was owned or rented) was prevalent in a wide range of regions.[24] In small and medium-sized family farms in sandy regions, the oldest son who aspired to be a farmer was likely to buy his own farm with the financial assistance of his family. A younger sibling, who was expected to take care of elderly parents, took over the home farm and moved into the farmhouse many years later. In other cases, men would take over farms owned by their in-laws if there were no male heirs or no one else interested in farming.

Who inherited a farm also depended on the scale of the farm enterprise and how many children wanted to continue farming. In the case of large, heavily-capitalized arable farms in Friesland, the oldest son of a wealthy owner-operator (*hereboer*) invariably succeeded his father, up until the late twentieth century. It was expected that the rest of the children would become well educated, so that they could enter various professions or become clergy. Nevertheless, many wealthy Frisian farmers also set up other offspring on farms in the newly reclaimed Wieringermeerpolder, on the other side of the Ijssel Lake, beginning in the forties. Even this option was not sufficient to accommodate all those who wanted to farm. Around the middle of the last century, it was difficult for young people to start farming regardless of the economic status of their parents. In all regions of the Netherlands, the lack of farmland was the main reason for emigrating.

[22] This was mentioned in several interviews, including one with someone not from that province.

[23] Interview (FS) with Matthew Gaasebeek, in Dundas, 27 November, 1989 (a lawyer who had dealt with rural Dutch immigrants).

[24] This was mentioned in interviews with people who grew up in different parts of the Netherlands.

Another topic that merits more detailed examination is the gender division of labor in rural areas. Although it was generally expected that the Dutch farm wife would be an efficient homemaker, as illustrated by the attendance at the *huishoudschool* (home economics school) by most rural girls, women participated in various aspects of production. The exact division of labor varied by class and by region, but also depended on such factors as the proportion of male and female children, and whether or not they were brought up in other regions prior to marriage. An example of variation in the gender division of labour among various regions is who milked the cows.[25] In specialized dairy farms on *laagveen* grasslands, particularly those in Zuid Holland, women generally did not do the milking. If a dairy farmer employed workers from outside the family, it was the hired man who helped with the milking, while a girl was more likely to be taken on as a *dienstmeisje* (girl servant), to assist with domestic chores. Likewise, in the Haarlemmermeer polder, where cash-crop farmers brought cows from other regions into their barns during the winter, milking was exclusively the task of permanent hired hands, all men, to occupy them during the off-season. In contrast, on mixed farms located in sandy regions, regardless of region or religious affiliation, the task of milking was divided equally among males and females. Bigger farms might also have a full-time *melkmeid* (girl who did milking). On smaller farms in such regions, where the man might have an off-farm job, the women sometimes took full responsibility of the cattle, including the milking. This was particularly prevalent in Noord Brabant. Inhabitants of the predominantly cash-crop regions on *dalgrond*, where some families kept a few cows for home use, also told me that men and women were equally likely to do the milking.

THE HISTORY OF DUTCH EMIGRATION

Prior to 1945, the number of émigrés from Europe to other parts of the world was proportionately lower in the Netherlands than other countries.[26] This can be explained in terms of the relatively high standard of living and economic prosperity. Indeed the economic development of the Netherlands, especially during its Golden Age, was in large part due to the strength of its agricultural production, as well as fishing. Nevertheless,

[25] I asked all of the men and women, who were old enough to remember, who did the milking in their region. Fifteen people who came from both areas of mixed farming and dairy regions provided detailed answers.

[26] The Netherlands ranked tenth in the proportion of people that emigrated overseas during the nineteenth century. See Swierenga, *Faith and Family*, 11.

people from the Dutch countryside moved to other parts of the world for a variety of reasons, and under different circumstances. Initially, Dutch settlers were brought to different parts of the world as minor components of commercial ventures associated with large trading concerns like the Dutch East India Company. Subsequently, groups of settlers left under the guidance of religious leaders.[27] This was the case of the nineteenth century Calvinist Dissenters who founded block settlements in Michigan and Ohio, in the United States. A similar block settlement in Canada is Neerlandia, Alberta, founded in 1912.[28] Group emigration and Dutch ethnic settlement also occurred after the Second World War, when postwar Dutch immigrants had the option of going to Brazil.

The second form of emigration is that of single individuals or families, each landing in different locations and various regions even if large numbers departed for the same country. That was the predominant pattern for postwar Dutch emigrants who ended up in Canada. However, massive emigration from the Netherlands, regardless which form it took, did not occur until after the end of the Second World War. Starting in the late forties, the push to emigrate intensified to the point where everyone was talking about leaving, and most of them wanted to go to Canada. This "immigration fever" lasted for almost a decade, although smaller numbers of people continued to emigrate in the sixties. In looking at this phenomenon, it is important to distinguish between farm and nonfarm migrants, although some of the causes and motives underlying this mass emigration were common to both groups.

When the German Nazis retreated with the arrival of the allied forces in 1945, the Netherlands was left in shambles. Roads and buildings throughout the country had been destroyed by a retreating German army and bombing by the allied forces towards the end of the war. People did not have enough to eat, and even in the period of recovery after the war there was insufficient housing and a high level of unemployment. Moreover, new regulations made it more difficult to run small businesses. Under such circumstances it was not surprising that so many people moved to other countries. The Dutch government encouraged emigration because they realized the urban industrial sector was unable to absorb the large number of young people looking for work. Ill feelings and disappointment resulting from the postwar political situation

[27] For a detailed historical account of these American settlements, see chaps. 3 and 4 of Henry Lucas, *Netherlanders in America: Dutch Immigration to the United States and Canada* (Grand Rapids, MI: Eerdmans, [1955] 1989).

[28] See Herman Ganzevoort, *A Bittersweet Land: The Dutch Experience in Canada* (Toronto: McClelland and Stewart, 1988), 29-31.

also motivated people to leave. Former members of the resistance movement noticed that erstwhile Nazi collaborators continued to hold public posts. People also emigrated because of the armed conflict between the Dutch government and independence fighters in the Dutch East Indies (now Indonesia), a struggle that continued until 1952. Many young men emigrated to avoid conscription. Others who had already served in Indonesia left their homeland again when, upon return, they found the Netherlands too small and confining. Still others left the Netherlands, starting around the mid-fifties, because they thought the Russians might invade Western Europe.

REASONS FOR AGRARIAN EMIGRATION AFTER THE WAR

On top of the difficulties experienced by everyone after the war, farm families faced unique challenges. Arable land was in short supply even under normal circumstances, but there was even less room for expansion because of flooding from damaged dikes. A botched military campaign by the Allies in Arnhem caused flooding in the low-lying regions between the Maas and Waal rivers, and the Wieringermeer polder was completely submerged in water because of the actions of vengeful German army commanders. Even after the damage was repaired, there was insufficient land for the many young men and women who had come of age during the war who wanted to marry and settle down on their own farms.

As in the past, the Netherlands continued to create new land after the war. The Noordoostpolder had already been pumped dry in 1942, but the process of land reclamation would not be completed until 1964. In the meantime, the preparation of a newly exposed, but soggy, marine clay bottom for cultivation provided paid work for numerous young men from farm families. They dug ditches by hand and helped prepare the land for the first trial plantings. These rural workers, most of whom had continued their agricultural education after the war by taking night courses, hoped that they might be allowed to remain and set up their own farms. Many learned new skills such as driving tractors while working on the new polder project. But the hopes of those prospective settlers, who already faced long waiting lists and a stringent set of requirements, were dashed when a disastrous flood in the province of Zeeland in 1954 made it necessary to give preference to flood victims who had lost land and homes. Quite a few Dutch immigrants who had worked in the Noordoostpolder stated that disappointment with this experience was their main reason for emigrating, when I interviewed them many years later.

In the rural Netherlands, young people who wanted to set up their own farms could not find land, even if they had the money to buy or rent.

In some regions, it was customary for farmland to be sold through a blind auction system, whereby the person who put in a bid closest to the market value would have the option to buy. However, an already established farmer who wanted that land could challenge a sale if he felt that the prospective buyer was not legally eligible under a complex system of land-use rights.[29] In other places, where members of local farm families could gain access to land belonging to a municipality, they would have to add their name to a list seven years before they could start farming.[30] Even for those willing to wait, there was no guarantee that enough land would be available for everyone who applied.

Family Dynamics

The shortage of land after the war, resulting in a strong desire to emigrate, had a direct influence on family dynamics. Farming around the time of massive emigration was labor-intensive and most farmers depended on family members to do chores. Moreover, rural families derived part of their income from children who worked off the farm and handed their earnings over to their parents. It was also expected that farmers would eventually help their offspring get established in farming. The decision by an older son or daughter to move to Canada could upset this pattern of reciprocity. The thought of their children departing to a faraway land could be particularly upsetting to older farm couples. Yet, these same parents realized that the only way their children could get ahead was by emigrating. One solution to this dilemma was for the entire family to emigrate together to stay together.[31] Sometimes an already middle-aged or older man or his wife would be the first to suggest that they emigrate "for the future of the children" as well as a way to fulfill a life-long dream. They might even send a son ahead to explore the chances of buying a farm, or the availability of jobs, later joining him with the rest of the children. In some cases, the person who initiated this move had themselves been thwarted in attempts to emigrate when younger.

In cases where a member of the younger generation wanted to emigrate, they were not likely to do so without the approval of their parents. Persuading their parents might require some effort. It was more difficult to leave, regardless of age, with a reluctant spouse. Older women were more likely to resist, although I encountered several cases where a woman

---

[29] Interview (FS) with Bill Joukema, July 29, 1992 (originally from Tjummarum, Friesland).

[30] Interview (FS) with Jack Van Geel, December 7, 1992 (originally from Hilvarenbeek, near Tilburg, Noord Brabant).

[31] I came across this scenario in many interviews.

was more interested in emigrating that her husband. Rarely did everyone get what they wanted. Another deciding factor was whether or not relatives, close friends, or neighbours had already emigrated. Few of those aboard the first ships that left the port city of Rotterdam knew anyone who had emigrated to Canada prior to the War. News, especially from those who managed to start farming on their own in Canada, made it more likely for people to leave their parents, sell a farm, and take the risks associated with overseas emigration. Having relatives abroad also meant a guaranteed network of support.

Money Matters

It was difficult to leave the Netherlands without money. Even when there was a job waiting, it cost money to travel, not to mention paying for shipping furniture, children's clothes, and other household goods. Such costs were partially covered by the Dutch government for those in need, increasingly from the mid-fifties on, but not everyone applied for such assistance. Yet, if someone did have enough cash, foreign currency restrictions imposed after the war limited how much money a person could take: each family was allowed to bring only 100 dollars per person. It was illegal to have funds transferred by other means while such restrictions were in place, and any surplus cash had to be left behind. Prosperous emigrants thus faced very different problems from those of their poorer neighbours. Someone with capital assets or other forms of property was not eligible for subsidies to cover the cost of emigrating, and had to pay hefty taxes on the sale of a farm. Any remaining money had to be deposited in a Dutch bank account, where it was kept until further notice. The only legal option was to buy as many consumer goods as possible and have them shipped over in a crate. One farmer shipped over the frame for a pre-fabricated house, and another a sugar beet wagon.[32] In contrast, many families incurred debts in order to emigrate.

Dutch emigrants found ways to circumvent foreign currency restrictions. People smuggled in several hundred dollars or even a thousand-dollar bill, sewn into women's undergarments or hidden in the back of a painting.[33] Farmers in the south who had good connections with Catholic clergy managed to smuggle in larger sums via Belgium. A more common method for transferring capital to Canada, especially for people who lived in close-knit Protestant communities, involved lending Dutch money

---

[32] Interview (FS) with Tony de Groot, at his farmhouse in Kinkora, 16 September, 1990.

[33] I encountered numerous stories about the smuggling in of *zwart geld* (black money).

to poor immigrants to be repaid in Canadian dollars. A person with capital, interested in buying a farm in Canada prior to emigration, would pay the passage and other costs of one or more families who wanted to emigrate. A minister's affirmation of honesty and a verbal agreement was sufficient. The lender then sent over a close relative, usually his oldest son, to collect the money owed, in instalments. Borrowers were usually able to repay their debt within a year, even if they were working on Canadian farms, since they immediately hired out their children. In some cases, the lender appeared on their doorstep several months after immigration.[34] People who loaned money in this manner told me it was extremely risky, since there was no legal contract and not all loans were repaid.

Why Canada and Alternative Destinations

Canada was the first choice for most Dutch postwar farm emigrants for several reasons. Many families and individuals would have gone to the United States, like many others prior to the Second World War, or Australia or New Zealand. However, the United States had more restrictive admission requirements, and Australia and New Zealand were much further away. The best option seemed to be Canada, and reportedly there were lots of jobs on Canadian farms, plus opportunities for buying affordable farms. For those who still dreamed of becoming Americans, Canada was a good second choice. Moreover, the Dutch had a good impression of Canada because its soldiers had liberated the Netherlands. Once emigration to Canada started to accelerate, favourable reports filtered back from relatives, friends, and neighbors and the big move across the ocean became less menacing. However there were other possibilities after the war.

In the late forties and fifties, Dutch farmers could also go to France or Brazil, countries that were recruiting Dutch farmers. The Dutch farm immigrants I interviewed frequently mentioned one or the other destination. In the case of France, it turns out that some farmers who ended up in Canada had first spent time there or at least checked it out. The reasons given for changing their minds were the fact that they could not buy their own farms, greater difficulty in learning French over English, and the fear of political instability. Dutch farmers were worried that the powerful French Farmers' Union might veto any approval of rental agreements if it decided French farmers wanted that land. One farm immigrant

---

[34] I heard such stories from several Dutch Protestants, who either recounted how they managed to finance their trip or how other people they knew did so.

complained that the only land in France easily available for rental was of inferior quality. The case of Brazil was somewhat different, since Dutch emigrants went there as colonists. The advantage was that they could start farming on their own right away, and would not have to first work as farmhands.[35] Dutch colonies established in Brazil prior to the War sent representatives to Holland to recruit. They promised the prospective colonists farm equipment and cattle. However, many people initially interested in Brazil backed out when they realized they would have to pool their money prior to emigration and farm as a group.[36]

Canada continued to be the preferred choice for Dutch farm emigrants well beyond the period of postwar reconstruction. As early as the sixties, farmers were no longer motivated to leave because of the aftermath of war or economic necessity. However, they were still interested in accommodating offspring hoping to make a living from farming. Such continuities and differences in motivation and experiences between 1950 and 1970 can best be illustrated through a series of case studies presented in chronological order.

Case Studies

The first group of people to emigrate had the Second World War fresh in their minds. A clear case is that of Joe Dibbits. Prior to emigrating, Joe lived on what had been his father's farm in Almkerk in the Land van Altena region of Noord Brabant.[37] It was a 27-hectare mixed farm with orchards on river clay. His father died in 1932 while Joe was still living at home, and his mother died six years later. As the oldest child of six children, he assumed full responsibility for the household and continued to run the family farm even after he married and had children of his own. Joe decided to emigrate as soon as the war was over, but was not sure where to go. The impact of the War renewed a desire from his youth to leave a hometown that seemed to have no future. At the beginning of the occupation, the Nazis evacuated the village to erect a barracks, and left a mess. They later confiscated cattle. In the interim the Dibbits had up to forty civilians living in their house at any time, because of the destruction of other houses. Sometimes German soldiers were billeted there as well. The final straw was the Arnhem offensive which caused flooding of most of their farm. The

[35] Interview (MF) with Sidney and Margaret Zeldenrust in their home in Dunnville, December 15, 1993.

[36] This was the most common reason given to me by the people I interviewed in Canada, some of whom had relatives who did go to Brazil.

[37] Interview (FS) with Joe Dibbits, at his home in Wooler, 3 July, 1991.

compensation paid by the Dutch government partly allowed Joe to finance his trip to Canada.

After the war, a new bureaucracy was established to address land shortages. However, apart from regulating the transfer of property and supervising rental arrangements, little could be done to ease the land hunger of young couples wanting to start farming on their own. The only option was to emigrate. A typical example is Piet Bouwhuis and his wife Catrina (née Witteveen) who left for Canada in 1948. They had both been raised on dairy farms in Friesland.[38] Her parents, clay-land dairy farmers going back several generations near the village of Blauwhuis, rented their 35 hectare farm from the Catholic Church. Piet's father likewise rented a dairy farm in that region, but he was born in the Frisian sandy soil region of Gaasterland. Piet's father had already moved to a *laagveen* region close to Joure to rent his own dairy farm, where he met his future wife, whose parents also farmed in that region. When Piet was nine, his father rented a bigger 25 hectare farm near Heereveen, further to the east. The Bouwhuis family illustrates a tradition of migration in Dutch farm families where the move to Canada was but one phase.

Piet was born in 1918. While helping out on the family farm he studied two days a week at a non-denominational (public) *lagere landbouwschool.* For four years Piet pedaled his bicycle six kilometres to Heereveen, and received his diploma at the age of twenty-one. That same year, his father moved once again, this time to the Bolsward region, where, though connections, he rented land owned by the former president of a steel company. A year after Piet went to Bolsward to work for his father, the Germans occupied the Netherlands. While still living at home, he met Catrina and they married as soon as the war was over. Her parents had been well-to-do farmers in the twenties, with a full-time hired hand and a maid, but their income had declined greatly during the Depression. Like many generations before them, this young couple looked for a farm of their own and started off by renting a dairy farm in Blauwhuis owned by Catrina's cousin. They wanted to farm on a larger scale, and Piet's father was willing to help them with cattle and some equipment. Piet discovered that a religious institution owned the land he wanted to use. So he submitted three tenders to rent through an application to the Ministry of Agriculture. The ministry made their selections on a combination of need, qualifications, and farm experience. Piet did not hear anything for a long time, and at one point his hopes were raised when an official told him that his name was among only eleven remaining

[38] Interview (FS) with Piet and Catrina Bouwhuis, in their home in Grand Valley, 30 April, 1998.

from the original 360 applicants. They were eventually turned down. That is when Piet and Catrina decided to emigrate to either Brazil or Canada, whatever country would accept them first. In April of 1948, they received notice they had been accepted in Canada.

The same shortage of land motivated people to emigrate from practically every region in the Netherlands. However, it was not always easy to sever family ties. The story of Bastiaan Bos, who emigrated in 1949, illustrates the importance of at least tacit approval of spouse and parents before a young man could leave his homeland. Bastiaan Bos grew up in a polder region with permanent pastures, just west of Rotterdam, where the original family homestead dates back to the sixteenth century.[39] Bastiaan grew up on the home farm and attended primary school and the *middelbare landbouwschool* (equivalent to high school) in the area. The day after graduation he started working full time for his father, as did his brother. There was no talk of wages. At age twenty-six, he inherited the farm but had to share it with his brother and a sister, whom he could not afford to buy out. With slim prospects to advance, Bastiaan, still single, decided he had nothing to lose by checking out Canada. His mother objected and asked her brother how she could dissuade her son from leaving. He told her emigrating was the smartest thing her son could to. Thanks to the advice of that uncle, no one else raised objections and Sebastian was free to go.

The busy fifties

The pace of emigration picked up in the fifties, at a time when economic recovery was too slow to meet the demand for jobs and housing. While the proportion of city emigrants began outnumbering their rural counterparts, both prospective and established farmers continued to look to Canada as their only hope for a better future. The experience of the Van Veen family is a good example of multi-generational emigration. Jan Van Veen had to think hard about emigrating to Canada at the age of forty-six.[40] His ancestors had farmed in the region of Katwijk (Zuid Holland) since the fifteenth century. His grandfather had owned a 6-hectare mixed farm on good clay soil, but Jan's father had taken over his father-in-law's much bigger, rented farm, close to The Hague. However, only one member of the family could eventually assume the rental and it went to the youngest son. Jan, who was the fifth child and

[39] Interview (MF) with Bastiaan and Syma Bos, at their farm near Sheffield, 27 September & 4 October, 1993.

[40] The surname is fictitious; this case study is based on two separate interviews with a father and son (who wish not to be identified), in Southern Ontario, 8 May, 1997.

oldest son, had already married the daughter of a milk store owner and had entered that business. With the outbreak of the War, both he and his co-workers were conscripted into the army. Jan and his wife had several children, and their oldest son, Herman, badly wanted to farm, so they sent him to the *lagere landbouwschool* opened to serve the nearby Zoetermeer region. When Herman was called up to serve in the armed forces, he persuaded his father to apply to emigrate to Canada. His father, who had also always wanted to become a farmer, agreed with the idea of leaving the Netherlands, and the whole family, with five children, departed in 1950.

While there were many such cases of whole families, including parents and married children, moving to Canada, a more common scenario was the emigration of young couples. Hessel Baarda grew up in the twenties on the dairy farm his father rented from the city of Sneek, in Friesland.[41] He wanted to become a farmer, but his father had lost the farm during the Depression. When Hessel finished primary school at age fourteen, there was no steady work. He got a part-time job sampling and delivering milk, and occasionally worked as a gardener in the afternoons. The gardening eventually turned into a full-time job for fourteen and a half years; Hessel obtained credentials for grafting, pruning trees, and working with bees. Even after marrying, Hessel worked for long hours for his brother-in-law, and eventually, his wife suggested that they emigrate to Canada, where she had a cousin. Hessel was not interested initially, but a couple of years later, with few prospects for the future, he began to make inquiries about going to Canada. He was inspired by glowing letters from his sister, then living in Canada, in Niagara Falls. She wrote of the need for fruit workers on the farm where they also worked. Hessel and his wife left for Canada in 1951, where Hessel eventually became a dairy farmer, just like his father, but on his own land.

An example of a woman taking the initiative to emigrate occurs in a family from Brugge, Noord Brabant, which emigrated in 1952. Annie Hurkens (Vander Ploeg after her marriage in Canada) was fifteen years old when her parents left for Canada with eight children.[42] She was the second oldest and not interested in leaving the Netherlands. She clearly recalls that it was her mother who decided they should emigrate. Annie's father, who had bought his own milk truck at the age of twenty, had had his own route for twenty-one years until German forces confiscated

[41] Interview (MF) with Hessel and Wiepkje Baarda, at their home in Smithville, 23 February and 21 June, 1993.

[42] Interview (MF) with Martin and Annie (Hurkens) Vander Ploeg, in Drumbo, 5 October, 1993.

the truck. Her mother looked after a few cows and a vegetable garden. Times were tough, especially after the War, and Annie and her older brother not only helped out at home, but also worked on other farms and ran errands. Although her father would have been eligible to receive a pension in four more years, her mother insisted they move to Canada. She saw that her children were always doing chores and working for little money, and believed there was no future for them in the Netherlands.

Not all postwar immigrants were poor or landless. Sikko Oegema, born in 1911, had done well by Dutch standards prior to the war.[43] He was farming on two separate parcels of land, in Dedemsvaart, Overijssel, which made his total farm operation of twenty-seven hectares a slightly larger than average mixed farm on good quality reclaimed peat land (*dalgrond*). He even had a steady hired hand. Sikko was busy in public life, as a member of the local *waterschap* (water control agency) and an active member of his church (*gereformeerd*). However, he was worried that his children might not be able to follow in his footsteps. He and his wife raised seven children, and his boys, especially his oldest son, seemed to be interested in farming. When the war was over Sikko wanted to start farming in the Noordoostpolder, but his application was rejected, which increased his dislike of the Dutch bureaucracy. So he decided it was time to emigrate, but his wife wanted to wait until their oldest daughter was sixteen. By the time they emigrated, in 1957, their second oldest son, who also wanted to become a farmer, had just finished high school. Their hired hand emigrated at the same time to join a brother living in Brantford. By the time the Oegemas emigrated, currency restrictions had been lifted and they were able to bring money with them.

The Sixties

The number of farm emigrants leaving the Netherlands after 1959 declined rapidly, as economic conditions improved throughout Western Europe. Thus, those who choose to emigrate did so for somewhat different reasons and under other circumstances.[44] A new program of student exchanges promoted by farm organizations and agricultural schools, provided opportunities for young people to spend a year in another country to gain practical experience and learn about other systems of farming. Many of these students, especially those who had a chance to

---

[43] Interview (FS) with Tom and Heiko Oegema, at their farm office, in Talbotville, 19 February, 2001.

[44] Out of the 161 agrarian immigrants with whom I conducted in-depth interviews, sixteen came to Canada in the sixties and only two in the seventies. I also interviewed four farm immigrants who arrived after 1980.

visit Canada, eventually ended up emigrating. Another category of Dutch farm emigrants comprised those who had already sampled other destinations, and then chose Canada. For example, two of the postwar immigrants I interviewed had first lived in Luxembourg, where their respective parents had immigrated after the war. In the first case, a recently married couple that moved to Ontario in 1968 had not wanted to take over the husband's parents' rented farm in Europe because of uncertain rental arrangements.[45] In the other case, a young woman who did not like the country her parents had chosen as their new home married a Dutch agricultural exchange student she met when he was working in Luxembourg. That student, who then moved to Canada, had already spent nine months on a Canadian farm in 1958.[46] Older couples also continued to make the big move.

Piet Albers lived in Hatert, a village close to the city of Nijmegen.[47] The farm where he grew up consisted of seven different parcels of land, and Piet's father worked day and night, including at a part-time job as a gardener. Piet helped out at home and spent a lot of time at his maternal grandfather's farm where they needed extra help. In 1951, at age six, Piet saw three of his maternal uncles leave for Canada, and he told himself that one day we would go there too. As he grew up, Piet helped his older uncles, who were then partners with their father, to sort potatoes on the weekends until late into the night. After finishing grade school, Piet started earning money as a manual laborer, but most of his earnings went straight to his mother, who had to feed thirteen people. In addition, he would get up early to help with the milking, and picked up potato peels from a local factory for their hog operation. While still in school Piet became interested in farming, and one of his uncles gave him a pig to fatten. Piet's father, however, told him he would be unable to give him land, since there were too many children, and he would be better off learning a trade. So, instead of attending agricultural school after grade seven, Piet was enrolled in a trade school to learn tool and dye making. He quit after one year, more determined than ever to become a farmer, and returned to the family farm to help his father full-time. At age eighteen, Piet Albers had an opportunity to gain experience abroad through an exchange program with Canada. After the nine months abroad, he immediately wished to return, but he

---

[45] Interviews (FS) with Frank and Marie Ernst, at their farm near Mitchell, 3 March, 1989 and 12 October, 1992.

[46] Interviews (FS) with Ernest and Margaret Vanderschot, at their farm near St. Marys, 14 May, 1991 and 12 September, 1992.

[47] Interviews (FS) with Piet and Lenie Albers, in their farmhouse near Brussels, 27 January and 5 February, 1998.

did not have his parents' permission, and his father was very ill. Around that time Piet met Lenie, then working as housecleaner to help out her family. With the death of Piet's father three years later, in 1967, they married and emigrated. Within four years, they had their own farm. His wish had finally come true.

Not all farm exchange students were boys or even had a farm background. Henny Derks grew up as the only girl in a family of six children in the town of Druten in the Land Van Maas en Waal region of Gelderland. When she was eleven, the family moved to the village of Dreumel, where her father and four older brothers set up a metal shop. As soon as she graduated from high school, she started working in the office of a large metal company. One day, her sister, who worked in a rural extension office, showed her an advertisement for the young farmers exchange program. After much pleading, Henny convinced her parents to allow her to apply for work on a Canadian farm, even though she was only eighteen years old. Very few women applied for that program, and during one of her orientation sessions, a young man giving a presentation on his recent year in Canada, noticed the only woman in the audience. Little did she know that she would end up marrying him after returning to the Netherlands. They emigrated and she spent the rest of her life living on an apple farm in Canada.[48]

Not all of the Dutch farm emigrants in the sixties were former exchange students. Sidney Zeldenrust came from a farm family in Groningen.[49] His father had already moved from the small, thirteen hectare dairy farm he owned, to another, rented, farm near the city of Groningen. They emigrated to Brazil in 1949, when Sidney was fourteen years old. There he was to meet Margaret, a Dutch woman whose family had joined the same agrarian colony. The colony was located on land owned by a paper and pulp mill who had hired the Dutch farmers to produce dairy products for their company town. Sidney, who had completed his education in the Netherlands before their departure, returned to his native land for two years (1956-58) for his Dutch military service, and to work for an uncle, before returning to Brazil. There he and Margaret became engaged. Through Dutch acquaintances in Canada, his parents had discovered that farming there was good, and that there were many Calvinist churches in Canada. They thought an English-speaking country would be a better fit for them, so the whole family decided to move to Canada in 1963. Sidney, now married, made an exploratory trip to

---

[48] Interview (FS) with Henny Derks, in her home near London, 21 May, 1980.

[49] Interview (MF) with Sidney and Margaret Zeldenrust, in their home in Dunnville, 15 December, 1993.

Ontario, but learned that he and his wife would have to first return to the Netherlands before they could apply as landed immigrants. Four months later, they were back in Ontario, and the following year, his parents and the rest of the family joined them on their new dairy farm in Dunville, Ontario.

Emigration from the Netherlands almost came to a standstill by the mid-seventies, as Europe's economy boomed. There were no longer many reasons for people to leave for Canada, as both city and county people now enjoyed unprecedented increases in their standards of living, and social services were superior to those of North America. The Netherlands had also undergone a cultural revolution: lifestyles changed; the Dutch became more secular, and the influx of outsiders made the Netherlands even more diverse. The pillar system had also started to crumble, beginning in the mid-sixties. At the same time, Dutch agriculture, and even the landscape of the Netherlands, was experiencing a complete transformation.

THE DUTCH COUNTRYSIDE TODAY

With *ruilverkaveling* (re-division and consolidation of farm boundaries), rapid mechanization, and even more specialization, a postwar farm immigrant who has not visited his or her hometown would not recognize the Dutch countryside around the turn of the century. Like the rest of the economy, Dutch farming became totally integrated into the larger European Community with a common agricultural policy. Moreover, the old division between clay and sandy regions no longer define Dutch rural life. Farmers from Noord Brabant routinely buy land and set up dairy or hog operations in what were formerly cash crop regions of Groningen. Dairy operations that used to have twenty cows now have sixty on the same amount of land. However, the intensification of agriculture has not been without costs.

An exponential growth in cattle production, resulting in such environmental stresses as contaminated groundwater, has led to even stricter regulations of the use of land and water. Beginning in the nineties, for the first time in Dutch history, farmland is being reconverted to wetlands. The Dutch government has even decided to put the most ambitious planned land reclamation project, that of the Wadden Sea between the Northern coast and the string of islands jutting out in the North Sea, on hold. Simultaneously, commercial agriculture has become even more vertically integrated and technologically advanced than in North America. Yet, one thing has not changed. Dutch farmers and most of their children still want to farm, and today, they are again looking to Canada and other parts of the world, as an alternative option.

II. Ontario: A Province of Immigrants

Like the Netherlands, the Canadian province of Ontario is characterized by geographical and social diversity. Yet, while Ontario is twenty seven times bigger than the Netherlands in area, it actually has fewer people! It is therefore not surprising that Dutch immigrants thought that Canada had wide open spaces with lots of room for expansion. However, only a small proportion of the land in Ontario is suitable for agriculture, due to either soil conditions or climate, and most of its arable land was already occupied, admittedly not for as long as in Europe. Other groups of immigrants had cleared the land at least a hundred years before the arrival of Dutch immigrants. They established communities and social institutions quite distinct from those in the Netherlands. Yet in the period just after the Second World War many of their descendants were leaving the land, freeing up room for newcomers willing to take their place.

TOPOLOGY, CLIMATE, AND SOIL TYPES

Most of Ontario's land surface consists of a rugged Precambrian mass of mostly granite, rocks and bogs, known as the Canadian Shield. A lumbering and mineral extraction region, the Shield is not suitable for agriculture. Most of the best soil is located in Southwestern Ontario, between Lakes Huron, Erie, and Ontario; in the narrow Ottawa valley; and in the rectangular area where the Ottawa and St. Lawrence Rivers run parallel to one another. There are also several minor areas with arable land in Northern Ontario including the Thunder Bay lowlands along the northern shore of Lake Superior (see figure 5).

Apart from quality of the soil and topography, average temperature, measured in heat units for corn, determines the type of agriculture practised. The warmest spots are in Essex, on the western tip of Lake Erie,[1] and the Niagara fruit belt on the southern side of Lake Ontario. In contrast, the harsher climate of Northern Ontario, with its shorter growing season, limits the type of crops that can be grown even in its clay belt region.

[1] Most of Essex County, including the city of Windsor, across the Detroit River, is the only section of Canada that lies south of 49th Parallel (where one faces north to a section of the United States).

Figure 5. Arable Land: Northern and Southern Ontario.

The Landscape of Southern Ontario

There is considerable variation in topography and soil types, including the moraines and drumlin fields (the result of glacial deposits) that give much of Southern Ontario its hilly appearance. Other landforms include clay, sand, and limestone plains where melting glaciers once created lakes subsequently subject to sedimentation. The most striking landmark is a long escarpment running north to south from the tip of the Bruce Peninsula to Niagara Falls. The Niagara Escarpment divides Southwestern Ontario into two halves. Another way of looking at the geography of Southern Ontario is to focus on the river valleys that define watershed regions. To the west of the escarpment, the Thames River and its tributaries, which flow through the cities of Chatham, London, and Stratford, have their outlet in Lake St. Clair. The largest watershed west of the escarpment is the Grand River Valley. It drains an area of 6,734 square kilometers encompassing the cities of Guelph, Kitchener-Waterloo, Cambridge, and Brantford. The longest river and the largest watershed east of the Escarpment is the Trent River system that flows into the Bay of Quinte.

Figure 6 shows the location of these watersheds and the major landforms and soil types in Southern Ontario. As in the case of the Netherlands, the best lands for cultivation are associated with the clay lowlands and plains although some clay soils, such as those on the Niagara Peninsula, are of lower quality because of the rough texture of the soil or uneven drainage. Sandy plains and hills are also suitable for agriculture, but may be susceptible to wind erosion. Till plains also have highly prized soils. In contrast, the moraines and limestone plains have a higher proportion of soils suitable only for permanent pastures or hay. The rest of Southern Ontario, consisting of uplands, hills, drumlin fields, and various slopes and fringes, has a mix of soil types.[2] These geographical features define where farming is possible, as well as patterns of human settlement and economic activity predating the arrival of Europeans.

THE HISTORY OF RURAL ONTARIO PRIOR TO THE END OF WORLD WAR II

The economy and culture of native peoples remained largely intact in what is now Ontario long after the arrival of Europeans on the North American continent. In the sixteenth century, the Huron, Petun and Neutral peoples practised swidden (also known as slash-and-burn) cultivation in the area between Georgian Bay and Lake Ontario.[3] They lived in longhouses occupied by matrilineal kin groups. Like their Iroquois counterparts south of the Great Lakes, they combined hunting with the cultivation of maize, squash, and tobacco. Bands of Algonquin people hunted and trapped for furs in most of the rest of Ontario, including the Ottawa valley.[4]

---

[2] For a detailed coverage of fifty-two minor physiographic regions of Southern Ontario, including patterns of settlement and types of agriculture around 1960, see L.J. Chapman and D.F. Putnam, *The Physiography of Southern Ontario* (Toronto: University of Toronto Press, 1966, second edition). For a briefer, but more general overview of the physical environment, see L.G. Reeds, "The Environment," chap. 1 in *Studies in Canadian Geography* (Toronto: University of Toronto Press, 1972).

[3] See plate 33 (the St. Lawrence Valley, sixteenth Century) in the *Historical Atlas of Canada*, vol. 1, ed. R. Cole Harris (Toronto: Toronto University Press, 1987).

[4] At the time of European contact some Algonquian groups, like the Nipissing and Ottawas, cultivated maize in Central Ontario and on Manitoulin Island respectively. See Robert Leslie Jones, *History of Agriculture in Ontario, 1613-1880* (Toronto: University of Toronto Press, 1946), 5-9.

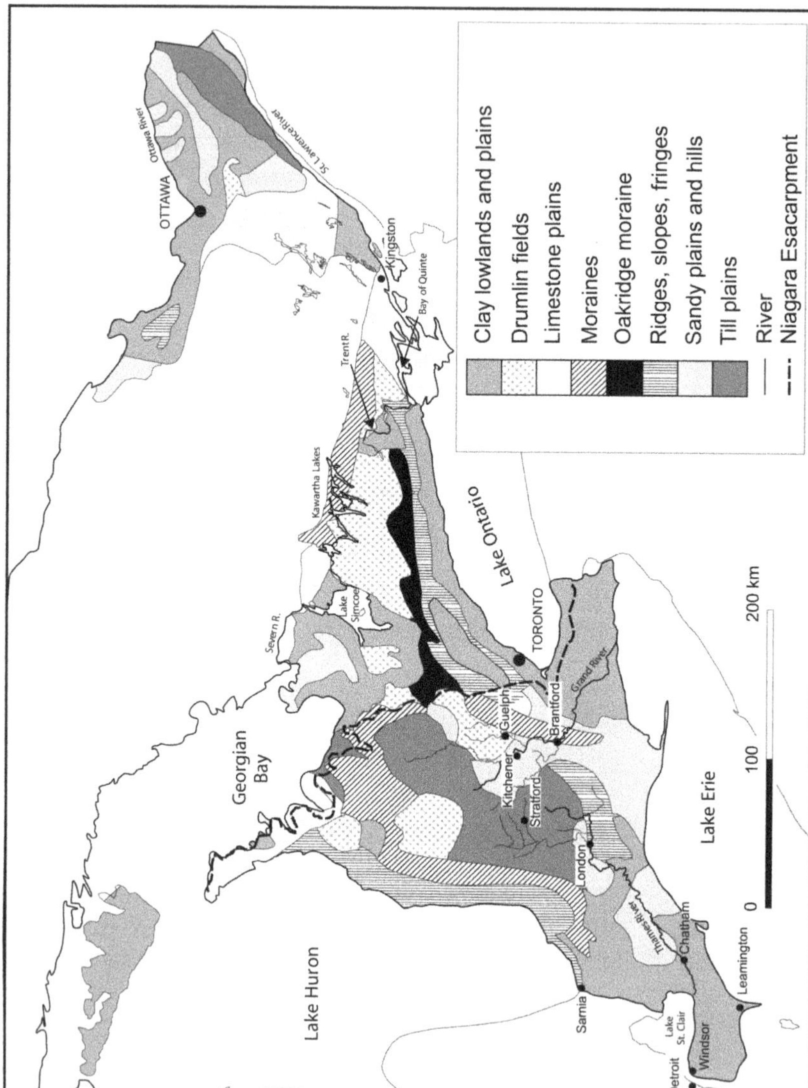

Figure 6. Landforms and soil types in Southern Ontario.

This cultural and economic contrast, predating European contact, parallels a regional distinction between Northern versus Southern Ontario that still exists today.

European Settlement in Upper Canada

A permanent European presence was slow to gain a foothold. By 1655, the region was abandoned by both Europeans and native people after attacks by the Iroquois with guns obtained from Dutch traders operating out of New Amsterdam (today Lower Manhattan of New York City). European settlement did not resume until after 1759, when the British captured Quebec City. From that point, what is now Southern Ontario became an appendage of the English-controlled colony of Quebec. But a significant influx of white people did not begin until after the American Revolution, when the British government invited people loyal to Britain to move to their Canadian colonies. Known as United Empire Loyalists, they included people of Dutch descent who settled near the Bay of Quinte. Native land claims were extinguished through purchase or treaty by the colonial government, who set aside native reserves. The rest of the land was divided into lots, concessions, and townships. Such cadastral surveying, completed in several stages between 1784 and 1850,[5] created a pattern of land use and administration that remains today.

The salient feature of Southern Ontario is the rectangular shape of fields and properties (lots), and the creation of townships, which paradoxically have no connection with nucleated settlements.[6] There were allowances for concession roads (also known as lines) at regular intervals and farmsteads were aligned along the fronts of concessions, with changing uses for the land moving from front to back.[7] Land maps today still shows the rectangular blocks or tracts of land within such concession roads, usually divided into 100-acre lots. The short end of each rectangular lot faces a concession road, which is identified by a numbers (see figure 19 of chapter 12), while the back ends of pairs of lots are separated by fences. Unfortunately, straight survey lines were to present challenges for future farmers in areas with hilly terrain or valleys, because roads and farm lines

[5] See map in C.F.J. Whebell, "The Politico-Territorial Structure," chap. 6 of *Studies in Canadian Geography*, ed. Louis Trotier (Toronto: University of Toronto Press, 1972), 166

[6] The original intent of the British colonial rulers was that the settlers of such townships, originally assigned numbers, not develop any sense of a spatial community or political independence. See Whebell, "The Politico-Territorial Structure," 110-112.

[7] See R.L. Gentilmore, "Settlement," chap. 2 in *Studies in Canadian Geography: Ontario*, ed. Louis Trotier (Toronto: University of Toronto Press, 1972), 24.

were often placed diagonally across drumlin fields and other hilly terrain. The irregular lay of the land also resulted in odd triangular lots.[8]

Initially, the Loyalists were settled close to military forts, such as Kingston, York (later called Toronto), and Niagara.[9] Another group of newcomers around the time of the American Revolution were German-speaking Mennonites (Pennsylvania "Dutch") who did not want to join the armed rebellion of the Thirteen Colonies that became the United States. In 1786, 500 people were brought to what is now Eastern Ontario directly from Glengarry in Scotland.[10] These early settlers cut down the forests of beech and maple, and used the logs to build cabins and cover roads in swampy sections of land. While the economic and political center of this British colony was in Montreal, in the French sector, a growing English-speaking population of the colony soon demanded their own system of governance based on English Common Law, rather than the French Roman legal system. In 1791 the colony was divided into Upper and Lower Canada, the former term referring to what is now Southern Ontario. Upper Canada received its own Provincial Assembly modeled after the British system. The early pioneer period ended around 1812, when war with the United States put an end to the influx of pioneers from there and abroad. When the Napoleonic Wars ended in 1815, a second wave of settlers came from the British Isles. They established farms under the supervision of people who had received large tracts of land and were in charge of building roads.[11]

The colonial period saw a gradual transformation of crops, farm animals, buildings, and tools. The pioneer farmers of the Loyalist period were able to grow enough food to feed their families. However, they needed money to buy supplies, pay taxes or cover other expenses. The sale of wood and potash was the main source of income. Once the land was cleared, wheat became the principal cash crop. With the income from wheat, farmers converted their log cabins into stone or wooden

---

[8] See Chapman & Putnam, *The Physiography of Southern Ontario,* 193, 227, 230, 286.

[9] These military forts were located on the Great Lakes or the St. Lawrence River, the main means of transportation.

[10] An attempt by the Earl of Selkirk to set up a block settlement on a land grant he received in Kent County in 1804 failed miserably, since neither his colonists nor his local representative were familiar with the environment. See G.P de T. Glazebrook, *Life in Ontario: A Social History* (Toronto: University of Toronto, 1968) 21.

[11] The Amerindian population did not fare so well. While the Iroquois successfully replicated their agricultural way of life in the Grand River Valley, original native inhabitants saw their standards of living decline since their seasonal movements were severely restricted by European settlement.

frame houses. Horses replaced oxen as traction animals. However, apart from tobacco, which was grown in the counties of Essex and Kent, there was little diversification in agriculture. Farm families had cows but there was no commercial dairying. Wheat cultivation was more lucrative than any other form of farming, and livestock and cheese could be imported cheaply from the States.

The building of canals across the escarpment in the twenties to connect Lake Ontario and Lake Erie and the Rideau Canal in the eastern counties provided the infrastructure for further economic development. John Galt, founder of the Canada Company in 1824, oversaw a large settlement scheme in the Huron tract, which stretched east of Stratford to Goderich (then an abandoned Indian clearing). He also founded the city of Guelph where the Canada Company sold land to better-off settlers. Other entrepreneurs set up mills and taverns to service a growing population in other parts of the colony. By 1840 much of Upper Canada had been transformed into towns and farmland, although homesteaders in the Ottawa valley continued to derive most of their income from timber. Within the next two decades additional settlement filled up the rest the arable land in the interior with the building of the Garafraxa road (now highway #6) to Owen Sound, the Sydenham road to Shelburne, the Hurontario road (now highway #10) and finally, the road from Owen Sound to Southampton (Hw #21). Railroad lines connecting Toronto, Hamilton and Windsor in the 1850s facilitated greater integration of this section of Upper Canada.

The province of Ontario was created in 1867, with the union of the remaining British possessions in North America to form the Dominion of Canada. The two decades following Confederation saw a rise in immigration from Europe, as English-speaking people in Central Canada sold their farms to move to the Prairies in Western Canada. With increasing competition from this part of Canada, farmers in Ontario began producing more dairy products, pork, and speciality crops, while wheat production declined. At the same time, manufacturing of farm implements expanded to meet the growing demand of wheat farmers in the Canadian prairies.[12] In Northern Ontario, the economy developed along very different lines. When a railway line along the Ottawa river turned west at Mattawa, a group of settlers moved into the clay belt regions, first around Lake Nipissing, then the Sudbury area, and finally along

[12] Reaman, *A History of Agriculture in Ontario*, 93, 144. The biggest farm equipment firm was the Massey Ferguson Company, which eventually became a major international corporation.

the north shore of Lake Huron. However, the movement of people to these clay belt regions did not result in communities revolving around full-time farming. Most migrants were more interested in cutting timber than farming, and many did not stay once forests had been stripped.[13]

During the first decade of the twentieth century, people throughout the eastern half of Canada continued to shift to the West, but fruit growing expanded in the East. Apple production for export to England became particularly important in Southern Ontario. The warmer region comprising Essex and Kent counties specialized in a variety of cash crops, including alfalfa and sugar beets.[14] The 1920s saw a major shift in tobacco production, as burley tobacco from Kent County was replaced by flue-cured tobacco in the sandy Norfolk plain, particularly in the area around Delhi. At the same time there was a gradual shift from the use of horses to tractors, although most farmers continued to use horses.[15] Motorcars became more widespread, but few farm families had modern amenities. After the stock market crash of 1929, conditions in rural Ontario deteriorated rapidly.[16] By 1941, only 16.8 percent of Ontario farms had gas or electricity; most people continued to use kerosene lamps, just as they had in the nineteenth century.[17]

AGRICULTURAL REGIONS

While it is possible to generalize to a certain extent about Ontario agriculture, each region and sub-region has unique traits. It should now be obvious that most of the far north never developed viable farming, although Dutch dairy farmers carved out a niche in fertile pockets in the Lakehead region, which includes the Rainy River district and Thunder Bay (see chapter 7). The section of the Canadian Shield that extends into Southern Ontario is also not suitable for agriculture, and few Dutch farm immigrants ended up in this part of the province. Therefore this chapter will provide an overview only of the agricultural regions in the arable parts of Southern Ontario, to provide a better picture of the different

---

[13] Several government-sponsored settlement experiments to turn returning war veterans into pioneer farmers after the Boer War and the First World War, plus a relief land settlement plan during the depression, were failures. Sinclair, "The North and Northwest", in Ian M. Drummond et al., *Progress without Planning: The Economic History of Ontario from Confederation to the Second World War* (Toronto, University of Toronto Press, 1987) 86.

[14] Drummond et al., *Progress without Planning*, 35.

[15] Drummond et al., *Progress without Planning*, 40.

[16] By 1992 real net income per occupied farm was a third of what it had been in 1939. Drummond et al., *Progress without Planning*, 39-40.

[17] Drummond et al., *Progress without Planning*, 42.

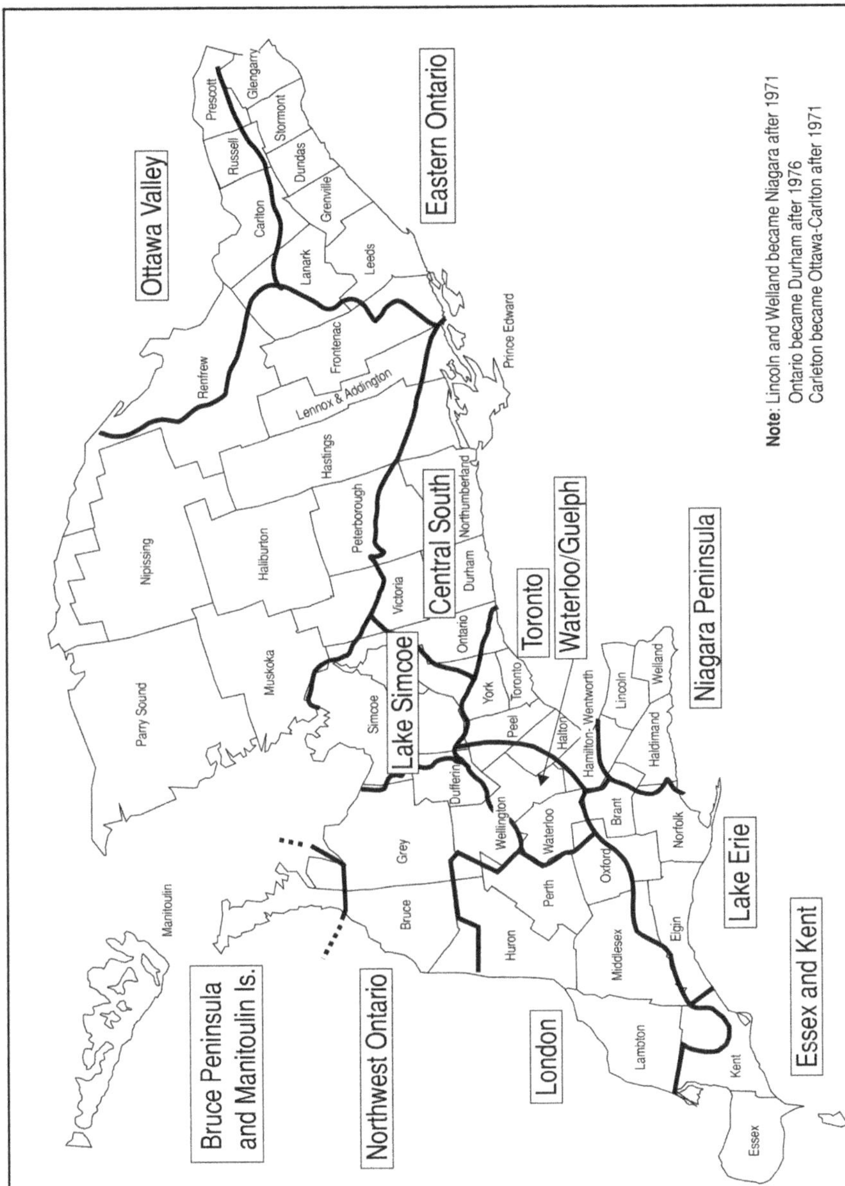

Figure 7. Agricultural Regions and Counties in Southern Ontario.

farming systems Dutch farmers encountered around 1950. Subsequent chapters will refer back to these agricultural regions. Figure 7 shows thirteen such regions, as well as the counties existing at that time.[18]

Essex and Kent Region

Encompassing nearly the entire Clair Plain (see figure 6), this region specializes in cash crops. Horticulture is practised in two fertile sandy areas, including the Leamington area where tomatoes are currently grown in greenhouses. The most prominent tomato producers are of recent Italian descent. In 1960, 10 percent of the farms were dairy operations and sugar beets were cultivated in the Chatham area. By 1990, most of the dairy farms and all of the beet producers were gone. The cash crop mix has shifted to predominantly small grains such as millet, spelt, barley and sorghum. The border city of Windsor (across from Detroit) is the most significant urban center, followed in order of decreasing size by Chatham, Wallaceburg, Leamington, Essex, Tilbury, and Blenheim. Today farmers of Dutch descent live in the areas surrounding these towns, particularly Chatham and Blenheim (see chapter 12). The only other significant ethnic presence is people of Flemish (Belgian) descent, who used to work in the tobacco fields around Blenheim.

Lake Erie Region

Situated along the northern shores of Lake Erie, this sandy soil region includes the former county of Norfolk. I have included the Mount Elgin ridges, located in the southern fringes of Middlesex and Oxford counties, plus a section of the Bothwell sand plain. In 1960, this region specialized in flue-cured tobacco, while the ridges were characterized by a combination of dairy and beef farms. By 1990, we find the same small grains as in the Essex and Kent region, combined with either cattle or tobacco. Towns and villages include: Brantford (Brant); St. Thomas and Aylmer (Elgin); Tillsonburg, Norwich (Oxford); and Delhi, Simcoe, and Waterford (Norfolk). Farmers of Belgium descent have a strong presence

[18] To formulate these regions, I have modified and simplified the fifty-two microregions for 1960 listed in Chapman, *The Physiography of Ontario*. My extrapolations for 1990 are based on a combination of my own observations and a map in Philip D. Keddie and Julius A. Mage's, Agriculture in Ontario, in *Ontario: Geographical Perspectives on Economy and Environment*, chap. 7, Bruce Mitchell (University of Waterloo: Department of Geography, 1991). I also consulted a series of maps by the same two authors in *Southern Ontario: Atlas of Agriculture: Contemporary Paterns and Recent Changes* (Guelph, Department of Geography, 1985).

in the area around Delhi and Simcoe,[19] while Dutch-Canadians constitute a significant proportion of farmers in the vicinity of Aylmer, Jarvis, and Brantford.

The Niagara Peninsula

This increasingly more urbanized region, which borders on the suburbs of Hamilton, includes the cities of Niagara Falls and St. Catharines. It consists of two distinct, but closely interconnected, sub-regions, both of which have a strong Dutch presence. The "fruit belt", a narrow plain sandwiched between the south shore of Lake Ontario and the Niagara Escarpment, was, and continues to be, the principal fruit and vegetable region in the province. Small towns include Grimsby, Beamsville, Jordan Station, and Niagara-on-the-Lake. Between 1960 and 1990, its vineyards and wineries evolved to the point where they now have an international reputation. This region, which has 200 greenhouse operations dedicated to cut flowers, floral crops, and nursery stock, is North America's third largest greenhouse floral producer. Their owners today are mostly of Dutch descent.

The Haldimand clay plain in the upper area of the Peninsula along the north shore of Lake Erie features a horticultural island of light soil around Font Hill, which specializes in peaches. The rest of this clay plain, with soil deficient in organic matter and minerals, once supported general farming with an emphasis on livestock, and dairying closer to Hamilton. Today, the outskirts of Hamilton have an equal mix of dairying, market gardening, nurseries, and orchards typical of regions located close to larger urban centers. Towns and villages in the upper section include Dunville, Caledonia, Hagersville, and Cayuga (in Haldimand), and Smithville and Wellandport (in the Niagara Regional municipality).

The Toronto Region

The region of urban development that runs along the western end of Lake Ontario is known as the Golden Horseshoe. I will use the term "Toronto region" to characterize only the section of the Horseshoe from Hamilton to Bowmanville, plus its rural fringe. In terms of physiographic features, this region consists of the Flamborough and Peel Plains, parts of the South Slope and Oak Ridges Moraines and a section of the narrow Iroquois coastal plain. By 1960, the apples and pears that used to

---

[19] See Joan Magee's, *The Belgians in Ontario: A History*, chap. 6 (Toronto: Dundurn Press, 1987).

be grown in this section of the plain had almost disappeared. The Peel area then became part of the milk shed for the city of Toronto. Since that time, urban expansion has reduced viable farmland, but the agriculture that remains is quite mixed. Smaller urban centers not yet absorbed into metropolitan areas include Uxbridge in the east, Caledon in the north, Milton, Acton, and Campbellville in the center, and Copetown and Freelton to the west of Hamilton. Today, it is difficult to find Dutch dairy or cash crop farmers in this region, although the majority of nurseries and many garden centers are in Dutch hands.

The Waterloo and Guelph Region

The well-drained Waterloo Hills and the Guelph Drumlin Field are two physiographic regions west of the Escarpment that include the twin cities of Waterloo and Kitchener, and Guelph respectively. Together with nearby Cambridge, which is a fusion of Hespeler, Galt, and Preston, the three cities form another center of urban growth. In the rural areas beyond this triangle, there are many towns and villages, especially in the Mennonite country north of Waterloo. In this region, intensive livestock was the dominant form of farming in 1960. The Waterloo area had more hogs, while a tendency towards beef cattle was typical of the Drumlin Field, with dairying well developed near Guelph. A mixture of livestock types and poultry has continued to be dominant, but the spread of small grains means that the southern part of the region now has the same combination of livestock and small grains as in London. There are Dutch farmers, but not to the same extent as in other regions.

The London Region

The London region is the part of Southwestern Ontario located between Lake Huron and the Waterloo/Guelph region. It encompasses the Huron Tract mentioned in the history of Upper Canada. Apart from Sarnia, located at the southern tip of Lake Huron, and the city of London, this region is predominantly rural agricultural, and includes the following towns and small cities: Stratford, St. Marys, Mitchell, Milverton, and Listowel (in Perth); Exeter, Hensall, Seaforth, Clinton, Goderich, Brussels, and Wingham (Huron); Woodstock, Thamesville, and Tavistock (Oxford); London, Strathroy, Mount Brydges, Park Hill, Ailse Craig, and Lucan (Middlesex); Sarnia, Grand Bend, Forest, and Wyoming (Lambton); and Thamesville (Kent). Dutch farmers are well represented over this whole region. In terms of physiographic divisions, the region consists of the Bothwell and Caradoc sand plains; almost the entire Stratford and Oxford till plains; half of the Teeswater Drumlin Field; sections of the St. Clair and Ekfrid clay plains, and the Huron Slope and Fringe; plus the western arm of the Horseshoe Moraines, which run parallel to Lake Huron. There are several marshes in the Huron Fringe that were drained prior to the Second World War. That micro region of muck farming specializes in onions and celery (see chapter 7). London has seen a gradual shift in the type of farming, with the northward and eastward expansion of small grains, which have become more important than the field crops common in the sixties. Since the early twentieth century, farmers have derived more income from livestock farming than from cash crops, except for the southwest portion where, by 1990, the opposite was true. A small area, just to the north of Thamesville has the same combination of small grains and tobacco that we saw in sections of the Lake Erie region. The area around Woodstock has a more developed diary industry, while the northern portion of Perth has more hogs. The area east of the city of London, which was a dairying region prior to 1950, saw an increase in market gardening and apple orchards, starting in 1960. By 1990, these specializations were overshadowed by small grains. Turkeys, grown mainly on the Lambton clay plain in 1960, are now found in most of this region.

The Northwest Region

This part of Southern Ontario has a shorter growing season and less land suitable for cultivation than fertile regions further south, such as London or Waterloo/Guelph. It consists of the central and upper portions of the Horseshoe Moraines, while the Huron slopes define its western flank. Other physiographic sub-regions include the Arran Drumlin Fields, the Saugeen Clay Plain, the northern half of the Teeswater Drumlin Field, the high Dundalk Till Plain, and a narrow extension of the Stratford Till Plain. I have included the northern side of the highly indented section of escarpment along the southern shore of Georgian Bay, including the township, but not the town, of Collingwood.[20] Small cities and towns include: Owen Sound, Hanover, Durham, and Meaford (Grey County); Walkerton, Paisley, and Chesley (Bruce); Orangeville, Shelburne, and Grand Valley (Dufferin); Mount Forest, Clifford, Palmerston, Harriston, and Drayton (Wellington). Despite the rugged terrain and swampy sections of most of the Northwest Region, there are many pockets of viable agriculture, including the section of the Dundalk

[20] This small area consists of the Cape Rich Steps plus the Beaver and Big Head valleys. In 1960, the loamy terraces near Meaford, the clay loam valley plain west of Thornbury, and the Cape Rich Steps were known as the "Georgian Apple belt".

Till Plain north of Shelburne, where potatoes have been the main crop since the sixties. All of the fertile areas, especially in the Drayton area and other parts of the southern fringe of this region, have concentrations of Dutch farmers. In contrast, one finds few Dutch farmers on the less fertile sections of the Dundalk Plain, characterized by a pattern of hay and beef typical of more marginal areas.

The Bruce Peninsula and Manitoulin Island

The Bruce Peninsula, with its rocky outcroppings, is not really an agricultural region. Scattered farms were established in the aftermath of the lumbering era, but the area did not, and still does not, have much beyond some beef farming. The Peninsula serves mainly as a recreation area, much like the cottage country of central South Ontario. Manitoulin Island, which can be reached by a ferry departing from Tobermory, at the tip of the peninsula, is similar to parts of the mainland portion of Southern Ontario, although much of it is still forested. Its soil is poorly drained, but it has some diary farms and beef cattle farms. There are not many Dutch farmers on the Island.

The Lake Simcoe Region

The region around Lake Simcoe, which corresponds closely to the county of the same name, lies to the east of the Niagara escarpment and along the southeastern shoreline of Georgian Bay. I have included all of the Simcoe Uplands and Lowlands, including those parts that spill over into neighboring counties, and the Schomberg Clay Plain. Barrie, the only city in the region, and the town of Orillia, are both located on Lake Simcoe. Other towns and villages, from north to south, are: Collingwood and Beaverton (in Victoria); Stayner, Midland, Elmvale, Alliston, Beeton, and Bradford (Simcoe); and Newmarket and Schomberg (York). This region is characterized by a complex pattern of soils and farm types, including both the best and the most marginal. The lowlands northeast of Lake Simcoe and the upland region in the eastern half of the Penetang Peninsula combine dairy and beef cattle, just like much of the Northwestern Region. The swampy strips of lowland on the west and east shores of Lake Simcoe are also agriculturally poor. In contrast, Holland Marsh at the south end of the lake, which was dyked and drained in the 1930's, has become a prime horticultural area of Ontario (see chapter 3). On the west side, close to Georgian Bay, the Essa flats near Alliston are an important potato-growing area. The clay soil of the lowlands around Elmvale and Stayner, where livestock was still predominant in 1960, is today a mixed livestock/small grain region. The Schomberg Plain is a well-drained silt clay loam area. It had always been diversified and is now indistinguishable from the rural fringe of the Toronto region. All of the fertile areas within the Simcoe region have a Dutch presence, especially the Holland Marsh.

Central South Ontario

This region is located below a jagged line falling from the northeast corner of Simcoe County to the point where the Canadian Shield approaches Lake Ontario, above Wolfe Island. This line, marking the edge of the Shield, cuts across many counties. Geographically, it consists of several elongated bands of soil and terrain types, from west to east. The main city is Peterborough, and secondary centres include Lindsay (Victoria), Hastings, and Campellford (Northumberland); and Lakefield, Frankford, and Stirling (Hastings). In the south, the narrow Iroquois shore plain up to Brighton is an extension of a mixed agriculture, also found in the Toronto region, and several areas of apple orchards, including Colborne, which advertises itself as the "oldest apple town in Ontario." At Carrying Place, this plain turns northward and fans out towards the east edge of Rice Lake. This is a sandy section, which became an outlier of the tobacco belt. Next, going from south to north, the South Slope is also suitable for agriculture. After the original settlers cleared its stony soil, it evolved from wheat to mixed farming (including canning crops), to eventually become part of the dairy shed of Toronto (the situation in 1960). The higher lying Oak Ridges Moraine, with its sandy hills and boulder clay, has cattle on sparse pastures, although potatoes and rye were once cultivated near Pontypool. Hobby farms and reforestation has since transformed most of the Ridges. Better known for its scenery than good soil, the proportion of people who live in small hamlets and country residences is greater than those who operate working farms. Only in a section near Uxbridge has the livestock that was dominant in the sixties been supplemented by small grains. Today, there are some Dutch farmers in all these sub-regions. The Napanee Plains at the east end has more farmers, including Dutch-Canadians, than the rest of the region. Most of these Dutch farmers own dairy farms. However, in 1960 20 percent of farm operators had full-time jobs elsewhere, signalling a trend that was to become more prevalent in other parts of Ontario. Currently this environmentally sensitive area, especially the Oak Ridges Moraine, is subject to a huge development debate.

Prince Edward Peninsula

This peninsula consists of a limestone plain with sections of shallow sandy and clay soils. Barley, once the main crop, was replaced with cheese production well before 1960, although the Peninsula already had some horticulture and produced eggs. Since 1970, dairying, cattle, and a variety of crops in that order have been the three major agricultural activities. The two main towns are Picton and Bloomfield. This region attracted many Dutch rural immigrants, many of whom live there today.

Eastern Ontario

Eastern Ontario, which has several concentrations of Dutch farmers, refers to the part of Southern Ontario that lies east of the Canadian Shield. Its southern boundary is the St. Lawrence River, and it borders on the Ottawa Valley. I have included the Leeds Knobs and Flats, Smiths Falls Limestone Plain, Edwardsburg Sand Plain, North Gower Drumlin Field, Winchester Clay Plain, Glengarry Till Plain, and Lancaster Flats. Howe and Wolfe Islands with their clay flats, also belong to this region. This almost completely rural agricultural area has the following small cities, towns,and villages: Cornwall (Stormont); Morrisburg, Winchester, Chesterville, and Iroquois (Dundas); Gananoque, Brockville, Lansdowne and Athens (Leeds); Maitland, Kemptville, Alexandria, and Lancaster (Grenville); Carleton Place, Perth, and Smiths Falls (Lanark); and North Gower (Ottawa-Carleton). This region is best known for its dairy industry. Beef cattle is also common, but prevalent only in the less fertile Smiths Falls Limestone plain. Small grains are increasingly becoming important cash crops, but not to the same degree as in Southwestern Ontario. The best farms are found on the Winchester Clay Plain (most of which is in the northern section of Dundas), where the soil is ideally suited for growing silage corn and alfalfa. That area has the highest concentration of Dutch-Canadian farmers.

The Ottawa Valley

The Ottawa Valley, located along the Ottawa River, stretching from Hawkesbury in the southeast to Petawawa in the north, consists mainly of a narrow clay plain with various other types of soil and terrain mixed in. Sand plains dot the northern sections of Prescott and Russell United Counties, while on the opposite side the valley ends in the Petawawa Sand Plain. The middle part is located within the boundaries of Ottawa-Carleton Regional Municipality, which contains Canada's capital city, Ottawa. From north to south, other populated centers are: Pembroke,

Renfrew, Russell, Bourguet, Alfred, and Van Kleek Hill. With the exception of the Petawawa sand plain and the Muskrat Lake Ridges in Renfrew (neither of which has had agriculture of any importance since the turn of the century), dairy farms are prominent in the whole of the valley. Crops in this region (including tobacco) have varied over the years, although the clay soils are not as fertile as in other parts of the province. Currently, the circular rural belt outside the city of Ottawa has the same mix of livestock and small grains as in nearby Eastern Ontario. A smaller number of Dutch farmers are scattered throughout this Ottawa Valley region.

RURAL ONTARIO: THE SOCIAL DIMENSION

Postwar farm immigrants who settled in all of these regions had to adapt to more than just soil conditions, markets and some mix of crops and livestock. They encountered ethnic and religious diversity, lack of overt status distinctions, and a political system that was quite different from that of their homeland.

Ethnicity, Race and Religion

In North America, ethnicity has shaped social interaction and group formation in much the same way as religion has in the Netherlands. Religious differences can also take on an ethnic dimension. The history of Canada has been one of conflict and accommodation among the Catholic French, Protestant English, and the native populations. The development of agriculture in Upper Canada took the form of block settlements: German-speaking Mennonites in the Kitchener-Waterloo region and the Irish in Peterborough. In some counties, this pattern of spatial segregation was replicated at the townships level. In Simcoe, the Roman Catholic Irish settled in the townships of Vespra, Flos, and Medonte, while Protestants from the North of Ireland moved to West Gwillimbury, Tecumseth, Mono, Essa and Innisfil.[21] Over time, these ethno-religious differences became less significant as a single English Canadian culture and identity emerged. However, part of that culture, as a form of shared discourse, includes the recognition of ethnic background. Linguistic enclaves where French, German or other languages were spoken survived up to the sixties and beyond. For example, when postwar Dutch immigrants started farming in the vicinity of Bornish, in McGillivray Township

[21] Jones, *History of Agriculture in Ontario*, 50.

(London region), they discovered that their neighbors were descendents of people who came from the Hebrides (Scotland), some of whom still spoke Gaelic.[22]

Another basis of group formation is associated with racial categories. Forms of social interaction based on such categories were less salient, but not absent, even after Confederation. The impression of Dutch immigrants was that the inhabitants of Ontario were racially homogenous. They originally expected to see "red skins," as portrayed in books and movies, but soon received the impression that were no "Indians" (native peoples) left in Southern Ontario. Only in places like Onondaga, which borders on the Six Nations Reserve near Brantford, were the Dutch aware of the presence of an aboriginal population, which only constitutes less than two percent of the population of the province.[23] Moreover, Dutch immigrants never encountered, much less heard about, the descendants of slaves who once lived in several rural townships in Southern Ontario. By 1950, most of the Black settlers in Kent County, and Maryborough Township (in the Waterloo-Guelph region) had long ago departed because of the hostility of their neighbors.[24]

Religion also shaped social life and attitudes in Ontario. During the first half of the nineteenth century, the Church of England (Anglican), with strong Irish, as well as English connections, was not only the largest denomination, but received disproportionate favor from the government in the form of land grants called clergy reserves. Anglicanism was symbolically associated with Great Britain and a hierarchically ordered society, in contrast to what was perceived to be a disorderly republican United States. Another ideological current with religious connotations was hatred of the Pope and Roman Catholicism, associated with the Orange Lodges found in most towns in Ontario. Originally introduced by Protestant Irish immigrants in the eastern counties, this institution was associated with a broad range of Protestants across Ontario.[25] Such lodges were still strong at the time postwar immigrants started arriving, and the members of such lodges were surprised to learn about Dutch

[22] Interview (FS) with Jan Hendricks, in Park Hill, 16 December, 1988.

[23] According to the 1996 national census of aboriginal peoples conducted by Statistics Canada.

[24] For historical references to the history of Blacks in rural Ontario, see Jones, *History of Agriculture*, 40, 53; Glazebrook, *Life in Ontari*, 116-17. Postwar farm immigrants might have heard racist comments about Black people, while simultaneously hearing declarations that "Canadians are not racially prejudiced".

[25] Glazebrook, *Life in Ontario*, 64: Donald H. Akenson, *The Irish in Ontario: A Study in Rural History* (Kingston: McGill-Queen's University Press, 1999), 277.

Catholic "Orangemen" (see chapter 8)! Yet, there was no sectarian violence even in places like Lansdowne, with Irish farmers from both groups.[26] Nevertheless, in many townships, Catholics and Protestants did not intermarry, and Protestants refused to sell land to Catholics well into the second half of the twentieth century.[27] There was also prejudice against French-Canadian Catholics. Dutch immigrants came across the expression "frogs", a pejorative label applied to French-Canadians.

Class and Social Hierarchy in Rural Ontario

The nature and extent of inequality in rural Ontario has changed over time. With the creation of Upper Canada, land grants varied from fifty acres for private soldiers to 5,000 acres for officers.[28] This was based on a hierarchical system in Britain that included absentee landowners, landed gentry, and tenants. However, the British agrarian structure, based on large estates, could not be replicated in North America with its ever expanding frontier. Over time, the predominant pattern was that of the 100-acre lots of most surveys and owner-operated farms. Farms were intended to remain in the family and even today "century farm" is proudly displayed to indicate that a farm has remained in the family for over a hundred years. In the fifties, the inhabitants of rural Ontario used the family name of previous owners to identify individual farms. Often a postwar Dutch immigrant would arrive at "the Harrison place" or "the McDougal place," even though they might be working for a Mr. Jones, who owned both of them. These names lasted for at least one more generation.

Farmers in Ontario are generally perceived as equals even when one farmer owns more than one farm. This image largely corresponds to the distribution of landownership; by the turn of the century and beyond, the proportion of farmers who did not own their own land was small, although there were, and still are, various forms of short-term share farming, especially in tobacco and other cash crops. Farms owned by families have in fact continued as the predominant pattern of land tenure, although bigger farms are increasingly incorporating to take advantage of tax write-offs, including salaries for family members. Today, farm corporations that control over 1,000 acres are still owner-operated, with

---

[26] Akenson, *The Irish in Ontario*, p. 280.

[27] I heard stories about Protestant farmers not wanting to sell land to Catholics from several people I interviewed in both the western and eastern halves of Southern Ontario, although several Dutch Catholic postwar immigrant farmers did end up buying such land.

[28] Glazebrook, *Life in Ontario*, 28.

family members or other relatives running their various operations. However, some family farm corporations have expanded to the point where they employ professional managers, who are not family members. Yet, even such farms cannot be included in the same category as a large corporation with industrial assets, which holds land for speculative purposes.

Another type of social hierarchy is that between farmer and farm worker. This distinction has never been rigid in Ontario, since many hired hands were family members or friends. A hired man, whether the son of a poor farmer or newly arrived immigrant, could end up setting up a farm of his own. Throughout the twentieth century, successive groups of foreigners have been the seasonal labor force in agriculture until returning to their countries of origin, moving into more lucrative occupations, or starting farming on their own. There has always been a shortage of seasonal workers, which became acute during and after the Second World War. This shortage was addressed by hiring, first Polish war veterans, then Hungarian and German-speaking refugees (often displaced persons) from Eastern Europe. Beginning in 1947 Dutch immigrants became the next source of labor.

Government Policy, Farmers' Organizations and the Politics of Agriculture

Since the establishment of Upper Canada, successive governments have promoted or shaped the development of agriculture. The most dramatic policy change to affect Ontario farmers was the adoption of marketing boards, whereby farmers had to co-operate in negotiating for farm prices, and, in the case of supply management, restrict output. In 1937, after holding plebiscites, the first four marketing schemes were established, including the Holland Marsh Vegetable Growers' Marketing Plan. The continued decline in farm output during and after the Second World War, resulted in additional marketing schemes, and by 1966, nineteen more plans came into being. The implementation of all of these post-war programs coincided with a period when rural Dutch immigrant farmers were starting to buy farms in Southern Ontario. In 1967, the Ontario Milk Marketing Board became the most comprehensive supply management scheme, although dairy farmers operating under the protection of the Milk Marketing Board still had to become more productive in order to stay in business.

Throughout the second half of the twentieth century, the Ontario Ministry of Agriculture and Food (OMAF), later called OMAFRA (adding on RA for "Rural Affairs"), played a much larger role in the regulation and promotion of agriculture than its nineteenth century predecessor. That ministry continued to be closely linked to the University of Guelph, with its various programs of ongoing research, rural extension, and the education of aspiring farmers. One of the OMAF programs, which made a difference for many rural people in Ontario, was the Farm Enlargement and Consolidation Program, initiated in 1966 to alleviate rural poverty.[29] Originally designed for Eastern Ontario, it expanded into other regions of Ontario, and became a source of credit for a variety of farmers, including Dutch-Canadians. Prospective young farmers received financial assistance in the form of Junior Farmers' loans. By the year 2002, various OMAFRA programs and most of the marketing boards were still operating, although the 1996 North American Free Trade agreement (NAFTA) between Canada, the United States, and Mexico marked a shift in policy that is still a topic of discussion and debate among farmers within the various organizations.

Farmers' Organizations and Rural Institutions

In the second half of the nineteenth century, various professional farmers' organizations, as well as horticultural societies, shared information and promoted new technological advances. By the turn of the century, many of the producers' associations that still exist today were in operation, including two types of dairymen as well as various cattle breeders and poultry associations, plus vegetable, vine, corn and fruit growers' associations.[30] Their respective members were later to vote on establishing marketing schemes. Women's Institutes, which organized speakers and offered leadership training courses and course in home economics, were found in almost every township in rural Ontario both before and after the Second World War. Many Dutch immigrant women came into contact with them.

Another organization found in every farming community today are the 4-H clubs (standing for "Head, Heart, Hands and Health"). Originally known as Boys and Girls Clubs, they date back to the turn of the century. These clubs not only became an important part of the education of the children of Dutch farm immigrants, but in some cases were taken over, and largely run, by their parents (see chapter 8). Dutch farm immigrants also became involved, though to a lesser extent, in another well-established rural institution, United Cooperatives of Ontario (UCO), whose 'Co-op" stores, fertilizer plants and fuel stations can be seen all

[29] This program was part of a federally sponsored initiative called the Agricultural Rehabilitation and Development Act (ARDA).

[30] Another, more general, farmers' organization closely linked to the Extension branch of the Department of Agriculture is the Ontario Soil and Crop Improvement Association, founded in 1938.

over the province. Founded in 1914, UCO began as the wholesale supplier for farmers' buying clubs and independent co-ops, which were increasingly affiliated directly with UCO. Another institution that became significant for farmers were co-operative financial institutions, known as credit unions. Ontario, unlike Quebec or Western Canada, does not have a strong tradition of co-operative institutions. In the forties, people in the English-Canadian heartland of Canada were still leery of credit unions; some saw them as an extension of trade unionism, while others equated them with the Catholic Church.[31] The credit union movement, which grew very slowly prior the Second World War, was invigorated as a result of the arrival of immigrants from different parts of Europe after the War. In rural Ontario, Dutch farm immigrants led the way in setting up many rural parish credit unions, as well as establishing their own, ethnic-based, co-operative institutions (see chapter 8).

The Farmers' Movement

The institutions presented so far do not for the most part have an explicit political agenda. In contrast, all-purpose or general farmers' organizations, which originally started out as social clubs or educational bodies, did become involved in lobbying on behalf of farmers. Currently the largest general farmers' organization is the Ontario Federation of Agriculture. In the past, such organizations could become quite powerful and influential politically. In 1919, farmers spoke with a single voice when the United Farmers of Ontario formed a government with the Independent Labour Party. The farmers' movement was less active during the twenties, as in earlier periods of relative prosperity, but it was revived in a new form during the Great Depression.

Renewed attempts to organize farmers politically resulted in the formation of the Ontario Federation of Agriculture (OFA) and its subsequent merging with another organization, the United Farmers, in 1943.[32] The OFA was a strong proponent of producer marketing boards, and collaborated closely with the co-operative movement to set up alternative insurance schemes and leadership programs. However, despite its successes, market forces and broader political concerns created new challenges and conflicts. Ontario farm output fell in 1943, even though the provincial government provided subsidies for hog and cheese producers. Then, as the war came to an end, increasing competition among farmers

---

[31] See Ron Kenyon, *To The Credit of the People* (Toronto: The Ontario Credit Union League, 1976), 16.

[32] Reaman, *A History of Agriculture in Ontario*, 243-45.

in the province, and overproduction, posed threats to price stability.[33] Farmer militancy flared up in 1952 when the Ontario Farmers' Union was founded as an alternative to the more established Ontario Federation of Agriculture. This occurred a year after a dramatic decline in farm prices, around the same time that Dutch immigrants were moving into rural Ontario by the thousands.

A decade later, at the height of its popularity, the Farmers' Union, whose leaders included Dutch immigrant farmers (see chapter 9) organized protest marches and tractor rallies in Ottawa, the nation's capital. The most controversial political issue related to farming in the sixties was the attempt by the then provincial agriculture minister, Bill Stewart, to form a General Farm Organization by amalgamating the OFA and the Ontario Farmers' Union. The more radical Farmer's Union was adamantly opposed, and the proposal did not pass. However, unlike the case of militant farmers' organizations in the prairie provinces, the influence of the Ontario's Farmers' Union was short lived. Its membership had dwindled to insignificant numbers by the early eighties, although in 1984, farmers rallied again in Ottawa on tractors to protest the high interest rates that were hurting farmers heavily dependent on credit.

Despite the disagreement between the department of agriculture and the Farmer's Union, Bill Stewart, himself a farmer and founder of the Ontario Milk Marketing Board, was a strong supporter of supply management. Farmers of different political persuasions today reminisce about how his government paid attention to farmers and remark that agriculture has now became a low priority. Dwindling numbers have reduced the political clout of the farm sector in both absolute and relative terms. Yet, this decline, along with increasing competition, has spawned a number of new organizations, including several with religious affiliations. The bishop of the Catholic Diocese of London started the Catholic Rural Life Conference in the eighties as a forum for discussion and self-help. Many of its more active members are Dutch-Catholics (see chapter 9). Another farmers' organization based on religious principles, which gained momentum in the nineties, is the Christian Farmers Federation of Ontario (CFFO). Dating back to the fifties, this newcomer to the world of farm politics was formed, and is still to a large extent controlled by postwar Dutch-Canadian immigrant farmers who belong to various Reformed denominations (see chapter 8).

---

[33] Kenneth J. Rae, *The Prosperous Years: The Economic History of Ontario, 1939-1975* (Toronto: University of Toronto Press, 1985), 137.

THE TRANSFORMATION OF RURAL ONTARIO AFTER WORLD WAR II

The era of postwar immigration coincided with one of growth and prosperity. As early as the end of the fifties, most farmhouses had electric washing machines and refrigerators, although many lacked running water and indoor toilets.[34] Rapid urbanization provided new markets for agricultural products, but high wages in the cities resulted in a five-fold increase in the cost of labor,[35] which made farmers more dependent on family members. At the same time, the introduction of electricity allowed for the near universal use of milking machines. Dutch immigrants witnessed the introduction of bulk tankers for milk transportation, which enabled farmers to increase output without increasing labor. The total agricultural output increased, but occupied farmland decreased between 1941 and 1971, as a result of urban expansion and the abandonment of marginal farmland. The number of farmers also declined rapidly, while average farm size went from 126 to 169 acres. At the same time, farm specialization increased. Between 1940 and 1971, total gross income from farming for all farmers combined, controlling for inflation, remained the same, yet the yearly net income for those who continued to farm doubled.[36]

The Futher Transformation of Agriculture

The past three decades have witnessed a transformation of agriculture throughout North America, similar to what also happened in Europe. Starting in 1970, Ontario agriculture underwent the same kinds of restructuring. Commercial farmers with bigger operations became increasingly tied into agribusiness, through vertical integration and contract farming. However, some already sell-developed forms of contract farming disappeared with the closing of small canneries. At the same time, the increase in standards of living and alternative employment elsewhere meant that sectors of farming still dependent on seasonal labor, such as tender fruits and vegetables, had trouble finding workers. Toward the end of the twentieth century, a new source of seasonal workers from abroad emerged. Since the mid-eighties, people from rural regions of the developing countries have come to Canada to work the long hours for minimum wages.

[34] Glazebrook, *Life in Ontario*, 267.
[35] Rae, *The Prosperous Years*, 136.
[36] Rae, *The Prosperous Years*, 134.

This also applies to the Netherlands, where workers from northern Africa and the near East have become the latest seasonal labor force in agriculture. In the case of Ontario, a special program for offshore labour brings men and women from the West Indies and Mexico in ever increasing numbers. These seasonal contact laborers are flown to Canada for periods of four to eight months to work in greenhouse operations, vineyards, and orchards. Nowadays it is not uncommon to see Mexican or Jamaican workers riding bicycles along country roads or congregating in supermarkets throughout Southwest Ontario during the summer. Recently, another source of farm labor has appeared on the scene: Mexican Mennonite families.[37] These new, mobile labor forces are a prime example of new forms of globalization.

The early eighties was a difficult period for farmers, when interest rates climbed to 20 percent. With increased dependence on credit, many farmers were unable to stay in business. The remaining farms became more specialized and even larger. By the nineties, it was not unusual for business-oriented farms in Ontario to be well over 1,000 acres in size. Farmers with small operations relied on part-time work and the income of other family members to make ends meet. Land speculation near urban centres gobbled up land at an unprecedented rate. While those who sold out could buy new farms outside of these zones of urban expansion, the land vacated was too expensive for most farm operations. In the fifties, one could see dairy farms on both sides of highway 25, all the way from Bronte to Milton. In 1997, only one was left.[38] On the other hand, the development of new varieties of soybeans and other small grains allowed those cash crops to expand into regions with lower average temperatures. Dutch farmers not only witnessed, but were active participants in, these changes, as we shall see in part two.

People from the Netherlands have settled in what is now Ontario since the time of the first European arrivals to the New World. However, contact between the Dutch and the native population or with settlers from the British Isles was, at best, sporadic. With the exception of the arrival of a handful of Dutch-American Loyalists from New York during the American War of Independence, and the immigration of families from the Netherlands during the first part of the twentieth century, most

---

[37] For a short treatment of Mexican Mennonite women handpicking tomatoes, see Ellen Wall, "Agribusiness and Hired Farm Labour in the Ontario Tomato Industry," PhD diss (Department of Sociology, McMaster University, 1992), 133-36.

[38] Interview (FS) with Herman Windmoller, in his home in Milton, 25 May, 1997.

people originating in the Netherlands came to Canada between 1947 and 1957. These postwar immigrants and their descendents were to become an integral part of rural Canada, especially in the province of Ontario. However, before telling the story of the Dutch farm immigrants who came to Ontario during the second half of the twentieth century, I will provide an overview of the Dutch presence in Canadian agriculture, including provinces other than Ontario, prior to 1947.

III. Dutch Farm Immigrants Prior to World War Two

We have seen how the Dutch presence in North America goes back to the period of early contact between Europeans and native people. However, few people of Dutch descent came to Canada prior to 1947. Even the Dutch immigrants in the period between the First and Second World Wars were numerically insignificant in comparison to the influx of people from the Netherlands who came after the end of Word War Two. Nevertheless, the Dutch had already come to the attention of people in Ontario prior to 1945 because of their close association with the Holland Marsh, a market-garden center located close to the city of Toronto.

HISTORICAL PRECEDENTS

The first Dutch families to settle in what is now Ontario were United Empire Loyalists (see chapter 2) although little evidence of that early presence remained at the time of postwar immigration. Later American Dutch newcomers also blended in. For example, Dutch immigrants who came to Wolfe Island in the early 1950s would not recognize that the original name of the Pike family (originally Snoek) had long ago been anglicized.[1] The only village in Eastern Ontario that has a Dutch name is Vankleek Hill, named after its founder, a United Empire Loyalist of Dutch descent who was granted a 500-acre parcel in 1790 in what is now Prescott and Russell United Counties.[2] However, that village, which later became predominantly French-Canadian, has no people of Dutch descent today.[3]

Dutch Immigration prior to World War One

Unlike the case of other European countries, there was no mass emigration from the Netherlands to the New World during the nineteenth century.

---

[1] Interview with Len and Lientje DeBruyn, at their farm in Wolfe Island, June 2, 1991.

[2] Albert Van der Mey and Marten Mol. *The Dutch Touch in Ontario* (Toronto: Marten A. Mol & Associates Ltd., 1997), p. 217.

[3] The percentage of people of Dutch descent in West and East Hawkesbury townships (Prescott) in both the 1986 and 1990 census was less than one percent.

The only notable emigration to the States consisted of Calvinist Dissenters who settled in Michigan and Ohio around the middle of the century, where they founded the Christian Reformed Church.[4] These Reformed Dutch-Americans gradually spread to other parts of the western United States, and eventually into Western Canada. Individuals of Dutch descent also showed up in Upper Canada. Anthony Van Egmond, a veteran of the Napoleonic Wars who first emigrated to the States and later moved north, received a contract from the Canada Company to build the Huron road (now highway 8). He moved to the Huron tract in 1828 and received 13,000 acres of land where he put up a large house and an inn. Van Egmond joined a revolt led by William Lyon MacKenzie and ended up dying in the Don Jail in Toronto. His sons set up businesses, including a sawmill, a brewery and a gunsmith shop in the village of Egmondville, in Tuckersmith Township (London Region).[5] Although several historically-minded citizens formed a Foundation to commemorate colonel Van Egmond in 1971, the postwar Dutch immigrants who bought farms in this area in the 1950's did not become aware of this Dutch connection until much later, when a historic plaque was erected in this village.

Toward the end of the nineteenth century the opening of the West to farm settlers became a magnet drawing immigrants interested in owning their own land. In 1893, the newly formed Christian Emigration Society in the Netherlands arranged for the emigration of unemployed farm labourers from the Dutch province of Friesland to Winnipeg, with the idea of then setting them up in homesteads in Yorkton, Saskatchewan. They did not fare well and the project was abandoned in 1897.[6] Like so many other immigrants, these prospective settlers worked as hired farm hands and for the railways. Those who stayed in, or returned to, the city of Winnipeg made a living peddling fresh vegetables. By 1905 some of these peddlers were growing their own vegetables on rented land on the outskirts of the city.[7] There were also block settlements by Dutch-Americans, including a group from Iowa and Montana who had originally emigrated from Nijverdal (Overijssel). They founded a village

---

[4] For a brief overview, see Schryer, The Netherlandic Presence, pp. 20-22.

[5] See John Martens, *Making Waves* (Waterloo: Waterloo Printing, 1988), pp. 23-35; G.H. Needler, *Colonel Anthony Van Egmond* (Toronto: Burns and MacEachen, 1956).

[6] Herman Ganzevoort, *A Bittersweet Land: the Dutch Experience in Canada, 1890-1980* (Toronto: McClelland and Stewart, 1988), 9-10.

[7] The most successful and best known of these vegetable growers in Winnipeg, who won the title "Cauliflower King" at an agricultural exhibition in the United States, was Klaas de Jong. Ganzevoort, *A Bittersweet Land*, pp. 19-20. His autobiography was published in 1973. Cauliflower Crown: The Cauliflower King of North America (Saskatoon: Western Producer Books, 1973).

in the Nobleford area of Alberta in 1904 and named it after their Dutch hometown. Their prior experience with dry-land farming in the American West enabled them to successfully adapt to farming in Canada. In 1912 a group of Dutch immigrants previously living in the city of Edmonton founded a similar settlement in a fertile area north of Edmonton. They called it Neerlandia. Members of the Christian Reformed Church, they were motivated by a desire to maintain their religious values. They set up a cooperative that bought bulk supplies for the community and marketed eggs and butter to Edmonton. By 1916 there were 60 families.[8]

A different kind of settlement scheme involving government assistance brought a handful of Dutch to the Maritime Provinces, on the opposite side of Canada. The Department of Agriculture of Nova Scotia, concerned about the abandonment of agricultural land, recruited Dutch farm immigrants who were placed on abandoned farms in 1914 and again in 1915. The first group of five families had all been farm workers with little experience of running their own farms, while a second group consisted of young farmers who hoped that their children would have a greater chance of becoming farmers in Canada. By 1920, they had abandoned their farms and moved to other parts of Canada due to the poor quality of the soil, a lack of other employment opportunities, and the size of their debt to the provincial government.[9] Other pockets of homesteaders between 1890 and 1914 were the result of the recruitment efforts of emigration agents working on commission for transoceanic transportation companies (Holland-America Lines, the Red Star Line) and Canadian railway companies (Canadian National and Canadian Pacific Railway). However, a series of scams and ill-conceived colonization projects, including one involving a Dutch priest, discouraged further emigration of prospective farmers.[10]

DUTCH IMMIGRATION PRIOR TO WORLD WAR TWO: THE "OLD TIMERS"

Dutch immigrants continued to come to Canada between the First and Second World Wars. Postwar immigrants later referred to all of these earlier Dutch immigrants as "old timers." Even more than was the case before World War One, land sharks and emigration agents enticed people from a wide range of occupations with false promises of plentiful jobs

---

[8] Henry S. Lucas, *Netherlanders in America* (Grand Rapids: Eerdmans Publishing, 1989), pp. 463-64; Ganzevoort, *A Bittersweet Land*, p. 30.

[9] Ganzevoort, *A Bittersweet Land*, pp. 27-29.

[10] Henri van Stekelenburg, *Hier is Alles Vooruitgang* (Tilburg: Stichting Zuidelijk Hisorisch Contact, 1996), pp. 81-88.

and cheap land. These agents duped the Canadian government, whose restrictive immigration policy allowed only agrarian immigrants; they told numerous Dutch people with no farm background to claim they had agricultural experience. The most deceitful recruiter, Willem Van Ark, a Dutch-Canadian from Neerlandia, had dealings with both Catholic and Reformed emigration societies. Another infamous pair of recruiters was the Cox brothers who lured Catholic colonists from Noord Braband to a marginal settlement north of Winnipeg under false pretenses.[11] These men appealed to religious values as a guise for land speculation. Yet, echoing the Dutch pillar system (see chapter 1) Dutch Calvinists, Roman Catholics, and those affiliated with other Churches or without religious affiliation had no contact among each other once they came to Canada. This was evident to those who ended up in Ontario.

Dutch Roman Catholics

Between 1918 and 1940, Dutch Roman Catholics originating from the Dutch provinces of Noord Brabant and Zeeuws Vlaanderen were part of a larger group of Dutch-speaking, Flemish farm laborers from Belgium with whom they shared a common language and religion. These farm laborers used to travel back and forth between Europe and Southwestern Ontario by boat as part of a 'swalllow migration'. Prior to the Second World War these Dutch Catholics were absorbed into existing parishes dominated by the Flemish in places like La Salette, Delhi, Leamington and Blenheim (Lake Erie and Essex-Kent regions).[12] Most of them worked in sugar beet and tobacco fields but returned to the Netherlands during the Depression. Few Dutch Catholic immigrants ended up in other parts of Ontario during this period.

As in the case of Western Canada, there was a connection between English Canadian Catholic clergy and the Canadian Pacific Railroad (CPR). Their regional representative was Father R.H. Dignan of Chatham.[13] Father K.E. Morrow, who lived in the Lake Simcoe region, became an agent for the Cunard and Red Star Lines. He set up the Catholic Settlement League of Brechin, Ontario, in Mara Township, with the blessing of the archbishop of Toronto, to promote the settlement of Dutch farmers, and he established contacts with the Catholic Emigration Organization in the Netherlands. As a result of his recruitment efforts, nine Dutch farm boys were placed on Canadian farms in his home

[11] For a more detailed account of these and other crooked recruiters, see Herman Ganzevoort, "Sharks in Wooden Shoes", in *Dutch Immigration to North America*, edited by Ganzevoort and Mark Boekelman (Toronto: Multicultural Historical Society of Ontario, 1983), pp. 147-65; van Stekelenburg, *Hier is Alles Vooruitgang*, pp. 92-107.
[12] Schryer, The Netherlandic Presence, pp. 150-51.
[13] Ganzevoort, *A Bittersweet Land*, p. 39.

region in order to learn English and Canadian methods of farming, with the idea of eventually buying farms. Further attempts to recruit farm settlers, involving Father Dignan of Chatham and Van Ark, the crooked agent from Neerlandia we met earlier, did not bring concrete results. Of the eleven immigrants they recruited not one was a farmer, nor did any ended up working for farmers.[14]

Dutch Calvinists (Reformed)

Dutch Calvinist immigrants, most of whom joined the Christian Reformed Church, came from various parts of the Netherlands. Like their Catholic counterparts, they worked in the sugar beet district near Chatham, Essex, and Sarnia, and were employed in vegetable, fruit, and dairy farms in the Hamilton area. These prewar Calvinist immigrants set up congregations in all of these places. Like many families, they started working for wages as agricultural laborers or menial labourers, although some set up farms on the outskirts of small cities or in nearby villages, such as Blackwell, near Sarnia (London region).[15] In fact, one newcomer, Leonard Mol, even bought a farm within a year of coming to Canada. This man and his family are a good example of a small group of better-off pre-war Dutch rural immigrants whose previous experience, plus access to capital, allowed them to get ahead quickly.

Leonard Mol was born in St. Philipsland in the province of Zeeland, the son of a cash crop farmer. Leonard was a farmer in the Netherlands and married a woman whose parents had gone to the States.[16] Prior to emigrating, Leonard spent three years in Wieringermeer, where he co-ordinated a 1000-hectare flax growing operation and did custom work for farmers who rented plots of land from the government. There were no economic reasons to emigrate since he was earning good money and owned three cars. However, several people who worked for him had been in Ontario as farm laborers and they asked him for money to return, on more than one occasion. Leonard finally decided to emigrate himself because he was unsure of being able to rent his own farm once the polder was completed. They were the only Dutch family aboard the ship in April of 1940, just before the Nazi occupation of Holland. Leonard did not speak English, but his sons had studied English and were able to help him. In Toronto, they contacted one of the few Dutch people in the city, the owner of a small factory, who used his connections to

---

[14] Van Stekelenburg, *Hier is Alles Vooruitgang*, pp. 105-6.

[15] Interview with Marinus Mol, in his home in Ayton, 12 May, 1991.

[16] Their story was recounted to me by Marinus Mol (one of his sons), in Ayton, 12 May, 1991.

exchange their guilders (not easy as a result of the Nazi occupation of Holland). That man recommended the Mol family take a train to Chatham, where they stayed the first night in a hotel. They looked up Dutch names in the phone book, and discovered there was a small Christian Reformed Church. They soon found a farm for sale for $5,200.00, with a brand new house and a standing crop of white beans, alfalfa, and grain. In a few years, the Mols were farming close to 300 acres with a couple of part-time helpers and a sharecropper who grew burley tobacco.

Other Groups

Not all of the Dutch immigrants who landed in Ontario before the war were Roman Catholic or Calvinists, and frequently Calvinists joined other churches in Canada. Jacob Vandenberg[17] who came to Canada at the age of twenty-one in February of 1930, arrived with twenty-five dollars in his pocket. His first job was with a dairy operation in the Waterloo-Guelph region, in Galt (now part of Cambridge) for thirty-five dollars a month, plus room and board. He was soon able to buy an old Model T Ford Coup for $150. Jacob's schooling in Holland was useful for working out feed rations. He enjoyed a good relationship with his employer and the family, but nevertheless, when the Depression deepened he worked for only room and board. Two years later, he met a farmer from Newmarket (Lake Simcoe region), who wanted to hire someone to manage his dairy herd. While still living on the farm, Jacob courted the daughter of a nearby farmer. It took them three years to persuade her father to allow her to marry an immigrant. Together they saved up money to buy a farm, while building up their herd to fifteen cows. When a 100-acre farm with a good house and barn came up for sale in a nearby village, they did not have the $250 down payment plus initial operating expenses. Her maternal grandfather lent them $1,000, and they began farming on their own. Jacob Vandenberg and his Canadian wife gradually built up a successful farm business, which thrived after 1940. They rented five farms to supply cheese and bacon for the war effort, but Jacob found it frustrating that he was not able to keep up with news from home; the Nazis censored all of the letters from his family. Nor did he have contact with any other Dutch people in Canada. He was the only Dutchman in the county, and he had been the only Dutchman in Newmarket. He joined his wife's United Church and served as superintendent of their Sunday school.

[17] This man, who was interviewed in 2001, did not want his real name to be used. Any references to his present location have also been removed.

Another case of a prewar immigrant who married a non-Dutch spouse and joined a church with no Dutch connections is Jan Verdun. The second oldest of twelve children, he emigrated to Canada in 1930, and started working as a farm labourer. That same year he was admitted to the Ontario Agricultural College in Guelph, where he earned extra money working as both a janitor and a headwaiter.[18] After graduation, he became a salesman for a fertilizer company and married a woman who taught in a one-room school near Strathroy. Jan started his own business in 1942 after buying a feed mill in Aylmer (Oxford County, in the London region). Like Jacob, Jan combined running a farm with farm-related businesses.[19] He was also completely integrated into Canadian society, his sole Dutch links being his birth family, until his contact with postwar Dutch immigrants increased, as many settled in the Aylmer area.[20]

A handful of Dutch immigrants not of farm background arrived in Ontario in 1939, just narrowly avoiding Nazi persecution in Europe. Walter Bick, a Dutch Jew from Amsterdam, had apprenticed in banking, and his father owned a textile factory. Recently married, and twenty-two years old, he came to Canada with his parents and several siblings. They bought a 106-acre mixed farm in Agincourt, near Toronto (Golden Horseshoe region), with a modest milk contract, and started growing cauliflowers and cabbages.[21] His mother had an egg route. They then taught themselves how to make pickles. Walter became the main exporter of pickles in Canada, where Bick's Pickles is still a household name. Another Dutch family from Amsterdam, the Redelmeiers, also arrived in the Toronto region just before the outbreak of the war. Francis Redelmeier who later became the first Dutch-Canadian member on the Milk Marketing Board (see chapter 9) was then sixteen years old. His father was a banker who wanted to farm, and they, too, bought a farm close to Toronto, in 1940.[22] Like Jan Verdun before him, Francis started studying at the Ontario Agricultural College in Guelph. During the summers he worked on a farm owned by a pre-war Dutch immigrant who

[18] Van der Mey and Mol, *The Dutch Touch*, 149-50. Information in this book was obtained from an interview with Jan's son, Bob Verdun. I also learned about Jan in a 1989 interview with Jan's younger brother, Arie Verduijn. See also Arie Verdijn's *Sojourners: A Family Chronicle* (Burlington: self published) 119, 122, 392.

[19] Jan Verdun, who died in 1996, became well known as the "Jug Milk King" because he designed a three-quart jug and opened Canada's first milk convenience store. He started breeding Charolais cattle and founded the Ontario Charolais Association. See Van der Mey and Mol, *The Dutch Touch*, 150.

[20] Unlike Jacob, Jan Verdun was not from a farm background, although his parents and grandparents were involved in farm-related businesses such as a flour mill and a seed business. See Verduijn, *Sojourners*, 109-115.

[21] Van der Mey and Mol, *The Dutch Touch*, 22-23.

[22] Interview (FS) with Francis Redelmeier, at his farm, 30 January, 1990.

had been in Canada for several decades. Francis later took over his father's farm and became the main advocate, and later president, of the Channel Islands Breeds Milk Producers.[23] However, most inhabitants of rural Ontario at that time would not have heard of the Netherlands, nor were they aware of the diverse, though tiny, presence of Dutch farmers in their midst.

These two cases illustrate how some prewar Dutch immigrants not only became farmers, but branched out into other business ventures. However, the vast majority of prewar Dutch farm immigrants, with few or no resources, were never able to buy farms, much less find sufficient work to maintain their families. During the Depression these families became dependent on relief work and welfare, with a real risk of being sent back to Europe. For such people, a settlement scheme in Holland Marsh, started prior to 1940, represented not only an opportunity for work but the possibility of becoming independent farmers. The success of Holland Marsh, which coincidentally had a Dutch-sounding name, brought Dutch farmers to the attention of a larger public, at least in the Toronto area.

The Holland Marsh

Holland Marsh, just south of Lake Simcoe, was once a swampy area considered unsuitable for agriculture. This marsh, and the nearby town of Holland Landing, were both named after Samuel Holland,[24] the first surveyor of British North America. Prior to 1930, the only way one could earn money in the marsh was by mowing marsh hay used for stuffing mattresses in the nearby town of Bradford. The idea of draining the marshes of the Holland River, then covered with brush, and converting them into fertile farmland, was conceived and promoted by William Day, a former lecturer in physics at the Ontario Agricultural College (OAC). With the support of the townships of West Gwillimbury, and the city council of Bradford (in the Lake Simcoe region), canals were built, and the work was completed in 1930. That year, Professor Day reported a $27,000-profit on his first 37-acre crop of vegetables.[25] However, the "highland" farmers on both sides of this low-lying area, accustomed to dry land agriculture, had no interest in muck farming.

---

[23] He is mentioned at least four times in Everett Biggs's book, *The Challenge of Achievement: The Milk Marketing Board's first 25 years of operation – 1965-1990* (Mississauga: Milk Marketing Board, 1990), 92, 167, 178, 181.

[24] Captain Samuel Holland was born in the Netherlands of British parents and his first wife was Dutch. However, most of his military career was performed in British colonial possessions.

[25] "Holland-Bradford Marsh," information bulletin from the Ontario Ministry of Agriculture of Food, Farm Products Inspection Branch, June, 1976, 2.

Several outsiders tried their hand at clearing sections of the Marsh, including Dutch immigrants. For example, three men from Hamilton, who had learned about Holland Marsh when they were passing through Bradford, jointly purchased 175 acres of uncultivated land at the west end from a highland farmer. One of them, whose father owned a transportation business in Holland, had worked for a farmer after moving to Canada, but his main job had been as a night clerk in a Vancouver hotel. He and the second partner, originally from the southern part of Limburg, started clearing their land in 1932. The third partner, a farmer who continued to rent land in Aldershot, near Hamilton, hired someone else to look after his section of land in the Marsh. His partners, who lived in the highlands on its edge, were among the few who continued to farm there for at least another decade. A dozen other farmers who bought or rented land in the thirties, quit after one or two years.[26]

A noticeable Dutch presence Holland Marsh did not become apparent until after 1934, because of the efforts of Mr. John Snor, an antique furniture dealer living in Hamilton and an official representative of the Netherlands Emigration Foundation (*Stichting Landverhuizing Nederland*).[27] Mr. Snor, a former bulb grower from Holland, was already acquainted with the Marsh, and knew at least one Dutchman who had been a pioneer there in 1928.[28] Snor persuaded the Dutch foundation to send an agent to the Netherlands to recruit immigrants with enough capital to start farming in the Marsh, but they were unsuccessful.[29] In addition to economic difficulties connected to the Depression, they encountered legal difficulties associated with land disputes among the highland farmers, a syndicate set up by Professor Day, and the municipalities that had contributed to the cost of drainage. Snor was discharged as a representative of the Netherlands Emigation Foundation in 1934. Unwilling to give up, Snor approached several businessmen in Hamilton, and together they promoted the idea of making the Holland Marsh a practical business undertaking. They envisioned a settlement of Canadian farmers and Dutch immigrants already living in Ontario, who would clear and cultivate the newly drained land.[30] Snor bought 125 acres of land in the Marsh

---

[26] Albert Van der Mey, *And the Swamp Flourished: The Bittersweet Story of Holland Marsh* (Surrey, B.C.: Vanderheide Publishing, 1994), 9-12.

[27] See article on Holland Marsh in *The Spectator*, 25 July, 1970, p. 29.

[28] That other "old-timer" had managed to clear and make a living on a plot of land rented from a lumber dealer in Bradford. In an article in *Maclean's* (15 September, 1953, p. 82) Marjorie Wilkins Campbell refers to him as Peter Rolls. A Mr. Sytema, who started working in the Marsh in 1932, refers to this man (who had three sons working with him) as Rol in an interview with Albert VanderMey (personal communication).

[29] Ganzevoort, *A Bittersweet Land*, 53-54.

[30] See *Bradford, 1857-1957*, compiled and arranged by Ina and Stewart McKenzie (editors and publishers of the *Bradford Witness*), 1957, 49-63; and first of a three part

from the Canada Land Company, and paid to have some of the land cleared. However, he was only able to sell land to six families. Close to bankruptcy, Snor negotiated a joint venture with the federal and provincial governments, and the government of the Netherlands, to establish a settlement of people on relief on the land he owned in the Marsh. The Dutch government retained Snor as its representative to oversee the project for ninety dollars a month.

The settlement project, partly financed through a Relief Land Settlement Agreement passed in 1932, provided $600 a month to each settler for the cost of building a house, buying supplies to start cultivating the land, and living costs for twenty months. Once they started producing crops, the settlers would be able to pay back the $500 cost of their land, including a house plot, in easy instalments. Snor signed up seventeen families.[31] Initially the men commuted from their homes to clear their allocated five acres, with plans to build houses in the Marsh as soon as possible. This project marked the beginning of a process that resulted in Ontario's most prosperous market-gardening region.

Ansnorsveldt and Springdale

The Dutch-Canadian settlers who started clearing land in the Marsh in 1931 came mainly from Hamilton. Most were unemployed and dependent on welfare. Nonetheless, some had had business experience; Jan Rupke, who operated a hatchery in Burlington, provided the chicken coop that served as temporary accommodations for the men who commuted to the Marsh. In 1934, seventeen families, fifteen of them Dutch-Canadian, one English, and one German, moved into small wooden houses built under the supervision of a Dutch carpenter. That carpenter married one of the Rupke girls a few years later. The new settlement, named Ansnorveldt after Mr. Snor, consisted of a single row of houses at the eastern end of the Marsh. They pooled their resources to build a wooden church in 1935. These pioneers, who performed hard physical labor, endured harsh winters, long treks to nearby Bradford for supplies, and summer dust storms. No one owned a car, but they occasionally caught rides from someone, who, independently, had bought ten acres of land just north of the settlement. That person, a Dutchman who had operated a delivery route in Hamilton, brought over his truck as soon as a dirt road was built. He slept in one half of a double garage,

---

article by Lia Wolters, "Holland Marsh verdient bijnaam 'saladekom van Ontario,'" in *Hollandia News*, 19, 1976, 2.

[31] See Van der Mey, *And the Swamp Flourished*, 16.

until his wife joined him once he built a second story.[32] In emergencies, Snor, who lived in Bradford, also provided rides. Everyone helped each other and members of the new settlement developed a strong sense of unity. However, relations between the settlers and Snor, and between Snor and the Dutch government, deteriorated. His position was terminated in April, 1939.[33]

Between the time of their arrival and the start of the Second World War, these settlers grew potatoes, carrots, and lettuce, but township taxes and land payments were always in arrears. They had to plead with various levels of government to supply them with adequate roads and better pumps. Another problem was lack of adequate marketing and storage facilities for their perishable produce. To address these issues, the ratepayers of the Marsh set up a drainage commission, which later became the Commission of Marsh Land Growers.[34] The Dutch settlers also took the lead in setting up the first producers' co-operative, the Dutch Growers' Organization, headed by John Rupke. Although all of these farmers started off with equal shares, some were ultimately more successful. Several members of the Rupke family were able to buy additional land. Others, such as two brothers from Chatham who arrived later, began with greater resources and had no dealings with Snor. The Verkaiks bought privately owned land, and initially spent only their summers in the Marsh. They then established a permanent residence in the western half of the Marsh and put up storage facilities. In contrast with the settlers in Ansnorsveldt, who had had five acres each, the Verkaiks soon owned several hundred acres of fertile land. This family later branched off into real estate and construction in nearby Bradford. One of the Rupkes, who moved to this part of the Marsh, became a vegetable storage owner.[35]

In 1946, the year highway 400 cut across the far west side of the Marsh, the Verkaiks and several other families adopted the name Springdale to set themselves apart from Ansnorsveldt. In 1950 these families built their own church, and later a parsonage that became part of a cluster of houses and barns that eventually included a small store.

---

[32] Van der Mey, *And the Swamp Flourished*, 27-28.

[33] Ganzevoort, *A Bittersweet Land*, 55.

[34] See "The Holland Marsh Story," in *Trade and Transportation*, Toronto, 1949, 25-63.

[35] Non-Dutch owners of agricultural-related businesses in the Marsh lived in Bradford or on Highway 11. All these private entrepreneurs competed with the Bradford Co-operative Storage Limited, set up in 1945.

Religious and cultural diversity in Holland Marsh

Most of the early settlers in the Marsh were Dutch Calvinists who belonged to the Christian Reformed Church. Over time, Dutch Calvinists living just beyond the geographical boundaries of the Marsh also joined the congregations established in Ansnorsveldt and Springdale. Each had societies for young people and other associations. In 1942, they established their own separate Christian school, at a time when a shift from Dutch to English was well under way. However, one of the original Marsh settlers, William Valentyen, was a Dutch Roman Catholic.[36] He gravitated towards people who were not Reformed, including an increasing number of immigrants from a variety of backgrounds moving into the area: Polish, Ukrainian, and German-speaking Yugoslavian. Many were Catholics, like Valentyen. Some of these newcomers, who also settled in the Marsh just before and during the War, came to work for the Dutch farmers. Others were themselves immigrant farmers who came to clear other sections of the Marsh. Valentyen became the main proponent for building a chapel where they could worship, but this project did not come to fruition until after the war.

The hamlet of Ansnorsveldt became more ethnically homogeneous when the only immigrant from England (a war veteran) and a German tailor, who had joined the original settlers, moved away. While the children had freely mingled, the parents did not have much in common with their Dutch Calvinist neighbors. Thus, while Ansnorsveldt and Springdale became completely Dutch, the Marsh as a whole, as well as the surrounding highland area, became ethnically more diverse. However, social ties and religious bonds among Catholic farm immigrants of different nationalities are not the only example of intra-ethnic collaboration. The executive board of one of the growers' associations, the Holland Marsh Cooperative, encompassed both Dutch and non-Dutch farmers. With an increasing demand for labor, starting in 1940, growers of diverse ethnic backgrounds and religious affiliations co-operated to establish a farm work service camp in Ansnorsveldt to house seasonal workers during the summer months. An example of still closer collaboration, involving even kinship ties across ethnic lines, was a business partnership created by one of the leading Dutch Calvinist farm families.

[36] Members of another Catholic family, the Brouwers, came on their own. Originally from Zeeland, they moved to the Marsh in 1936, where they assembled a house from two construction shacks, which had been used by workers building a bridge across Highway 11, which touches the southern portion of the Marsh.

Holland River Gardens

The largest privately owned Dutch-Canadian enterprise that emerged in the Holland Marsh was a partnership of three brothers, the Horlings. Not only did this extended family manage to buy up a considerable amount of land, one of them, George Horlings, started buying wholesale, vegetables for packing and shipping. Together with several Canadian (non-Dutch) partners, they established Holland River Gardens Ltd, which at its peak incorporated more than 300 acres of marsh gardens. The non-Dutch partners included a former highland farmer, who had been reeve of West Gwillimbury, and two former Dominion grocery store employees, with extensive management experience in sales and cold storage. One of the latter, who became the plant superintendent, was a brother-in-law to the Horlings. By 1950, their packing plant supplied ice-packed fresh celery and lettuce to the growing population of Toronto and also (via rail-links) to markets as far away as Halifax, and Vancouver and cities in the U.S.

The Horlings brothers, like the Verkaiks, arrived in the Marsh from Chatham. In 1924, two brothers had emigrated together from Nieuwe Pekela (in Groningen), where they were farmers. They had been recruited by the Dominion Sugar Company to work in the sugar beet industry. A year later they rented a 30-acre farm close to the town of Chatham. In the meantime, two other brothers, three sisters, and their parents also joined them in Canada. In 1926, they rented a 200-acre farm, but a bad harvest forced them to abandon it. Opportunities for making money deteriorated during the Depression, but one of the Horlings, who had bought his own truck, continued to haul sugar beets for the sugar company on commission. When he learned about the new settlement scheme in the Holland Marsh, Walter Horlings bought fifteen acres with $300 down and moved there on his own in 1935, using his chicken coop for temporary accommodation, also on the west side. He borrowed twenty-five dollars from Frank Flach, another Dutch immigrant, who married one of the Horlings sisters, enabling him to secure additional credit from his bank in Chatham to continue farming. Brothers George and Harry soon arrived with additional money for discounted land acquired by paying the back taxes of bankrupt plot holders.[37] The brothers lived very modestly since they were all still single. In 1936, a contract with a local branch of the Dominion grocery store chain gave them access to further credit, and business picked up during the war years. By 1942, they

[37] Based on the rough notes of an interview with Walter Horlings, 26 October 26, 1981 and two interviews with Frank Flach, on Sept. 29, 1989 and on October 26, 1981. I was given these notes by Ger Graaskamp.

owned ninety acres and, in 1945, they organized and incorporated the Holland River Gardens.

The success of the business enterprises run by the Horlings, the Verkaiks, and the Rupkes illustrates the importance of strong family ties among Dutch immigrants who had a previous exposure to farming. The ability to recruit family labor, combined with thrift and determination, gave these immigrant farmers a competitive edge. These cultural traits enabled Dutch-Canadian farmers in the Ansnorsveldt settlement, and those who entered the Marsh on their own initiative, to get established and stay in the risky business of farming. Prior business experience helped some of these pioneers to expand their operations beyond the family-farm scale, although exceptional growth depended on close collaboration with siblings and in-laws. We shall see later that the postwar Dutch immigrants who succeeded had the added advantage of more education and an even more favorable economic climate.

The name Holland Marsh continues today to be closely associated with Dutch farmers, in the minds of the general population. However, the vast majority of the large cohort of postwar Dutch immigrants settled in other parts of rural Ontario. Their arrival marks the beginning of an era when Dutch farmers were to have an even greater impact on the development of agriculture and the social landscape of rural Ontario. From then on, the Dutch became noticeable in almost all the major agricultural regions of Ontario.

PART II – DUTCH FARM IMMIGRANTS AFTER WORLD WAR II

IV. The Netherlands-Canada Settlement Scheme

Prior to the outbreak of the Second World War, Canadian immigration policy had been quite restrictive.[1] At the end of the war, politicians were worried that more immigration would jeopardize the economic security of returning soldiers. There was no objection, however, to the entry of farm workers, who were given the same preference they had enjoyed before 1940. The Dutch government at that time was anxious to promote the emigration of rural people. In 1946, the Dutch government sent A.S. Tuinman to Ottawa as its agricultural attaché and to negotiate the admission of Dutch immigrants. The outcome of these negotiations was the Netherlands-Canada Settlement Scheme, which allowed Dutch farm workers to come to Canada providing sponsors make available a guaranteed job and lodgings for at least one year. Initially the Canadian government anticipated no more than 500 single men per year. Yet by the end of 1947, close to 3,000 Dutch immigrants, including whole families, had landed in Canada.[2] From 1949, they became highly visible, arriving at train stations throughout Southern Ontario.

The agreement established the respective government departments responsible for the movement of Dutch farm people to Canada. However, the day-to-day business, including searching for sponsors, was handled by various immigration organizations. These organizations, which had Dutch counterparts, were organized along the same denominational lines as the Dutch pillar system (see chapter 1). The Christian Reformed Church, which had its headquarters in Michigan, was the first to set up a Canadian Immigration Committee, in 1947. It functioned through a network of home missionaries, who had helped settle prewar immigrants.

---

[1] Herman Ganzevoort, Sharks in Wooden Shoes, in *Dutch Immigration to North America*, H. Ganzevoort and Mark Boekelman (Toronto: Multicultural History Society of Ontario, 1983), 148. A restrictive immigration policy was the outcome of conflicting views, resulting in a compromise solution. Industrialists and English Canada were in favor of increasing immigration, while French Canada and labour unions were opposed. See William Petersen, *Planned Migration: The Social Determinants of the Dutch-Canadian Movement*, chapter 6 (Berkeley, University of California Press, 1955).

[2] Herman Ganzevoort, *A Bittersweet Land* (Toronto: McClelland and Stewart, 1988), 68.

Dutch Catholic immigrants were initially channeled through the Catholic Immigration Aid Society, established in 1950. The smaller Immigration Committee of the Reformed Church of America was also started that same year.[3] All of these associations worked closely with the Dutch embassy in Ottawa, which took care of citizens who had other or no religious affiliations.

By the late forties, when arrivals already numbered in the thousands, the annual influx of people from the Netherlands grew to tens of thousands, peaking at over 20,000 in 1952 and 1953.[4] They landed in all parts of Canada, although more than half came to Southern Ontario or soon ended up there.[5] Rural schools were swamped with Dutch-speaking students, often outnumbering their non-Dutch counterparts. After 1951, the Canadian government opened the doors to a broader range of immigrants, including craftsmen and urban workers. Consequently, the proportion of Dutch immigrants working in agriculture declined from almost 100 percent to just under 20 percent.[6] By the mid-fifties a variety of occupations were represented, and most Dutch immigrants no longer started off on a farm. The flow of Dutch immigrants declined during the rest of the decade, peaking again at close to 12,000 in 1957, and diminishing rapidly throughout the 1960s.

FROM THE NETHERLANDS TO ONTARIO

The paperwork required to emigrate was relatively simple. However, the logistics of shipping thousands of Dutch families to numerous destinations in Ontario presented frequent challenges. During the first couple of years, ships that had been used to carry troops to the former Dutch East Indies (Indonesia) were refitted to transport immigrant families, with separate quarters for men and women. In Canada, immigration organizations hired full-time fieldmen who assisted newcomers after they disembarked in Halifax. Each Dutch immigrant family left the Netherlands with a letter indicating a final destination, including which farmer they would work for. However, people sometimes discovered they no longer had a sponsor, which necessitated a last-minute replacement. Volunteers

---

[3] Ganzevoort, *A Bittersweet Land*, 70-71.

[4] See chart shown in Albert Van der Mey, *To All Our Children*, Jordan Station, Paideia Press (1983), 52-53.

[5] See Anthony Sas, "Dutch Immigration to and Settlement in Canada: 1945-1955." PhD diss., Department of Geography, Clark University, MA, 1957, 63-68.

[6] The percentage of Dutch immigrants who found jobs in industry rose from a low of 4.8 in 1948 to a high of 32.8 in 1954, then tapered off to 25.8 in 1955. This pattern of proportional increase holds for the two other major occupational categories – construction and services. See Sas, "Dutch Migration to and Settlement in Canada," fig. 9, 52.

did much of this groundwork, but there were occasional breakdowns in communication. Some immigrants arrived after a long train journey from Halifax or Montreal to find no one waiting for them, or to discover that their luggage had ended up in the wrong place. There were misunderstandings between the three immigration societies until a single a co-ordinating body, the Canadian Netherlands Immigration Council (CNIC) was set up in 1955. Not all Dutch immigrants knew where to go when they stepped off the train, and the fieldmen, Dutch immigrants themselves, picked up stranded newcomers and provided temporary lodgings.[7]

Treatment of the newcomers varied widely, depending on where they landed and for whom they worked. The agreement stipulated that Canadian farmers would provide room and board, plus a monthly salary in return for the labor of the Dutch immigrant for at least one year. However, the immigrants usually did not find out about salary, living conditions, and conditions of employment until they met their employers. Indeed, the contract was subject to considerable variation, depending on the nature of the farm work, the solvency of the farmer, and the size of the Dutch family. Although they preferred to hire single men, Canadian farmers soon recognized the advantages of sponsoring big families, especially those with grown children, able to help out. However, whether or not husband and wife, or even older children were all expected to work was not always clear, and subject to different expectations. The language barrier and different cultural values created misunderstandings, especially if sponsor and farm worker were poorly matched. The experiences of both Canadian employers and Dutch immigrants varied greatly, depending on their respective personalities and economic backgrounds.

THE SPONSORS

Some farm sponsors had prior experience with foreign workers, although they had never had to share their homestead with whole immigrant families. Starting in 1949, and for about a decade, Ontario farmers became increasingly dependent on Dutch families. A farm in Bronte, just south of Milton (west of Toronto) can be used to illustrate the rapid transition from non-Dutch to predominantly Dutch immigrant farm workers between 1945 and 1950. Its owner, George Atkins, a man who later founded the Developing Countries Radio Network, represents one type of farm sponsor

---

[7] This was a common theme in many of my interviews. See also Van der Mey, *To All Our Children*, 129-152; Margaret Vanderschot and Mary Smith, "Netherlands Research Project: A Study of Postwar Immigration from the Netherlands to the St. Marys area", A History of Ontario's People project report for St. Marys Museum (1991), 1-2.

with whom many Dutch immigrants came into contact: a well-educated, progressive farmer, who also had sufficient capital to run a commercially viable operation.

Woodland Orchards

George grew up on a farm called Woodland Orchards Ltd. He was four years old when his parents bought a 330-acre farm that included forty acres of orchards.[8] The property included three extra houses for the hired hands, all bachelors. George, who studied animal husbandry at the Ontario Agricultural College, took over the management of the farm after he graduated in 1940. When George married, he moved into one of the houses. He started building some of his own equipment when he saw the need for further mechanization, but he was still short of workers. As soon as the war ended, George used his connections to contact farm worker placement officers. The first available farmhands in 1945 were Polish war veterans, who had fought for the allies. For two years, an unmarried man worked on George's farm. Their next Polish worker, an army officer during the war, boarded in George's house, followed by a Hungarian war refugee who only stayed a short while.

Between 1950 and 1955, George Atkins had four Dutch families working for him.[9] The first Dutch family to arrive were the Van den Heuvels: a middle-aged couple with six children, including three boys old enough to do farm work. They had lived in a small farm in Veggel, in Noord Brabant, where the father had worked in a needle factory. Their family was too poor to own a horse, but they all knew how to milk cows.[10] This couple paid their own passage, leaving them with little money to buy items to ship to Canada.[11] Once in Canada, the father and his oldest sons performed a variety of chores. George remembered that they were hard workers. They, in turn recall that George was a demanding, but fair, employer who was always busy doing something. The next big

[8] There were ten acres of Northern Spies and one year the family exported fourteen railroad cars full of these luxury apples destined for export to Great Britain. Interview (FS) with George Atkins, in his farm in Bronte, 29 April, 1997.

[9] Interview (FS) with George Atkins, in his farm in Bronte, 29 April, 1997.

[10] I spoke to these two older sons, now living in Milton, by telephone a few days after I had interviewed George Atkins. They both ended up as urban workers. George had fond memories of this family and had kept in touch, but he did not know anything about their background or whether they had become farmers in Canada.

[11] George Atkins had written them prior to their departure, asking the father to buy a methylene explosion lamp for scaring off birds in his orchards, for which he would be reimbursed. George knew that a particular model produced in Amsterdam was more efficient than an American one he owned. George showed me that lamp, which he had kept, during my interview.

farm family George sponsored were the Vanderzanden. They came from the same Dutch province, from the village of Mariahout. The father, who was forty-nine, was an experienced, full-time farmer.[12] They moved into the house vacated by the Van den Heuvels. The oldest daughter worked in the Atkins household, helping both George's wife and mother, who also lived on the homestead. Another son (age sixteen) went out to work for a nursery in Oakville. A year later, this family moved to Stoney Creek, where several of them got jobs in town. George Atkins later took on two more Dutch families, who had already been in Canada for several years.

George preferred families over single men because they didn't need board, and he also got more helpers. He was amenable to a child working elsewhere, as long as at least one son was available to help on his farm when necessary. Only one sponsored man got "under his skin" at times for being too opinionated and insubordinate. Nevertheless, George considered that this third Dutchman, who later successfully managed a farm for one of his neighbors, to be quite capable. When George later accepted a full-time job as a radio commentator for the CBC in 1955, he did not hesitate to rent out his farm to yet another Dutch immigrant who cultivated green beans.

George Atkins was better off than most Canadian farmers who sponsored immigrant farm workers, yet he only had one family working for him at a time. In contrast, farmers who operated larger, or more labor-intensive, operations would have several Dutch families working for them at a time. A good example is Lawrence Kerr of Chatham.

Kerr's Farms in Chatham (Essex/Kent Region)

Lawrence Kerr's ancestry is Protestant Irish and Highland Scots through his father and mother respectively. Lawrence graduated in 1929 and worked for five years as government agriculture representative before returning to the home farm.[13] Still single at the age of twenty-eight, he bought and moved onto his own farm of 145 acres near the city of Chatham. He gradually expanded his farm holdings and, in 1944, married a high school assistant principal in Blenheim. That same year, Lawrence bought his third farm. By the time he took on his first Dutch families, he had diversified cash crops and beef cattle operations. He also grew vegetables on consignment for processors such as Libby's and

[12] After selling their Dutch farm, they had money left over and were able to smuggle in enough money to buy a used car soon after their arrival. I did extensive interviews with several members of this family, the Vanderzanden (including Freda Leenders, nee Vanderzanden), most of whom became full-time farmers.

[13] Interview (FS) with Lawrence Kerr (then in his nineties), at his home in Chatham, 25 June, 1998.

Cairns. Like many other farmers in Ontario, Lawrence first employed a young Polish man who had served in the English army. That man boarded with the Kerrs for six months. Between 1946 and 1951, Kerr's farms had workers of various nationalities, including a group of Japanese-Canadian women, but he did not sponsor any farm immigrants from the Netherlands. Then, in the fifties, he invited several Dutch families, who had already been in Canada for one or more years, to live on his farm.

The first Dutch farmer who came to work for the Kerrs had learned English in previous jobs. He knew how to drive a tractor and was a good mechanic, and knew how to plant corn. The Dutch man's son, then attending high school, also worked on the farm for a while. His wife babysat on numerous occasions when Mrs. Kerr entertained guests. This Dutch family, who lived in a spacious, remodelled brick farmhouse, stayed for four years before moving to Thamesville to start their own farm. In 1952, a new family from the Netherlands moved into another house owned by the Kerrs. The Twynstra family, originally from the Dutch province of Friesland, had already worked for a Canadian sponsor in Wallaceburg for a year, when Lawrence Kerr took them on. Mrs. Twynstra especially liked their living arrangements. Compared to the damp shack where they lived during the first year, she was amazed to find central heating and a modern bathroom in their new home.[14] They had three sons, the oldest of whom had attended the *lagere landbouwschool.* He too became one of Kerr's employees and handed over his earnings to his parents. The second oldest, then twelve years old, continued his schooling in Canada. But he also spent a lot of time on the farm. Henry Twynstra, the oldest son, recalls that during the busiest time of the year, there were twelve Dutchmen working for Lawrence Kerr, although the foreman was an Englishman. Henry, who spoke Frisian at home, had already picked up some English in Wallaceburg, but improved his Dutch through interaction with the other workers at Kerr's farm.[15]

Lawrence knew that immigrants from the Netherlands were unlikely to stay, since they wanted to become independent farmers. Yet, he was willing to help out his Dutch workers and even drove them around to look at farms to buy. The men and boys who lived on his farm also learned a lot about beef farming and general farm management. For example, Peter Twynstra, Henry's younger brother, learned about tractors on the Kerr farm. He recalls how one day Lawrence invited him to

---

[14] Interview (FS) with Peter Twynstra at his home in Southcott Pines (Grand Bend), 18 July, 1997.

[15] Interview (FS) with Henry and Shirley (Sjoukje) Twynstra, at his home in Nairn, 13 June, 1997.

drive a TD6 Crawler tractor, which pulled a four-furrow plough, home along the road.

Elliott's Place (Lake Simcoe region)

Earl K. Elliott, who also sponsored a Frisian-Dutch family, typifies a young Canadian who became a successful farmer without formal training in agriculture. Earl served in the navy during the Second World War, where he received most of his technical training. When he returned to Canada at the age of twenty-six, a year before the end of the war, he worked for two years in the mines of Northern Ontario, where he earned enough money to buy a farm in Elmvale, northwest of Barrie, in 1948. At that time, most farms in the area were 100-acre beef operations unable to compete with Western cattlemen. Most of their owners either sold their farms or rented out their land. Earl took the risk of setting up a large scale enterprise, and within a few years, he bought several more 100-acre farms for a combined milk and hog operation. In 1952 he bought a small dairy in Elmvale.[16] When Earl started farming, he had only one hired hand, a French-Canadian, but around 1950, he realized that he was going to need more help. Through a letter back home from one of a neighbor's Dutch workers, Sijmen Kloosterman, then a 21-year-old single man working as a farmhand in Garijp (Friesland), first heard about his future employer in Canada. In the spring of 1951, Sijmon boarded a propeller plane and flew to Montreal, via Scotland and Gander, Newfoundland.

When Sijmon stepped off the train station in Toronto, the hometown friend who worked for Earl's neighbor was the first to greet him. He then met Earl, whom Sijmon found to be younger than he imagined.[17] Sijmon boarded with Earl's household and was put in charge of milking twenty-six cows. He also helped the French worker in the fields. Unlike the majority of Dutch farm immigrants, Sijmen continued to work for Earl Elliott for more than a decade before buying his own farm. His experience illustrates the incentives Canadian farmers used to keep experienced and loyal farm workers. When Sijmen married Yt, his hometown sweetheart in 1954, Earl let the young couple live in a house

[16] The history of his farm career, and his farm, was described in a report written by a Dutch agricultural student who worked on his farm in 1961. "Verslag van de praktijktijd, doorgebracht on het bedrijf van Earl Elliott, Elmvale, Ontario, Canada, gedurende de maanden juli en augustus, 1961," by Abraham Kloosterman (University of Wageningen).

[17] The story of this man and his family was written down in an unpublished manuscript of over one hundred pages, which he wrote after he retired. See *Memoires, opgetekend door Sijmen Paulus Kloosterman* (circa 2000).

in 'Jameson's place', a hundred acre lot that formed part of his 450-acre farm. As Earl got busier in his milk delivery business, he gave Sijmen more responsibilities and put him in charge of a third hired hand. Sijmen started receiving a third of the profits from a dairy operation that had grown to eighty cows. His boss also helped him out when Sijmen decided to buy his own farm.[18]

Earl Elliott represents an ambitious modern Canadian farmer whose operations were later emulated by Dutch-Canadian farmers. The Dutch people he sponsored liked and admired their boss. In turn, Earl, like so many other Canadian farmers, appreciated their entrepreneurial spirit and willingness to work hard, essential to the success of his own farm business. However, not all Dutch farm immigrants had such a favorable impression of their sponsors, as we shall see later.

Other Sponsors/Employers

People with a wide range of occupations, backgrounds, and needs sponsored or employed postwar Dutch immigrants. Of the Dutch immigrants I interviewed, one reported working for the owner of a feed mill in Milton and another for the owner of a lumberyard in Cambridge. Physicians, veterinarians, and lawyers who owned farms also sponsored Dutch immigrants or took in Dutch families, who had already been sponsored elsewhere. These businessmen and professionals turned over a lot of responsibility to Dutch immigrant farmers once they realized they were not only hard workers, but had also received agricultural training.

Not all Canadians who needed Dutch workers were in the prime of their lives, as in the case of an elderly couple that operated a farm in the Lindsay area. While not impoverished, they both suffered from arthritis and could no longer look after their farm by themselves. Upon the recommendation of a neighbor, they took on a young Dutch couple that had just had their first baby. The elderly people moved into the back kitchen and let the Dutch couple occupy the rest of their house, which had hydro and water. The elderly couple paid $125 a month for the Dutch immigrants to look after their farm.[19] Another kind of sponsor was the prewar Dutch immigrant farmer. One such farmer, who was completely integrated into Canadian life with no connections to a Reformed church, or Dutch friends, nevertheless sponsored several Dutch farm immigrants after the war. His contact with other Dutch-Canadians then increased significantly.

---

[18] Sijmen, whose wife encouraged him to become an independent farmer, said, "It is better to be a small boss than a big hired hand (Beter om kleine baas te wezen dan grote knecht)." Cited in Sijmen Kloosterman, *Memoires*, 149.

[19] Interview with Martin and Wilhelmina Verkuyl, 23 November, 2000.

Leonard Mol, whom we met in the last chapter, assumed a dual role as fieldman and employer. When postwar immigrants started arriving in the Chatham area in 1947, he and his son employed four young Dutchmen. Their farm became a favorite gathering place, they drove people to church on Sundays, and Leonard accompanied other members of his congregation on expeditions to other regions looking for suitable land for newcomers to settle.[20] Farmers in the Holland Marsh, both big and small operators, also sponsored Dutch relatives and strangers alike. Whether they regarded them as convenient sources of labor or fellow countrymen requiring assistance varied from person to person.

A farmer who sponsored a Dutch family was expected to provide housing and at least one full-time job for a year. However, many Canadian farmers could not afford to keep their Dutch farm workers on the payroll during the winter. Nor were they always able to provide them with a separate house. In some cases, Dutch families occupied entire floors of their homes, which could become quite cramped. The lines between employer and employee often blurred, especially with Dutch-Canadian farm sponsors, who themselves had earlier immigrated to Canada. Everyone has his or her own story of coming to rural Ontario as a sponsored farm hand and what it was like to live and work on a Canadian farm. While subject to the same legal agreement, men's experiences differed from women's, and we cannot ignore the memories of children, especially those six to twelve years old, whose first contact with rural Canada was the one-room schoolhouse.

SALARIES AND WORKING CONDITIONS

People who came in the late forties received from forty to eighty-five dollars a month, plus a house to live in. The figure commonly quoted for a couple with school-age children was sixty-five dollars a month. These wages increased in the fifties, and older children or a spouse were sometimes paid extra. In 1951, when George Atkins of Bronte sponsored his second Dutch family, he paid seventy five dollars a month to the father and sixty-five dollars a month to his oldest son, then seventeen. A daughter who helped in the household received another sixteen dollars. When the head of the household was also the farm manager, the wages were higher; one man started with $120 a month in the mid-fifties and this amount was raised to $200 two years later.[21] A Dutch immigrant might also earn additional money during periods of peak labor demand. A laborer in the Lake Erie tobacco region in 1955 received

[20] See story and pictures in Van der Mey, *To All Our Children*, 346-47.

[21] All of this information was extracted from my fieldnotes, which I typed up after each interview.

a bonus of fifteen dollars a kilo during the harvest, garnering him an extra $900 during one season.[22] In the Niagara Peninsula the going rate for farm work in the early fifties was from fifty-five to sixty cents an hour. Some farmers permitted members of a Dutch family to work for a neighbor or customer on days when they were not needed. Whether or not they were paid extra depended on the sponsor. One man in the township of Garafraxa, in Dufferin County (Northwest region), never knew what he was to expect when he was "borrowed". If his boss did not like the other farmer he would say "charge him a dollar an hour, since he is not a good customer."

In addition to wages and accommodations, farmers often provided benefits in kind. An employer might give his Dutch farmhand a calf or allow him to graze sheep on his land. Many farmers gave permission to the families they sponsored to take the eggs or milk they needed, or to pick tomatoes for their own use. Even with such perks, they still expected their farmhands to work every day, with little time off. Dutch women of all ages became housekeepers as part of the sponsorship agreement; bachelor farmers expected wives and daughters to cook and clean on top of their other farm tasks. Canadian farmwomen might ask Dutch women to undertake tasks beyond officially described duties. Farmhands worked from 6:00 A.M. until 6:00 P.M., although in one case, work started at 7:00 A.M. and finished at 5:00 P.M. In contrast, a dairy farmer in Woodstock made his helpers work from 5:00 A.M. until 7 P.M., including Sundays, when they were allowed to break for only part of the morning or early afternoon to attend church. One of his workers considered himself lucky when he was given every second Sunday off at his next job.[23] Apart from back-breaking work, some immigrants faced such health hazards as exposure to toxic chemicals while spraying orchards.

Other sources of income

Dutch immigrants had other ways to earn money. In regions with cash crops, whole families harvested sugar beets, tobacco leaves or potatoes for their sponsors and other people. A farmer who required seasonal workers would send a Dutch farmhand to recruit members of other immigrant families. Married women with older children to look after the younger ones did housekeeping and babysitting for Canadian neighbors. Dutch women of all ages and backgrounds worked in the fields. One young, recently married, woman whose husband worked for a farmer

---

[22] Interview (MF) with Theo Dykstra, 24 November, 1993.

[23] Phone interview (FS) with a Dutch-Canadian farmer who now lives in Innerkip, 8 January, 2000.

in the Niagara region picked fruit even though she was not from a farm background. Another Dutchman acquainted with her husband suggested she might not be able to work as hard as the other women. In fact, she picked three times as fast! In other parts of Southwestern Ontario, Dutch teenagers also worked in the sugar beet fields. In 1952, the going rate on one farm in Kent County was fourteen dollars an acre for blocking beets, three dollars for the second hoeing and fifteen dollars an acre for harvesting.[24] In some cases, Canadian sponsors not only allowed the older children of their sponsored Dutch families to work off the farm, but even encouraged their farmhands to take jobs in the city during the winter. Peter Ellens worked for several years for a fruit farmer in the Niagara Peninsula, pruning, picking, and spraying, and eventually becoming the farm manager. The first year was a bad year financially for fruit farmers, so Peter's boss suggested he apply for a factory job in nearby St. Catharines during the winter. He told Peter he would be unable to continue paying his salary, but allowed the family to continue living rent-free.[25]

Helping a sponsored Dutch worker find a job in the winter not only saved sponsors money, it encouraged them to stay on. However, it did not take long for Dutch immigrants to find out that they could make more money off the farm, in canneries, factories, or construction. One man, who earned sixty-five dollars a month on a farm in Lambton County (London region) in 1949, soon started bringing home $110 every two weeks from a factory job in Sarnia. Another man, who had been a carpenter in the Netherlands, went from a forty-eight-dollars-a-month salary on a farm, to an hourly wage of $1.35, and then $2.30, framing houses.[26] This more than tripled his income. Another interviewee went from 33 cents an hour for picking fruit to a dollar an hour as a welder. Cutting wood was also a way for men to earn extra money in the winter, although that often involved "living in the bush" for months at a time. Dutch men sometimes also traveled to Northern Ontario to work in the mining industry, in places like Sudbury.

Dutch farm immigrants, especially those who had had their own farms, were initially hesitant to work in cities or mines because of the stigma attached to any form of factory work. Even poor part-time farmers in the Netherlands looked down on factory workers and miners, because of strong status distinctions there (see chapter 1). However, they quickly overcame their reservations when they realized such distinctions were

[24] Interview (MF) with Jake Oosterhoff, 13 October, 1993.

[25] Interview (MF) with Peter and Ann Ellens, in their home in St. Catharines, 30 November, 1993.

[26] Interview (FS) with Gerry Van Bussell, in Lucan, 9 July, 1990.

meaningless in Canada, and that their earning potential was greater outside of the agricultural sector. Many farm immigrants and their children, continued working in nearby factories even after buying a farm. In the mid-fifties, half of the approximately one hundred people working for Watford Wireworks, located in Middlesex County (London region) were Dutch immigrants.[27] Some were married men who worked on their own farms in the evening, but most were young men still living at home. Their wages helped sustain the household while the father ran the farm. However, despite the good income, they disliked the monotony and lack of fresh air, not to mention occupational hazards. One Dutch immigrant, who headed north to work in the Copper Cliffe mines, experienced a nose bleed on the first day, from sulphur fumes. He also witnessed the deaths of three men who were badly burned when molten metal spilled on them. This Dutch immigrant, who developed an ulcer and lost thirty pounds, did not last very long.

Single men were more adventurous in finding ways to earn money. One farmhand from a dairy farm in Zuid Holland started working for a diary farmer close to London. He was an experienced cattleman, but had few other skills. His sponsor, who served on the township council, was rarely home, so he looked after the farm. In his second year, the Dutchman started venturing out with a friend. During the winter of 1954, they worked in a feed mill in Chatham, hoed sweet corn in the spring, then worked in the tobacco fields, and finished by picking cherries and apples at the end of the summer. In between, they lodged on a farm where his friend's parents were working. They helped with groceries but could not afford to pay much rent. After a year of "chumming" around, this man found a steady job with another dairy farmer. He stayed there for another year and a half and met his future wife. She coaxed him to leave the dairy farm to work for her father, who was a tobacco sharecropper near near Brantford. This pattern of alternatively working on farms and holding a variety of other kinds of jobs was not uncommon.

Exploitation

Just as there were cases of exceptional generosity on the part of Canadian sponsors, there was also blatant exploitation. A case in point is a couple with ten children who emigrated in 1952. The father grew up on a farm near Leiden, but had moved to the city. Their second oldest son, who reluctantly accompanied them, recounted their story in the presence

[27] Interview (FS) with Martin Strijbosch, in his home just outside of Strathroy, 20 October, 1990.

of his now elderly mother. They all started working in a dairy and cash-cropping operation in Linwood, near Kitchener. That farm was one of ten owned by an executive of a major U.S. airline. The owner, who was a land speculator, kept his farms in minimal working condition. The Dutch immigrant family, with five able-bodied workers, received $150 a month and none of the extra benefits originally promised. Not only were they cheated by the farm manager, but the fieldman who had found their sponsor skimmed off ten dollars a month. They left the farm before the end of the year, and moved to Wingham, where the father and sons worked in factories. The only daughter found a job working in a store. The following year, they moved to Plattsville, closer to the Kitchener job market, and the father resumed doing odd jobs for local farmers.[28]

Another case of outright swindling involves a family sponsored by a farmer in Lucan, near London, in 1950. When their crate of personal belongings arrived, the farmer noticed a large number of carpentry tools. From that day on, instead of working in the fields, the Dutch immigrant and his sons were forced to build a circular stairway in the farmer's home, and a new barn. Then he hired them out to various neighbors. The farmer charged a dollar an hour for their skilled labor, but the family was still only paid the original forty-five dollars a month. Their sponsor would drive them to the work site, bring them home for lunch, and then drive them off to another shift, which often lasted until late. This situation came to an end when neighbors notified immigration authorities.[29] Other sponsors, while not that abusive, nevertheless cut corners or short-changed their Dutch workers. One farmer justified a monthly wage reduction from seventy five to seventy dollars because the wife of the Dutch farmhand was collecting a five-dollar baby bonus check. Immigrants in the Niagara Peninsula experienced similar stinginess or meanness; the name of a Canadian of German origin who owned several farms and greenhouses near St. Catharines was mentioned in several interviews. A Dutchman who spent part of a season on one of his farms, together with his father and a couple of brothers, recalled how at one time there were workers of four nationalities working there: German, Dutch, Polish, and Italian. They were put together so that they would not be able to communicate with one another. The foreman would yell at them to hurry up. The owner did not provide rides to church on Sundays because they could walk four miles to catch a bus. Even the permanent workers sponsored by this farmer lasted only a few weeks.

---

[28] Interview (FS) with her son in Brampton (in the presence of his mother), July 6, 1991.

[29] Interview (FS) with Gerry Van Bussell, on his farm near Lucan, 9 July, 1990.

Dutch immigrants who worked for old-timers in the Holland Marsh also preferred to find other employers as soon as possible. While many of the farmers who had settled in Ansnorsveldt in the thirties took in, or helped, relatives or hometown friends, most of the postwar immigrants who came to the Marsh worked for a few large growers in Springdale. They were housed in shacks and did not form close ties with their employers, in spite of the common bonds of shared language and religion. A former Dutch farmworker who first landed in the Holland Marsh, made the following comment: "Our bosses wanted us to learn English as quickly as possible when we joined their [Christian Reformed] Church, but they used to tell us off in Dutch when they thought we were not working hard enough in the fields."[30]

Resistance

The most common reaction of farm workers to sponsors they did not like was to vote with their feet. In some cases, a complaint to one of the immigration societies resulted in a transfer of a Dutch family to another sponsor. In such cases, someone would arrange to pick up their belongings and move them to another location. There was no penalty for quitting; many Dutch immigrants sought out other farmers willing to pay more money or provide better housing. Yet, some families needed some reassurance before leaving an employer. In one case, a Dutch farmhand refused to order his son, who was working in town for extra money, to return to work on the farm. The sponsoring farmer stopped paying their monthly wages even though the original contract did not include the son. Yet, this immigrant did not go to another farm until he was specifically told he was free to leave the farm.

Some farmworkers resisted in a more active way. A single man raised in Gelderland started working for a fruit farmer and vegetable grower in Trenton for forty dollars a month. When winter came, he helped to cut logs. But occasionally, the people who hauled away the logs with horses did not show up because it was too wet, and his employer would not pay him. So the Dutchman finally informed the farmer that, "you broke the contract, so now I'm free," and started working the night shift in a factory. One Dutch farmhand was more creative when his employer withheld his pay: on several occasions, the farmer had claimed he did not have the money, so one day this employee, a single man living in a shack on the property and working for straight wages, showed up at the front door of the farmer's house with two suitcases. He informed his boss that he was moving in until the wages were paid off through room and board. That farmer suddenly found the money.

[30] Interview, in Dutch, with a retired Dutch-Canadian farmer, March, 1989.

Sometimes resistance resulted in minor improvements that benefited other workers. One Dutchman, unhappy with no days off, insisted he should get at least one weekend free every fourteen days. The farmer relented, but then had to do the same for another long-time hired hand. That hired man, a Canadian not of Dutch background, later told the Dutchman it was his first day off. A similar incident involved a Dutch worker in Hawkestone (in the Lake Simcoe region) who spoke little English. When the Dutchman was made to work the first day he arrived, a Sunday, he said to another farmhand who was German that he would work this Sunday but not the next. When the farmer sent the Dutch worker off to the fields the following Sunday, he refused saying that since he didn't work on Sundays in Holland, he should not have to in Canada either. The German worker worked as usual. However, on the third Sunday, the German also confronted the farmer, declaring that he wouldn't work on Sundays either, since he didn't in Germany. However, such open defiance did not always succeed. Another Dutch farmhand, who was paid by the day, stopped receiving wages after three days on the job, although he and his family were allowed to continue living in the farmhouse they occupied. The Dutchman did not know how to express himself in English but managed to communicate, "no pay, no work." By noon, the farmer had thrown them out of the house and placed all their furniture by the side of the highway.

Dutch women could also say no. On one occasion the wife of a Canadian farmer told a married Dutch woman, who was already being paid to do housekeeping, to start painting the walls. The Dutch woman, who had never picked up a paintbrush in her life, and who wanted to spend more time with her children, refused. This did not ingratiate her to their employer.[31]

SOCIAL ACCEPTANCE AND LIVING CONDITIONS

As Dutch immigrants started arriving at train stations throughout Southern Ontario, they were objects of curiosity, frequently confused with displaced persons from Eastern Europe (DPs). The term "DP" evolved into a derogatory label applied to all postwar immigrants from continental Europe.[32] The individual experiences of Dutch immigrants were more positive. Canadians who had been in Holland during the War were especially friendly, and even generous; not infrequently Dutch immigrants

[31] Interview (FS) with Grace Joukema, in her home in Grand Valley, 29 July, 1992.

[32] Between April 1945 and March 1953, 165,000 displaced persons were admitted under a special program. See Petersen, *Planned Migration*, 153; and Jean R. Burnet and Howard Palmer, *"Coming Canadians": An Introduction to a History of Canada's Peoples.* (Toronto, McClelland and Stewart, 1988), 41, 74, 94.

were amazed when neighbors and even strangers stopped to help with a flat tire or spare gas.[33] Overall, the Dutch were well received, and less subject to discrimination, because they were regarded as White Anglo Saxon Protestants. One journalist commenting on the warm reception of the Dutch on Wolfe Island (in Central South Ontario) reveals the racial stereotypes then prevalent in Canada. He stated that the Dutch were "regarded by some immigration officials as the best immigrant stock of all Europeans."[34] However, those Canadian Protestants not fond of Catholics, were disappointed to learn that their new Dutch neighbours were 'papists' (see chapter 7). Nor did they appreciate the refusal of Dutch Calvinists to accept their invitations to join mainstream Canadian Protestant churches, establishing their own churches and schools instead.

Lending a helping hand

We have already seen examples of sponsors helping their Dutch workers. Neighbors also lent a helping hand. When one Dutch family arrived at a farm near Springbank (London region), with all of their belongings, their sponsor did not have their house ready. Yet, a neighbor immediately invited the whole family to share a meal and spend the night with them. That Dutch family also received a warm reception when they joined a small Presbyterian congregation in Springbank.[35] Friendly neighbors of Scottish descent welcomed another family who landed in Puslinch Township (Waterloo-Guelph region). Immigrants from other parts of Europe also helped their Dutch counterparts. The person most helpful to the Dutch farm family in Puslinch, was a bachelor in nearby Morriston, who had earlier come from Germany. The Dutch family, from the Achterhoek region of the Netherlands, near the German border, had no difficulty communicating with the German who would often visit, bringing his violin to provide entertainment. He was a good singer and they spent many enjoyable evenings together. When the Dutch family started looking for their own farm, he went with them and advised them against buying a farm near Morriston because it lacked water. Using his knowledge of local soil conditions, that man later helped them select a good farm. He then lent them money, charging only 3.5 percent interest. The Dutch

---

[33] These observations are based on numerous anecdotes recounted by my interview subjects, as well as my recollections of living as a young migrant in a rural area just east of Toronto in the late fifties.

[34] See Harold Hilliard, "How the Dutch Made Good on Wolfe Island," *The Toronto Star Weekly*, 23 February, 1952.

[35] These events were recounted by Dinie (Kuysten) Twyntra, whose parents emigrated in 1948.

family received further financial assistance from two other young single Dutch men, who used to visit their first home in Canada. These men handed over their money for safekeeping, but were repaid within a year.[36] Such reciprocity between Dutch families in Ontario and lonely immigrant bachelors was not uncommon.

Some immigrant families sponsored by farmers lived in large stone houses with bathrooms. These were, however, exceptions. For the most part, living conditions in rural Ontario in the late forties were primitive by Dutch standards. The Dutch family was often provided with accommodations in a small, uninsulated wooden-frame building. Even the wood siding was frequently left unpainted, and the houses seldom had plumbing or electricity. Isolation and lack of public transport were especially challenging. Dutch families, spread out over a wide area, depended on rides to church from neighbors or employers. Others had to venture long distances in rusty old cars over pot-holed gravel roads.[37] The first winters were particularly hard, but food was inexpensive. A typical experience of a postwar immigrant family is that of a young couple arriving on a large fruit farm near St. Catharines in the summer of 1953. They were lodged in a two-room cabin and had to share an outhouse. There was no heat or plumbing and only a bucket of water for washing up.[38] Another Dutch couple on a farm in Inwood (near London) lived in an eight cubic foot pantry, with access to a small upstairs bedroom. Most of their belongings were stored in their shipping crate, and they were charged for the wooden planks used to build shelving.[39] A family who started working in the Ottawa valley had no bedding prior to the arrival of their crate; they wrapped the pillows they borrowed with towels to hide their odour.

Poor living conditions were particularly hard on the women who washed, cooked, and cleaned in addition to working outside the home to earn money. One woman who had had a maid in the Netherlands now did all of her laundry by hand, including her husband's stained work clothes. One day a stranger appeared and sold her a used electric washer for fifteen dollars. Her workload was reduced but it offended her sense of respectability; she always shut it off when visitors arrived because it was so noisy. The experience of living and working in rural Ontario presented special challenges for young people still living with their parents. Some sponsors sent Dutch teenagers away for months at a time to board,

---

36 Interview with an elderly Dutch woman, living on her son's farm in Puslinch Township, 5 December, 1990.

37 One person recalled a large family's children crawling out a car's back window. Interview (FS) with Mrs. K. Rietkerk, Guelph, 7 January, 1992.

38 Interview (FS) with Peter Buis, at his home in Niagara-on-the Lake, 16 June, 1997.

39 Interview (FS) with John Van den Broek and his wife, in Wyoming, 17 May, 1991.

and work, with other farmers. They were paid separately, but the earnings belonged to the parents. One young woman of eighteen, the oldest daughter of a large farm family who came to the Kinkora area (London region) in the early fifties, disliked the first farm family she was sent to. However, the farmer forbade her to look for work elsewhere and threatened to cut off her share of their earnings if she quit.[40]

The experience of moving to rural Ontario was especially difficult for Dutch immigrants whose income and standard of living had been above average in Europe. The family of the young woman who landed in Kinkora had had a hired man and a live-in maid in the Netherlands. The house provided by their Canadian sponsor was big, compared to where they had lived in Holland, but they were unaccustomed to not having warm water or using an outhouse. However, that family managed to get ahead more quickly than less well-off Dutch immigrants, since they were able to ship over seven crates of belongings. They even shipped over a jeep. When they had their own farm, they used the jeep to carry everything, including animals, on their trips to town and back. Nevertheless, even for these better-off immigrants, life in Canada was always an adventure. Their oldest daughter recalls returning from Stratford one day with a calf in the back of the jeep. Whenever they encountered a police cruiser, she had to push calf's head down in between the groceries.

Going to School

Most rural children in the Netherlands started working on the farm or at home after finishing grade school. After the Second World War, more and more of them continued schooling by attending part-time evening programs in practical agriculture (the *lagere landbouwschool*) (see chapter 1). Indeed, despite the disruptions of Nazi occupation, the educational system in the Dutch countryside was superior to that of rural Ontario, where underfunded one or two-room schoolhouses were the norm. When the school board in the township of East Luther (Northwest region) appointed a Dutch farmer with school-age children to carry out a physical inspection, that farmer noted a lack of desks and books, and a polluted well. Even though the Canadian farmers on the board told the Dutchman that they had some money "for a rainy day," he had a hard time persuading them to part with some of it. They objected when he presented them with a bill for several hundred dollars for a load of sand.[41]

---

[40] From an interview conducted in Ilderton (near London), 29 April, 2002.
[41] Interview with Piet Bouwhuis, at his home in Grand Valley, 30 April, 1998.

Dutch immigrant children whose parents worked for different farmers might attend several such schools. One man, who came to Canada at age ten, had already attended four different rural schools before entering high school. He did not have happy memories:

> You know, kids were pretty tough when you came to these little public schools. You would introduce yourselves as the new kids on the block, with bright shiny faces, and the first thing you know you got a punch in the nose. Welcome aboard! There was no welcoming mat and you quickly had to learn to take care of yourselves.[42]

Children who had already attended school back home were particularly disadvantaged. One man, who was seven years old when his parents moved to Canada in 1950, had to repeat grade one and never advanced past grade six. When I interviewed him in 2001, he remembered it as a disastrous experience, in which he was never taught the proper foundations, especially in math. Like so many other Dutch children, he had no English, nor did he have a sufficient Dutch education to transfer to a Canadian setting. In contrast, Dutch children who had already gone to school for several years before coming to Canada were able to advance quickly once they learned English. They found that they were at least two years ahead in arithmetic.[43]

The older immigrant children who had already finished elementary school in the Netherlands, invariably started working in agriculture when their parents moved to Canada. While a lack of Canadian schooling and poor writing skills in English did not deter them from becoming successful commercial farmers, they were often at a disadvantage later in life if they were interested in becoming involved in farm organizations or running for political office. One farmer told me:

> I could have done much better if I had had the opportunity to go to school longer. I can read but have real difficulties with spelling and writing, which has been a real source of embarrassment for me, because I would have to ask someone to write things down for me. For example, when I was director for the marketing board, I realized I could never become the chairman because I would have to do office work and communicate better in a place like Ottawa.[44]

Regardless at what age they came to Canada, the children of farm immigrants who came to Canada prior to 1955 rarely studied beyond grade eight. However, many Dutch immigrants did not want their offspring to continue attending school even when this became the norm for their

[42] From a taped interview with a man at his home in Ancaster, 1994.

[43] This was also the experience of the author who attended a rural school in Ontario in 1957.

[44] Interview with a farmer in the Niagara region, 25 June, 1995.

non-Dutch counterparts. One man, who liked school and was doing well, had to help out on the family farm as soon as he had completed grade ten. In some cases, even well-established immigrant farmers, who had themselves received a good education, did not want their children to complete high school or beyond, for fear that "they might not want to become farmers."[45] The idea of continuing one's education was unthinkable, with few exceptions. This attitude started to change as immigrants realized the advantage of schooling. For example, several Dutch farmers in the London area who had studied at a *landbouwschool* changed their minds after a visit to Ridgetown College in 1964. One of them sent his third oldest son to study there, even though the son had worked on the home farm for five years.[46]

Getting Married and Gender Dynamics

Quite often the children of postwar immigrants were of marrying age. They frequently met other younger people also of Dutch background (see chapter 8), resulting in the first marriages of Dutch-Canadian immigrants. However, single people from the Dutch countryside also came to Canada on their own. Frequently, they were already dating or engaged to be married, and many young men returned to marry a hometown sweetheart before settling down in Canada. Occasionally a bride-to-be would join her future husband in Canada within a year of his arrival. If engaged couples emigrated together, they had to get married within thirty days.

In the fifties, Dutch parents exercised almost total control over their offspring prior to marriage, even when they were engaged. For example, Willie Knibbe, the eldest daughter of a large family, was not allowed to join her fiancé when he left for Canada in the spring of 1947. Her parents were worried about living conditions and primitive transportation, and they wanted him to go investigate first. He landed in the Holland Marsh, sponsored by a cousin of his wife's father. Ten months later, in March of 1948, she was allowed to leave for Canada. The following week, a Dutch pastor performed a double wedding ceremony in the Christian Reformed church of Ansnorveldt. The other bride was Willie's cousin who had traveled with her on the same boat.[47] Such marriages among people already connected through hometown links were not unusual. However, the more common scenario was for young single

---

[45] This attitude was reported in several interviews.

[46] This son, who was himself initially reluctant to return to school, told me this in an interview at his farm near Kerwood, in the 1990s.

[47] Interview with Martin Verkuyl and Wilhelmina ("Willy") (Knibbe) Verkuyl, at their farm home near Hickson, 23 November, 2000. This double wedding (and a photo) is also described in Albert Van der Mey's *And the Swamp Flourished*, 70.

immigrants to meet a future life partner from among other Dutch immigrants. This created a new dense network of friendship and kinship. I encountered several such cases in the region of London where two brothers married two sisters (see chapter 8). Such intra-ethnic marriages were more likely to ensure that family values and gender relations from rural Holland survived in Canada.

Gender roles in the Netherlands varied from region to region (see chapter 1). These roles were replicated, but also modified, in Canada. Gendered farm tasks in the Netherlands influenced the sexual division of labor in rural Ontario, especially when it came to milking. In one case, a woman not fond of milking, felt she had already done her share at the family farm in the Netherlands, particularly when her older brothers starting dating and working off the farm. Women doing the milking was typical of her home region. Once in Canada, she informed her husband she no longer wished to milk cows, except when absolutely necessary. Nevertheless, two decades later, her sons persuaded her to start milking cows every weekend! Women from Noord Brabant, more likely to work in the fields or in the barn than their counterparts elsewhere, continued to do so after emigration. However, in rural Ontario, even women from regions where this was not typical, performed physical labour. A woman from Friesland had been a maid in the Netherlands for one farmer and spent several years doing fieldwork for another. In Canada, she helped her husband with farming, even in extreme heat or well into the evening, when he was ready to call it quits.[48] In another family, originally from the region of Rotterdam, all five daughters did the same chores as their brothers. One of them operated the combine without break. She could even perform minor repairs herself. These young women also worked in the barn after school.[49]

While most spouses of Dutch farmers were of farm backgrounds, and became full partners in running a farm in Canada, they were generally taken for granted. Although some men recognized and even commented on their wives' contributions, the women themselves rarely mentioned this. When interviewed, many immigrant wives described their work as "just helping out." In conducting a study of Dutch immigrant farm women in Eastern Ontario, Elizabeth Allingham found it difficult to get these women to talk about their contributions to the farm because they downplayed it.[50] The contribution of farm women to agricultural production

[48] Interview with a retired couple in Guelph, 29 July, 1992.

[49] Interview with a farmer at his farm near Alliston, 30 May, 1991.

[50] Elizabeth C. Allingham, "Different Voices of Postwar Immigrant Women in Rural Ontario: victims of agrarian ideology and 'invisible' contributions to agricultural production", paper for a course at the University of Guelph, Spring, 2002, 12.

is undervalued because it is difficult to draw a clear line between household tasks, child rearing, and work in the barn or the fields. The significance, and, in some cases, indispensability of the farm work done by women became evident in several cases. In the Halton region (part of the Golden Horseshoe), one woman had sole responsibility for a chicken farm that was originally set up for her brother-in-law. After his departure, she tended 14,000 hens, paid the bills, and delivered egg to two towns. At one point she generated most of the household income, because her husband only worked part-time. After he developed asthma, she became the sole breadwinner.[51]

Women also did most of the farm work if a husband had a full-time job, got involved in politics, or was running a business. In one situation, this pattern was replicated in subsequent generations. One man, whose father worked in construction, told me his mother used to manage their pig farm. She did much of the physical labour and organized her sons to help. He recalls that she was enterprising and independent, often taking all of her children off for week long camping trips. The man I interviewed, became a schoolteacher, but also bought several farms. He married the daughter of Dutch immigrants. She also ended up doing most of the farming, including driving the tractor, like her mother-in-law. At one point, she was even farming full-time while raising their children. However, a woman's role in the running of a farm enterprise might be radically different if her husband was not only a full-time farmer, but had also attended an agricultural college. Such men, who saw themselves as professional farmers, were less likely to form equal partnerships with their spouses (see also chapter 10).

CONTACT WITH DUTCH ENTREPRENEURS

Almost all Dutch farm immigrants started off on Canadian farms prior to 1952, although most of them eventually drifted to nearby towns or cities in search of more remuneratively rewarding work. Here, they increasingly came into contact with other Dutch immigrants, including manual laborers or craftsmen with no farm or rural background. They were also apt to run into professionals or business people of Dutch backgrounds in an urban setting. These Dutch entrepreneurs in the fifties were well aware of the numerous immigrants living in the countryside. They also knew that farm immigrants aspired to buy their own farms and could well become future customers. However, it was not easy to

[51] Elizabeth Allingham reports on a similar case in Eastern Ontario. See "Different Voices," 13.

persuade farm immigrants to buy luxury goods when they were still struggling to make ends meet, or just starting out. To build up a clientele of rural customers of Dutch background, it helped to have some familiarity with farming. In the late forties, one insurance salesman covered a huge territory, from Niagara Falls to London, in an Austin Mini. In the Netherlands he had grown up on a farm owned by his father, who managed 200 farms for several big landowners. Moreover, this salesman himself was sponsored on a Canadian farm for a year. In an interview, forty years later, he recalled his insurance-selling techniques:

> I always took long rubber boots with me on my trips and would step right into the barn and help the farmer finish the milking or clean up. That way I could get him into the kitchen and talk about buying life insurance. I always persuaded them to sign the dotted line. But I would tell them I was not selling life insurance but a comprehensive policy, which meant that if he died, his wife would get all the money and even if he didn't, he would still get the money. Then the farmer's wife would say 'Ja, dat is een goed idee' and before you knew it, I had another customer.[52]

Other immigrants peddled Dutch household goods or delicatessen items. Piet Veurtjes, owner of a Dutch store in Ottawa, used to make personal deliveries with his van to farm households located off highway 31. John Niekerk, the owner of a delicatessen in Barrie, made the rounds of Flos Township. Similarly, there were Dutch bakeries and butchers in small towns and villages, whose customers included farm immigrants.[53]

Dutch farm immigrants came into contact with Canadian employers, teachers, and ethnic entrepreneurs. Like other immigrants, their first priority was to feed their families by working for wages. The next challenge was to create a home for the family or to start their own family. School-aged children had to find new friends. They all had to struggle to learn a new language and adjust to a new social and physical environment. Within a year, most of them were working in town or venturing into the bush in order to start saving some money. However, they never forgot why they emigrated – to become Canadian farmers.

---

[52] Interview with Bob Gosschalk, in Toronto, 10 November, 1988.
[53] See Schryer, *The Netherlandic Presence*, 223.

V. Becoming a Canadian Farmer

Becoming a full-time farmer took hard work and luck. Most immigrants also held down jobs in order to buy their own land, and continued to do so until they could make farming their sole, full-time occupation. Even when land was purchased within a year of arrival, family members still had to take on off-farm jobs to keep the farm going or to accumulate the capital necessary for further expansion. Prospective farmers who were timid, became ill, or did not have the full support of a spouse, did not stand a chance. While the majority of rural Dutch immigrants never achieved the goal of owning their own farms, it is still astounding how many did. For those who succeeded, the interval between landing in Canada and running a viable commercial operation varied from eight months to twenty years. Of the 162 cases examined in my study, thirty-two families were able to make a living on their own farm in under two years after arriving in Canada.[1] Fifty-seven families took between two and seven years to establish a farm, while seventy did not have a viable farm operation until after seven years. Five families studied never went beyond having a strictly part-time farm operation.[2]

GETTING STARTED

The greatest challenge for immigrants who wanted to buy a farm was coming up with enough money. Older couples sent their children off to work, who in turn handed their earnings over to parents, just like at home. Yet, while such savings might be enough for a down payment, additional money was required for operating expenses. Young people who came on their own, or recently married couples, found it more difficult to buy land, although there were alternative ways to start farming.

---

[1] In the case of couples who were born in Canada or came to Canada when they were children, I started counting the years after they finished their schooling.

[2] These figures do not include the farm immigrants who came to Canada after 1980, at which point farmers from the Netherlands started buying farms in Canada prior to immigrating.

Working for Shares

One option to start farming on your own was in a share-cropping or share-farming arrangement on a dairy farm. However, while the potential for a lucrative return was great, it was a risky undertaking. One postwar immigrant I interviewed knew a neighbor who had gone to Canada in the twenties to grow tobacco. That prewar immigrant ended up owing more than he earned and had to return to Holland. The story was almost repeated in the late forties, when another Dutchman took up tobacco growing on shares. He was barely able to scrape by and was left with little money after paying his debts to his Canadian partner. Other immigrants who entered into share-cropping arrangements had better luck, as in the case of Jan (John) Hendrikx, a farmer from Alfen, Noord Brabant. He landed in Blenheim (Kent County) in 1948 with his spouse, their four children, and a hired hand. Upon the recommendation of his Flemish uncle, a Capuchin priest working in Ontario, Jan immediately started working for shares for a farmer of Flemish background, who had come to Canada before the war. In his first year, Jan planted ten acres of tomatoes, ten acres of tobacco, plus some cucumbers. He made enough money to buy his first farm in Park Hill (London region), a year later.[3] The case of John Hendrikx is unusual because of his family connections and links with the Flemish community. Generally, Canadian farmers offered to enter into share-cropping arrangements only with Dutch immigrants who had already worked, or continued to work, for them. For example, Richard Van Keulen, who came to Canada in 1952, had already managed his sponsor's tomato production when he started growing his own tomatoes on a share basis in his spare time. Richard continued on as a full-time farm manager, on the condition that he tend his own crops only after 6:00 P.M.[4] His boss supplied the fertilizers, the seeds, and the land (twelve acres) but Richard took care of weeding, picking, and transportation. The tomatoes were shipped to Libby's, a nearby cannery, under the contract his boss had.

A Dutchman who landed in the region of Waterloo in 1947, entered into a somewhat different share-cropping arrangement. He first worked for a farmer in Cambridge who owned a lumberyard as well as a con-

[3] Jan rented out that farm and continued working for shares in Blenheim for another three years, as a favor to the widow of his sponsor, who passed away around the same time he bought his own farm. Interview (FS) with Jan Hendrikx in his home in Park Hill, 16 December, 1988.

[4] He later expanded his operation to twenty-five acres. Interview (FS) with Richard van Keulen, in his home in Blenheim, 29 May, 1997.

struction company. That farmer had five purebred Holsteins, twelve sows and 300 chickens, and his Dutch worker combined farm labor with chores in the lumberyard. After six months, the sponsor offered to let him run the farm part of his business on shares, but the Dutch employee was hesitant. He preferred to keep working for eighty five dollars a month. However, when he received a contract to build ten new houses, the Canadian farmer no longer had time to farm. The following spring, this immigrant became his own boss after all. His former employer provided the land and all supplies in return for half of the profits from the farm, although the Dutchman was also welcome to work in the lumberyard for an hourly wage at any time. The Dutchman and his family now had the brick farmhouse they had previously shared with the farmer all to themselves.[5] Such share arrangements worked out particularly well in the case of dairy cattle. Farm immigrants experienced in handling cattle could keep half the calves. Those with resources of their own could become partners in return for half of the milk, half of the bull calves and a quarter of the heifers. Such share farmers had to provide half of the inputs, including feed and their own labour. They were thus able to starting building up their own herd without owning land. But they would eventually still have to buy their own farm.[6]

Buying the First Farm

Another option was to continue working in town until one had enough money to buy a dilapidated or abandoned farm. "Bare farms" were readily available in the decade following the Second World War because many middle-aged and elderly couples, whose children were not interested in taking over, wanted to leave farming because low prices for agricultural commodities meant that farms no longer generated good incomes. Ontario beef operations, unable to compete with Western cattle businesses, were especially vulnerable. Some of these farms sold for as little as $2,000. The asking price for a 100-acres farm in East Luther (Northwest region) with a house, an empty barn, and not tiled, in 1951, was $3,600.[7] Farms in similar condition went for six or eight thousand dollars right up until the end of the fifties. A Dutch immigrant could buy such a farm for a down payment as low as $1,000.

Not all postwar Dutch farm immigrants bought marginal farms. As

[5] Interview (MF) with Hayke Haanstra, 7 September, 1993.

[6] I encountered only four cases of share-farming involving dairy cattle, but heard about other people doing the same; the terms and conditions varied from case to case.

[7] Interview (FS) with Piet Bouwhuis, in his home in Grand Valley, 30 April, 1998.

early as 1949, one Dutch farmer purchased thirteen hectares of good arable land near Kingsville (Essex), worth $16,000, to start growing vegetables. In 1952, another immigrant got sixty-seven acres of uncleared bog land in the Grand Bend Marsh, in the London region, for $5,000. During the fifties, a fully operating farm with equipment, plus cattle or fruit trees, was not out of reach for Dutch immigrants with modest resources accumulated over several years. One Dutchman bought a dairy farm worth $15,000 with a down payment of $3,000 in 1952. That farm, located in the Dunville area (Niagara region) consisted of sixty acres of land, horses, farm machinery, and seventeen cows plus milk quota. Someone else bought a beef farm "lock, stock and barrel" for $17,000 in Ailse Craig (London region). Dairy farms located close to large urban centres and cash crop or fruit farms in good locations were in the $30-50,000 range: a 200-acre full dairy farm with a milk contract in Waterdown (Hamilton-Wentworth) went for $40,000, with a down payment of $9,000 in 1955; a twenty-acre fruit farm in Niagara-on-the-Lake with a good house cost $35,000 (with a $6,000 down payment) in 1957; and a total operation on 130 acres in Blenheim (Kent), including 600 feed cattle, twelve acres of tobacco, and twenty acres of tomatoes, sold for $41,000. The top range of farms bought by Dutch immigrants with some capital included a 40,000 square-foot greenhouse operation on three acres of land in Grimsby (Niagara), which was worth $90,000 in 1954.

While some people bought land in established regions, where farm prices were higher, most Dutch immigrants working in more prosperous and fertile agricultural regions, such as Kent and Essex counties, had to go further afield. Dutch farm workers in the region of Chatham, where farms were beyond the reach of most immigrants, had to look around for land further north, in Huron County or in Middlesex or Lambton (all located in the London region). The stony soil in the higher, western side of the escarpment in the Niagara Peninsula was another possibility. The rule of thumb was that the value of farms in that region corresponded to that of the buildings, especially the main farmhouse. Indeed, up until the early sixties, a few Dutch immigrants were still able to trade a house for a bare farm. However, truly marginal farmland, such as that found in the middle portion of most of Central Ontario, was not likely to attract prospective farmers from the Netherlands, no matter how inexpensive.

In hunting for a farm, Dutch immigrants were not motivated by the sale price or a good deal alone. An important consideration was proximity to an industrial center within commuting distance where family members could earn extra income. They also had to take soil conditions and climate into consideration. Dutch farmers recognized the potential of wet, swampy land that only needed tiling. They also recognized that new varieties of crops with shorter growing seasons would enable them to develop successful farms in areas with lower heat units, hitherto only good for hay or oats. In some cases, the reason for choosing a location was the extent to which the soil resembled that of a home region. Someone bought a run down farm in Drumbo (part of the London region) because he noticed that the light, black, though stony, soil was similar to what he was accustomed to in the Netherlands. Two brothers pooled their resources for a down payment on a farm with mainly sandy loam located in Brachton (Waterloo region).

FINANCING A FARM OPERATION

While many farms sold at reasonable prices, it took ingenuity to buy, much less start, a full-time operation, if one did not have collateral or access to credit. Large families, who lived frugally, were able to save money by working at jobs in town. After buying a second-hand car, which was essential in rural Ontario, they would start looking for a farm that could also serve as a place to live – close enough to a town to commute to a job. Another possibility was to first buy a small house or cottage near an urban center, the sale of which could become the down payment for a farm. Dutch farm immigrants also pooled the resources of several families.

Partnerships and Family Dynamics

We have already seen how offspring working at off-farm jobs handed their earnings over to their parents. They continued to do so even once a family farm was established. Parents, in turn, were expected to set up their children in farming after they were married. Long-term reciprocity also took other forms. Sometimes an older son would emigrate first and then sponsor his parents, who would in turn look after the farm while he continued to work for wages. In other cases, older children and their mother would operate a newly acquired farm while the father worked off-site until the farm operation could accommodate all family members. Still other partnerships involved relatives and close friends. It was not uncommon for siblings to pool resources, as in the case of the two brothers already mentioned. After moving to the Waterloo region, the younger brother, a bachelor, continued to work for a farmer in another region while the older brother, who was married, lived on the farm they had bought together. After several months the older brother told the younger to quit his job and join them in Brachton. With the younger brother now helping out on their farm, the older brother was able to work the night shift in a nearby factory. There he earned the same amount of money his brother used to bring home, in half the time! The younger brother looked after a chicken operation while the older brother did fieldwork during his spare time. He worked off-site from August 1957 to the spring of 1958, and, a year later, they were both farming full-time. In another case, a group of brothers, all married, formed a partnership with their elderly father, only six years after arriving in Canada. Such close co-operation also took place between in-laws, although these arrangements did not always work out. One man, in hindsight, regretted that he had allowed his parents to convince him to unite with his sister and her husband, then experiencing difficulties and in need of additional labor and capital. An alternative form of collaboration among close family members or in-laws was to share farm machinery and labor.

Unrelated people were less likely to buy a farm together. Nevertheless, I came across several cases involving single men who had "ventured" together, pooling their resources to come up with a down payment. Bastiaan Bos, who came to Canada in 1949, at the age of twenty-six, had already worked for two sponsors when he teamed up with the three Vandenburg brothers, to look for jobs in Hamilton. There they met up with a Canadian tenant farmer who persuaded the four boys to take over a rental. The following year they pooled $4,000 dollars to buy a bare farm on the fifth concession of Waterdown (in the Toronto region). In 1951, after paying off the remaining $2,000, they invited a large Dutch immigrant family to share their house. The woman in charge of that household cooked and washed for the four bachelors. The partners stayed in that place for four and a half years. By 1954, Bastiaan had a wife who took over domestic duties,[8] and in 1955, they sold the farm.

The partnership of Bastiaan and the Vandenburg brothers lasted well beyond the sale of their farm. They were already renting another farm in Dundas when they used the proceeds of the sale of the Waterdown farm to buy a $40,000, 200-acre dairy farm in Cambridge (Waterloo region). This time they had to come up with a $9,000 down payment. During the next five years, one of the Vandenburg boys got married and moved off to another farm in Dundas, which was also jointly owned by the four partners. They moved cows and equipment between farms and another Vandenburg brother joined them a few years later. That left the youngest brother with the Bos family. The two groups contin-

[8] Interview (MF) with Bastiaan Bos, in his farm home near Sheffield, 27 September and 4 October, 1993.

ued to share equipment for another decade. The partnership ended in 1972, when Bastiaan and his wife, Syma, bought the Cambridge farm. Bastiaan died in 1995, but the tradition of co-operative farming continues: his four sons, their wives, and his daughter and her husband were all farming together in Cambridge (near Sheffield) in 2002.[9]

I came across several cases where people who were not related likewise helped each other. One young man with a truck working at odd jobs made a deal with a man who had already established his own farm. The farmer bought and butchered a horse, and the young man helped to bring it into the barn with his truck. Together they hung up the carcass, skinned it and brought horsemeat (*rookvlees*) around to all of the Dutch families who lived in the area.[10] In another case, a more permanent business partnership involved a young and a much older man. They met in Grimsby, in the Niagara Peninsula, in the early fifties and both aspired to owning a greenhouse business. The young man, Bill Vermeer, grew up in De Lier, in the heart of the horticultural Delfland region of Zuid Holland, where his father cultivated vegetables and fruit in greenhouses. The whole family emigrated together in 1952, except for a married daughter. Bill went off on his own shortly thereafter, once he had married his childhood sweetheart. He spent his first year working in a factory and then worked as a bricklayer for four years. In the fifth year, Bill launched the greenhouse operation with the older man, because the older partner was looking for a younger man with experience.[11] The case of the Vermeer family can also be used to illustrate changing family dynamics associated with immigration. At the age of eighteen, when he started thinking about moving to Canada, Bill worked for his father. Within a decade, the father was working in his son's new business in Canada.

Women in farming

Dutch farm immigrants were rarely able to start a farm operation and keep it going without the assistance of their spouses, as well as their older children. However, in some cases, a woman would end up running the entire farm operation on her own. This was especially common in the case of chicken operations. A good example is that of a Dutch immigrant woman in Eastern Ontario. In 1959, while still living in the

---

[9] Syma Bos provided an update to the earlier version of this story in a letter in the summer of 2002.

[10] Interview with Adrian Olsthoorn, in Ilderton, 29 April, 2002.

[11] Interviews with Bill Vermeer, in his home in Grimsby, 28 November, 1989 (FS) and 30 March, 1994 (MF).

Netherlands, she married a man just back from a trip to Canada. That same year they moved to Canada. He had already spent a couple of years in Canada on a dairy farm owned by his parents, in partnership with an older brother. However, when he returned to Canada, he got a good job in a cement factory and worked his way up to become its manager. The wife started off earning money by running a chicken farm across the road, but took it over when her daughter was five years old. The title to the chicken farm, which came with six acres of land, was in both their names but she was responsible for all decision making. He had nothing to do with the farm, nor did he know many farmers, except his brother who used to work their small piece of land. Instead, he channeled his energies into political life after joining the Liberal Party. In contrast, she was more interested in the chicken operation and gradually expanded it to the point where she had 10,000 hens. Five daughters became involved in collecting eggs and the many other requisite tasks. When her husband died unexpectedly, she continued to run the farm until she retired, when she sold it to one of her daughters and her husband.

Another story of a chicken farm run entirely by a woman comes from the greater Toronto area. In that case, her husband also had a full-time job, first working in a factory and later as the full-time manager of a credit union. They emigrated to Canada in 1958 with their three children. He had worked in the Netherlands as a claims adjuster for the Dutch railroad, with few connections to farmers. In Canada, they lived in a small rural town close to Toronto, which had few farmers even then. Although quite a few Dutch people from their home region also lived in that town, their only farm connection in Canada was his brother, who emigrated later, in 1971. In fact, it was that brother who suggested they buy a farm together, though when he moved, she ended up running the whole operation with the help of her children. This Dutch immigrant woman became the main breadwinner when her husband became too ill to continue working. The kids helped with grading eggs after school and she delivered eggs to various supermarkets.

Financial Assistance

Many Dutch immigrants would have been unable to afford even a bare farm without the assistance of a Canadian farmer willing to take the risk of lending them money. Such loans which usually took the form of a private mortgage at four or five percent interest, were unlikely to have occurred without the impression that Dutchmen were good and honest farmers, thus illustrating the importance of the positive stereotype. In some cases when a farmer did not have enough capital to invest, a local business owner might step in. In other cases, much bigger businessmen assisted large numbers of Dutch farm immigrants, though not for purely altruistic reasons.

Cattle barons

Dutch immigrants struggling to gain a foothold in farming were sometimes surprised to receive unexpected help from wealthy cattle drovers. Such men, also known as cattle barons, made their money buying and selling cattle on a regional, national, and even international level. Owners of thousands of acres of marginal land, these cattle barons would bring in beef cattle from Western Canada for finishing and sale in Ontario. In the forties and fifties this process involved a two-year cycle. The cattlemen would put the cattle in their own pastures in the spring and then look for local farmers willing to feed the cattle during the winter. The cattle would again be put out to pasture in the spring for finishing before being shipped to market. Feeding beef cattle for such barons became one way for Dutch farmers to start a farm and develop a dairy herd, since they were allowed to keep any heifers in calf. With the decline of the beef industry in Ontario, cattle barons were already diversifying their operations when they started providing credit to Dutch farm immigrants.

When Dirk Mous was living on Wolfe Island (in Central South Ontario), he worked for such a cattle dealer. He and his wife emigrated in 1948 and had just bought their first farm, although Dirk was still working at a job. The dealer initially gave Dirk ten dry heifers to feed during the winter with the option of buying some in the spring.[12] A year and a half later, the cattle dealer persuaded the Mous family, which now included several small children, to move to Brampton to take care of fifty grade Holsteins then under the care of a hired hand on rented land. The dealer could not send those cows to the United States, as he originally intended, because of an outbreak of hoof and mouth disease, so he sold the cows and a milk contract to Dirk for a $3,000 down payment. He installed a new milk cooler on the rented farm and let Dirk pay him off in instalments. Dirk rented an additional seventy-five acres to grow hay. The Mous family stayed in Brampton (undergoing urban expansion) for six years, then rented a bigger farm in Uxbridge, before ending up on their own farm in Smithville, on the Niagara Peninsula,

[12] He helped quite a few other Dutch Immigrant farmers in that way. Taped interviews (MF) with Dirk Mous, at his farm in West Lincoln, 26 October and 8 November, 1993.

in 1963. That farm had also been a beef farm before being converted into a Holstein-Frisian dairy operation by its new Dutch owners. When the Mous family moved to Smithville, they were not the only Dutch on the Niagara Peninsula. Many postwar immigrants had arrived much earlier, and also had received a helping hand, from a different kind of wealthy entrepreneur.

The millionaire and the visit of the Dutch Queen

The state of agriculture in Niagara was in dismal shape in the late forties, with most farms in poor condition. To make matters worse, owners of farms located on the brow of the escarpment shipped their manure to fruit farmers down the hill, thus neglecting their own soil. Many such farms were sold to urban investors, including the former president of a large rubber company in Toronto, who bought properties along the Ridge road in the Grimsby area. He built pig and chicken barns and installed Dutch families on the farms, hiring foremen to oversee his operations. This businessman also acquired what was then the largest range (40,000 square feet) of greenhouses at the bottom of the mountain, in addition to Grimsby Flour and Feed (an old mill) to provide feed for the farms on the Ridge Road. This man, remembered by many Dutch immigrants as "the millionaire", affected the lives of many Dutch people who landed in the Niagara region. A shrewd businessman with a temperamental personality, one man remembers him as a boss who was not unfair, but could at times be ruthless. For example the "millionaire" told his non-Dutch foremen that if the Dutch farmers he sponsored did not perform well, they should be kicked out. Wives and daughters had to help with picking fruit or tying grapes. When single Dutch workers arrived, the Dutch families living on his farms were expected to provide them with room and board. Dutch immigrants worked in other enterprises he owned as well, including a construction firm. He informed the families that if they stayed, he would help them get started on their own farms.

Bill Van Wely, who grew up in the Dutch province of Gelderland, arrived in Quebec City in June of 1949, six weeks after getting married. At the time, his brother-in-law was working for the "millionaire." Bill began looking after 600 laying hens on one of the farms on the Ridge Road and also worked in vineyards and orchards on other farms. One day, after a dispute with a Canadian foreman, the owner sent Bill to his feed mill operation. When Bill was transferred to the feed mill, he helped the manager, an Englishman, to clean up the place. It did not take long for the manager to realize that his Dutch helper, while still struggling to express himself in English, knew quite a bit about mixing formulas. So the manager gave Bill more responsibilities. Then, unexpectedly, the manager quit, and to his surprise, Bill became the new manager. A year later, Bill's employer offered to sell him the mill. He was initially reluctant and told his boss that he had come to Canada to become a farmer. Nevertheless, after making a down payment of ten dollars on the first of January, 1953, Bill became the sole owner of the mill. As part of the deal, Bill rented one of the farms on the Ridge Road, with an option to buy.

By the end of that year the "millionaire", who was then close to retirement, had sold all of his Niagara assets to Dutch immigrants and was collecting payments from the various agreements. The sudden decision to sell his farms disrupted some lives while benefiting others. One Dutch family was left homeless and jobless when the father, who was negotiating to buy, lost out to another Dutchman with a higher bid. The decision to sell his farms to Dutch immigrants may have been connected with a visit of Dutch Queen Juliana, in 1952. The queen and her entourage had just visited a local school when they stopped by for tea at the millionaire's mansion in Grimsby. Bill recounted how his boss had allowed him to pretend to cut the lawn, to have a chance to see the queen. Before he knew it, the queen stopped to say hello. She asked Bill why he had come to Canada, and what his expectations were, so he told her about his boss's promise to help his Dutch workers set up their own farms.[13] The next day, Bill received the offer to take over the feed mill. Around the same time, the investor started negotiating with immigrants who wanted to buy his farms. Ironically, the parents of a girl who presented the queen with flowers at the school were the same couple that lost their home as a result of one of those sales.[14]

Bill Van Wely and his wife worked hard on their rented farm, disking in early morning and sowing at night. Often he had little time to work the land after working at the mill. Some nights he did not go to bed. Instead of taking the option to buy that farm, Bill negotiated to take over another farm with a bigger house vacated by a bankrupt Dutch farmer. The Van Welys wanted to raise their children on that farm, but Bill channeled his energies into building up his feed business. He supplied the same farms he had when he was manager of the mill in Grimsby. Two years later, Bill moved his business to Smithville in the

---

[13] Interview with Bill and Audrey Van Wely (MF), in Vineland, 14 and 24 October, 1993.

[14] This last piece of information was obtained during another interview conducted with her father, who later bought a farm in another location.

upper part of the Peninsula, after buying another mill with money borrowed from the Canadian Development Bank. Many Dutch immigrants, especially dairy farmers from Groningen and Friesland (including Dirk Mous), had bought bare farms in that section of the peninsula. The feed business grew in tandem with the Dutch farms that were also starting to expand.

Other vendors

The people who helped Dutch immigrants were not all big businessmen or cattle dealers. Farmers on the verge of retirement were particularly motivated to sell their farms, and, in some cases, an elderly sponsor might propose this within the first year of hosting a Dutch family. However, a Dutch employee did not always accept such an offer. The man in Lindsay who sponsored a Dutch couple (mentioned in chapter 4), offered them his farm around the same time he arranged to get a job for the husband at General Motors in Oshawa. However, the Dutch farmhand declined because he did not like the soil conditions of a farm located "on the ridges" (the Oak Ridges Moraine in the Central South region). With such transactions, the terms were often flexible.

Elderly people who sold farms to Dutch immigrant families received non-material benefits as well as a stable source of income. One example is a seventy-year old bachelor farmer, originally from England, who was ready to quit farming in 1954. He was asking $16,000 for a beef farm of 300 acres that had two houses (one for a hired hand), sandy loam soil, and a good location, five miles outside of Alliston. The first Dutchman to express interest had come to Canada in 1950 with several sons, one of whom badly wanted to farm. He had already worked three years for a farm sponsor in Thornton, near Barrie, and his oldest son worked for another farmer, so they had already saved up some money. The Canadian farmer accepted their offer of $1,000 down payment and had a local lawyer make up a contract for a mortgage for the remaining $15,000. He also told the Dutchman that he could wait for payments. The only condition of sale was that the former owner be allowed to continue to live rent-free in the main farmhouse, which he did for another twelve years. The new Dutch owner and his family moved into the house vacated by the hired hand.[15] Sometimes people who were not farmers also provided financial assistance. One Dutch couple in Brooke Township, in Lambton County (London region) came up with the cash to buy a farm by selling a house to the manager of a

[15] Interview (FS) with a retired farmer, in his home in Alliston, 8 May, 1997.

small independent telephone company. The manager, an old man in poor health, moved into their old house and looked forward to visits from the Dutch couple when they brought him his monthly cheques. After several years he told them to forget about the principal and instead just to keep paying the interest, afraid that they would discharge the loan and no longer visit.[16] If that man had been unwilling to negotiate the monthly payments, the Dutch family might not have been able to get into farming.

Canadians sold farms to the Dutch for other reasons: to avoid bankruptcy, to finance a daughter's wedding, or to go into another line of work. Most often, there was no one to take over the farm. A man in North Dumphries (Waterloo region), whose two sons were not interested in farming, sold his 180 acre mixed operation to two brothers from Noord Brabant in 1957. The brothers, one of whom had recently married in the Netherlands, had been working for farmers over the previous five years, as well as in a plastics factory in Stratford. They were already acquainted with the farm because one of them rented a house on the property, and they also used to help out with haying. One day, the farmer unexpectedly asked them if they would be interested in buying the farm. The Dutch brothers offered to give the farmer the $1,000 they had saved up between them. However, the farmer said he would accept a down payment of $800 since they would need the rest of the money for other expenses.[17]

Once a farm was bought, coming up with the mortgage payments and covering operating expenses often required creative solutions, particularly if there were no other family members to work elsewhere. One possibility was to rent out, or even sell, part of the land. Another strategy, harking back to the pioneer days of Upper Canada, was to cut and sell logs on farms with extensive woodlots. In one instance, a Dutch purchaser worked without wages for two years in lieu of a down payment. He lived on his newly acquired farm, but did manual labor for the vendor, who continued to operate another farm.

LATECOMERS

After 1960, when most postwar immigrants had already bought some land of their own, a smaller number of Dutch immigrants interested in

[16] The Dutch family paid off the loan about a year before the old man died. Interview (FS) with John Jannsens, in his farmhouse near Watford, 11 November, 1988.

[17] Taped interview (MF) with Gus Donkers, on his farm home near Sheffield, 14 September and 8 October, 1993.

becoming farmers were still arriving in Ontario. The experience of these latecomers was not much different from those of earlier postwar immigrants. However, while no longer facing foreign currency restrictions, the rapidly rising price of land in Ontario presented new challenges, especially for younger people just out of school. Piet Albers, who ended up farming in Brussels (Huron), can be used to illustrate this particular cohort of immigrant farmers.

We met Piet Albers in chapter 1. Born in 1945, he was the oldest of fourteen children and grew up on a farm in Hatert (Gelderland). His father did not send Piet to the *landbouwschool* because he knew he would be unable to give his son any land. However, Piet did not like trade school and returned to work on the family farm. He was determined to become a farmer. At age eighteen, he spent nine months in Woodstock, Canada, as part of an exchange program, working three months for an uncle, and six for a Canadian dairy farmer. He liked the idea of making Canada his home. When he returned to the Netherlands, his father, then quite ill, would not allow him to emigrate. After his father passed away in 1967, he emigrated. In the meantime, Piet had married Lenie.[18] The Albers couple started off with a Mennonite dairy farmer close to Kitchener. They received $200 a month and occupied the second story of the farmer's house. Two years later, they moved to Brantford, where they worked in tobacco. After six months they moved to Fenelon Falls to manage a farm for a doctor in a share arrangement. Piet's wife was pregnant at the time. They lived in a trailer before moving into a house without running water or hydro. The deal fell through and they left within two years.

Piet was still interested in buying his own farm, but was unable to save enough for a down payment. Their next move, in 1971, was to a rented beef farm in Pfefferlaw. There, they were joined by Henk, one of Piet's brothers who had worked as their hired hand on the farm in Fenelon Falls. This time, Henk wanted to farm together with Piet. He received no wages since his work on the farm was considered his share of the investment. Piet then started working in an auto parts factory, while his brother looked after hogs in a leaky barn. Soon, Jan (John in Canada), another one of Piet's brothers, who had previously vacationed in Canada, came to board with them. He took a job and handed over his paycheck to Piet and his wife. In 1973, Piet bought his neighbor's cream quota and started milking cows on a small scale. He used a

---

[18] Interview with Piet and Lenie Albers, in their farmhouse near Brussels, 17 January 27 and 5 February, 1998.

vacuum pump that had cost him five dollars and plugged it into a light bulb socket with an extension cord. That same year they converted their cream quota into milk quota. Times were tough, especially for Piet's wife, who sometimes did not have enough money for groceries. Piet recalls the time a noisy truck arrived in their farmyard, delivering their first milk check for seven dollars. In 1974, the three brothers jointly purchased their first farm in Brussels with a down payment of $5,000. It was a bare, run-down farm with 125 acres of land worth $80,000. The Farm Credit Corporation (see next chapter) lent them the remainder of the money since they already had a tractor and a milk quota. They brought their thirty cows to the new farm, but Henk, who had just married, stayed in Pefferlaw, where he continued to look after their rented farm. That farm still had a crop in the ground and Piet did a lot of traveling back and forth because they had only one tractor between the two farms.

After buying their own farm, Piet and John continued to hold down outside jobs. Occasionally a calf would die when they were both absent. So Piet decided to stay on the farm full-time. In 1975, Henk, still living in Pefferlaw, wanted to buy into his wife's family's butchery business. Piet and John continued to work together and expanded enough to accommodate two families. They bought two additional 50-acre farms and, within the next five years, had paid off what they owed Henk. In 1983 John moved into a rented house. Since he was more interested in raising hogs, he was put in charge of a pig farm whose owner had gone bankrupt. The following year they jointly bought their own pig farm. The Albers brothers and their wives worked hard. As Piet told me, "We worked our guts out and did all the work ourselves."[19]

They went through a difficult period around 1986 when they encountered problems with the bank manager, but their luck held. Their children picked up stones in the fields and were soon milking as well. Even then, Piet was aware that children from both families might eventually want to become full business partners. To avoid future conflict, he suggested that the families split their joint enterprise. The lawyer they consulted was amazed they had run their partnership purely by verbal agreement. Even after the legal separation, the two families purchased new equipment together for several more years, until they could each buy their own fuel tanks. When I interviewed Piet in 1999, he and his wife owned three farms. However, for every farm couple like Piet and Lenie who established their own farm business, there were three others who could not make a living from agriculture.

[19] Interview with Piet Albers, in his farm near Brussels, January 27, 1998.

RATE OF SUCCESS

It is possible to get a rough estimate of the relative rate of success of Dutch farm immigrants by looking at a study conducted in the early fifties by Abe Tuinman, the Dutch agricultural attaché involved in the scheme to settle Dutch farmers in Canada after the war. According to that study, almost half of the Dutch immigrants of agrarian background, who arrived between 1946 and 1952, started their own farm operations. The average sale price was $7000.[20] Tuinman's success rate figures were adjusted in a subsequent publication, which indicated that about a quarter of the 16,500 Dutch farm immigrants, up to the end of 1955, had bought their own farms.[21] While these figures relate to Canada as a whole, I doubt that the success rates for Ontario, with the majority of postwar Dutch agrarian immigrants, would be any lower. On the contrary, given the more favourable conditions for farming in Southern Ontario, I suspect that the percentage would be closer to a third, and perhaps even higher, given that many farm immigrants who landed prior to 1955 did not start buying their first farms until well into the sixties. On the other hand, we cannot ignore the people who bought, but then in turn sold, their farms, or those for whom farming never became the principal form of income. Unfortunately, there are no statistics on these other trends.

The successful transplanting of postwar Dutch immigrants to rural Ontario must be seen in the context of social and economic conditions of both the country of origin and the country chosen for immigration. We saw earlier how, in the forties and fifties, the offspring of Canadian farmers were more interested in working in the city than going back to the family farm, and how prices for farmland were depressed. This situation was ideal for frugal and hard-working people who wanted to develop dilapidated, abandoned, or hitherto marginal farms with little capital investment. Dutch farm families were able to step in and occupy this niche at just the right time. The successful takeover of farms by these immigrants can be explained by the fact that they had a competitive advantage over their non-Dutch counterparts. Even those with few resources could save money, share resources, and keep labor costs to a minimum. However, once they bought a farm, they still had to learn how to farm in Ontario.

---

[20] Abe Sybe Tuinman, *Enige aspecten van de hedendaagse migratie van Nederlanders naar Canada* (S'Gravenshage: Staatsdrukkerij, 1952), 77.

[21] Abe Sybe Tuinman, The Netherlands-Canadian Migration, *Tijdschrift voor Economische en Sociale Geografie* 47 (8) (1956), 183.

VI. Learning and Getting Ahead

The ability of an immigrant group to adapt to a new social environment, and for its members to surpass those of other groups, is, in large part, a reflection of the level of education and training available in their county of origin. Stephen Steinberg demonstrated this thesis in a comparative study of various ethnic and racial groups in the United States.[1] While his research dealt with adaptation to an urban environment, his argument is just as applicable to a rural setting, especially one characterized by commercial, mechanized farming. However, averages gloss over individual variations. To what extent Dutch farm immigrants succeeded depended on their ability to transfer previous farm experience and their knowledge of plants and animals to Canadian conditions. Teenagers and young adults were able to apply prior education, but had less experience. Older Dutch farmers were more experienced, but set in their ways, making it more difficult for them to adapt to a new environment. Both groups had to learn how to farm in Canada. Those who learned well were able to get ahead more quickly.

PRIOR LEARNING AND DUTCH EXPERIENCE

Many farmers in the Netherlands received training in agricultural schools, even if only on a part-time basis. The emphasis of these schools was on practical applications. Those who attended the *lagere landbouwschool* in the winters, or at night, worked on a home farm or that of a neighbor. In the classroom, students learned how to recognize different soils types, the chemical composition of fertilizers, and animal husbandry. Several farmers recalled how their teachers used a chart showing a barrel with staves of different length and breadth to illustrate different proportions and amounts of nutrients needed to grow healthy plants. Depending on the prevalent form of agriculture in a particular region, greater emphasis might be put on a particular aspect of farming. Thus, in a school located in a dairy region, students learned more about feeding formulas for cows, different types of grasses and how much fertilizer per unit

[1] See his *The Ethnic Myth: Race, Ethnicity and Class in America* (Boston, Beacon Press, 1989).

of land might give a better hay crop. All programs included some bookkeeping. Few girls attended such schools, but rural women who grew up during or after the war studied home economics in a *huishoudschool.* At the more advanced levels (*middelbare* or *hogere*) of schooling there was more specialization, with separate *landbouw* (farming) and *tuinbouw* (horticultural) schools. Those schools, which included the same subjects covered in the Dutch equivalent of high school or college, were similar to agricultural colleges in Ontario. In contrast, few farm immigrants knew English; only teenagers who had gone to a secondary school could make themselves understood.

Older farm immigrants born around the turn of the century grew up in an era when attending a *landbouwschool* was a privilege reserved for the sons of the wealthiest farmers. Nevertheless, they might have certificates for grafting, beekeeping, or the care of fruit trees from short courses. Others picked up additional skills not directly related to agriculture, including carpentry, plumbing, and how to read blueprints. I have met immigrants who had learned how to weld, do electrical wiring, or had taken courses in business economics and/or auto mechanics. It was an official postwar Dutch policy to teach a trade to farm boys, in anticipation of an influx of young people from the countryside to industrial urban centres (see chapter 1). Such skills were a competitive advantage in Canada when these young people were looking for jobs in town, and they were also useful on the farm. One Dutch immigrant farmer in Ashfield Township (Northwest) became the farm machinery repairman for the whole region, since he was the only one with mechanical skills.

Dutch farm immigrants, even those with no formal training in agriculture or trades, brought with them a wealth of experience, although the nature of this prior experience varied widely. The form of agriculture found in clay regions on the old seabed was quite different from that of the sandy regions (see chapter 1). Farmers from the sandy parts of Noord Brabant routinely ploughed the land, usually with horses, while dairy farmers in the *laagveen* parts of Friesland or Zuid Holland were accustomed to having permanent pastures that were never ploughed. Few Dutch immigrants were acquainted with mechanized farming, but most would be familiar with portable milking machines that were rolled out to the fields. Moreover, just prior to and immediately after the Second World War, numerous young men received a greater exposure to mechanized agriculture when they were recruited to work in the recently created Wieringermeer polder or the Noordoost polder, then still under reclamation. Some men had also learned about different kinds of farming, or picked up other skills, through visits to France or Germany.

LEARNING HOW TO FARM IN CANADA

The process of learning about farming started within a day or two of arrival. Some of the things young Dutch farmers learned at school were applicable to the Canadian context, but soil conditions and climate were quite different from those in Europe. A sponsor like George Atkins (see chapter 3) would show his workers what to do and then let them practise on their own, but many Dutchmen had to learn completely through observation and imitation. Once Dutch workers had acquired some English, they could communicate more effectively. As immigration increased, the chances of meeting other Dutch immigrants with Canadian experience improved and more established Dutch-Canadian farmers acted as mentors to newcomers. Nevertheless, when they started farming on their own, most Dutch farmers had a lot to learn, including when to plough or to sow a new crop. Land drainage was also different.

An intimate knowledge of Dutch agriculture and entrenched work habits could interfere with learning how to farm in Canada. Much had to be unlearned. Farmers in the Netherlands were used to doing farm tasks by hand, from milking to digging up the soil. They meticulously ploughed into the corners of fields. In Canada, with plenty of land and a shortage of labour, such intensive forms of farming were not practical. A few immigrant farmers did implement such Dutch methods of farming, with disastrous results. However, the majority soon learned which methods just did not work in Ontario. Yet, in learning how to farm in Canada, Dutch immigrants did not accept everything they saw or were told. They appreciated such labor-saving devices as manure spreaders, harvesters, and combines, but their first impression of other aspects of Canadian farming was that it was hopelessly backwards. Dutch farmhands saw that most Canadian farmers were unaware of the chemical composition or formulas for different kinds of fertilizers. Learning how to farm in Ontario was thus largely a matter of trial and error.

Making mistakes

The most common error was to plough too deep. In most of the Netherlands deep ploughing (up to eight inches) brought fertile subsoil to the surface, which was not the case in Canada. Another common mistake was to start too early, especially on clay soils. Dutch newcomers would begin ploughing the land after the first rainfall in the early spring without realizing there was frost in the ground, and experienced difficulties. Worse, the soil would turn as hard as cement after a couple of days, especially if it was not tiled; Canadian clay was less porous than the sea clay deposits of the Low Countries. There was also more sunshine.

One newcomer, who started a greenhouse operation in the muck land near Grand Bend, bought glass frames from a tobacco grower in Norfolk County, only to discover that his first crops of carrots got burnt with the heat of the sun. In some cases, a Canadian farmer tried to warn a Dutch newcomer in a joking or indirect manner. One man told a Dutch neighbor who was ploughing too early that he would be better off going to sleep, mentioning that his own son had gone to town that day to play pool. Other Canadians, who thought the Dutch were driving up the price of farmland, deliberately said nothing at all, hoping that they would fail and leave. Nevertheless, those farmers willing to listen avoided the worst mistakes, including antagonizing their neighbors. Even if they were convinced that, for example, cutting grass for silage should be done earlier, they learned to be discrete and show by example. They recognized the benefits of combining the best of both systems.

Some immigrants had to learn the hard way. A good example is a Dutch family who got involved in the sod business. Initially, they applied too much fertilizer to stimulate grass growth and did not prepare the land with heavy rollers. The sod cutter kept on breaking down because the soil was too rocky. When they were ready to sell their sod, a buyer wrote them a bad check. They improved with their second attempt, although they still had a lot to learn. After ploughing and disking the land, they left a lot of twitch grass in the soil. Two years later, only seventy percent of the resulting mixture of bluegrass and twitch was fit for harvesting. Nevertheless, they paid off their debts and, in 1967, they bought an abandoned dairy farm to grow their own turf. Over time, their experience improved.

Other forms of learning

Learning to read English usually took several years and opened up an important source of information about Canadian agriculture. Dutch farmers subscribed to *Farm and Country*, the *Western Ontario Farmer* (later called *Ontario Farmer*), the *Farmer's Advocate*, plus more specialized publications such as *Holstein Journal*. With more difficult material and technical manuals, children attending high school in Ontario helped their fathers. Another source of information was representatives from feed or fertilizer companies, who organized day trips to farms. Dutch farmers also learned through contact with agricultural extension officers working for the Ministry of Agriculture and Food, who provided technical information and promoted new techniques, in co-ordination with the Ontario Agricultural College (OAC), now the University of Guelph. Another way to learn more about farming in Ontario was by attending meetings organized by farmers' organizations and Soil and Crop Improvement Association representatives.

In the late fifties, John MacLachlan, then president of OAC, promoted Dutch days for immigrant farmers at Guelph. During those visits, young people acted as interpreters for their fathers. Dutch farmers also benefited from research done at the university to develop new varieties of corn and other crops that could be cultivated in all counties in Southern Ontario, rather than just the southwestern corner of the province. The rapid expansion of corn into zones with lower heat units coincided with the influx of Dutch immigrant farmers into Huron, Perth, and parts of Grey counties as well as Eastern Ontario. However, it was to take another ten to fifteen years before OAC, and colleges such as Ridgetown and Kemptville, were to see a noticeable presence of the offspring of Dutch immigrants. Initially older children had to help out on the farm and work to supplement family income, even if they had not yet finished school (see chapter 5).

While anxious to learn as much as possible about farming, immigrants with formal education took the advice they received from experts in Canada with a healthy dose of scepticism. One Dutch farmer in Lambton County (London region) decided to experiment, when the recommendation of someone from Guelph involving nitrate application on potato plants did not work. Instead, he discovered on his own that it needed to be applied much earlier. Likewise, someone who landed in the Waterloo region refused to continue doing business with a salesman when he found out he did not recognize that a common problem with alfalfa was caused by a deficiency of boron. The Dutch farmer was appalled that the so-called expert did not know anything about micronutrients and trace elements. In contrast, immigrants without formal education in agriculture were more dependent on private and public extension efforts.

GETTING AHEAD

Regardless of how they picked up the technical aspects of farming, Dutch immigrants had to learn to deal with bankers and the agribusiness establishment. We have already seen how most immigrants started part-time operations. Their main problem was lack of access to credit. With little collateral, Dutch farmers were not eligible for bank loans. Even those who had smuggled in what they called "black dollars" did not have sufficient start-up capital to buy farm equipment, seeds, and farm animals, not to mention being able to pay workers to harvest labor-intensive crops. Nevertheless, a combination of savings and family labor enabled most immigrants to sow a crop before the end of their second year, while holding down a job in town. We have already seen how prospective dairy farmers continued to combine off-farm jobs with part-time farming, slowly building up their herds. The decision to become a full-time farmer was not easy, especially if it meant giving up a steady wage. Many people who came to Canada with the intention of becoming farmers ended up making good money working in factories. Quite a few nevertheless took the plunge and turned part-time farming into a full-time occupation. But this required more capital. One way to make a farm viable without holding down a job was to enter into a contract with a feed company to run a chicken, turkey, or a hog operation. The company would provide the feed and deduct the cost of these expenses at the end of the season, often to the farmer's disadvantage. Dutch farmers preferred to set up completely independent operations. But this meant knowing how to access, and negotiate with, banks or government agencies in order to obtain the money required for operating expenses.

Turning to financial institutions

Dealing with financial institutions in the immediate postwar era was simpler than today. Throughout the fifties, and even into the late sixties, a farmer could still submit a written proposal or show his financial plans on the back of a cigarette box,[2] provided he was seen as a good risk. It helped to be aggressive. One Dutchman, who combined a job as manager in a company at a Thornhill staircase manufacturer with farming in Pefferlaw, did something very daring. He had already had several run-ins with reluctant bank managers when he wrote a check to buy thirteen cows available at a good price. There was not enough money in the bank account to cover the check. However, the Dutch farmer then immediately went to see a new, young bank manager of the Canadian Imperial Bank of Commerce (CIBC) to borrow the money he needed. He told the bank manager, "I know I should not have written that check, but it was a good buy. If you don't help me I will have to sell the cattle next Monday, but still make a profit." The manager was hesitant but offered to visit their farm to inspect the cattle. He also met the farmer's wife and children. He granted the loan.[3]

---

[2] Interview (MF) with Bruce Dykstra, in his home in Ancaster, 15 February, 1994. Bruce, who was eleven when his parent immigrated, spent some time on his parents' farm in rural Ontario, but then studied to become a teacher and eventually ended up in banking. He was manager of a bank in Hagersville for 15 years and dealt with a lot of Dutch farmers.

[3] Interview with one of farmers mentioned in the previous chapter.

This occurred around 1970, when Dutch farmers were starting to be recognized for their ability in turning abandoned farms into viable operations. Their predecessors had to be even more persistent to obtain credit from a bank. It helped if an influential and rich Canadian businessman was willing to put in a good word. Dutch farmers still had to prove themselves. A Dutch immigrant who had already been in Western Canada for several years before moving to Ontario in the early fifties had just enough money to put a down payment on a farm, but not to pay for the Holstein cows he wanted. When he approached the Royal Bank, the manager laughed and told him to first prove to him that he could farm. The farmer had to return several times before receiving a loan for $500, and it took several more years before being granted a line of credit.[4]

Dutch farmers approached the local branches of most major banks, although the Royal Bank was mentioned the most in my interviews. This bank was said to have quotas in the fifties permitting managers in rural areas to assist farmers each month with a set amount. Farmers without collateral were lucky to encounter managers at a time when they had more power to make decisions and more connections to rural communities. Managers were also willing to take risks. Another source of credit was a series of government farm assistance schemes, including a Junior Farmers Loans program and the Farm Credit Corporation. The latter, initiated by the federal government under John Diefenbaker in the fifties, approved long-term loans. Funds were disbursed and managed by local banks with the backing of Ottawa. This program facilitated the paying off of private mortgages held by farmers who had sold their properties to Dutch immigrants. The Canadian farmer would get a lump sum payment while the Dutch farmer continued to make his monthly payments to the bank.

Obtaining operating capital from the Farm Credit Corporation could be just as difficult as borrowing from the bank. Mac Traas, who bought a farm in Campbellville (Toronto region) requested a line of credit, pointing out that he already had 10,000 recently transplanted apple trees worth $15,000. The Farm Credit Corporation had never dealt with fruit farms and turned him down. Mac needed the credit to raise hogs and grow crops until his orchards started producing. The only way to stay in business was to sign a contract with Maple Leaf Mills, a feed company willing to provide credit to raise chickens.[5] Many people initially unable to obtain credit ultimately not only survived but thrived. Some

---

[4] Interview (MF) with Gerrit Pot, in their home in St. Anns, 3 September, 1992.

[5] Interview (FS) with Mac Traas, on his farm near Campbelville, 26 January, 1997.

Dutch farmers, hesitant to go into debt further, borrowed only enough money to cover their own operating expenses. Others wanted to expand to remain competitive and help their children become farmers.

Improvements and Expansion

Long before they could make a living from farming, or even think about expanding, Dutch immigrant farmers started making improvements. Indeed, they seemed almost obsessed with giving their farms a neat, tidy appearance, sometimes to the chagrin of their more laid-back, non-Dutch neighbours. Dutch immigrants did not like rusty, broken-down farm machinery lying around, unpainted barns and houses, and weeds in their front lawns. As soon as they moved in, the new Dutch owners of these farms removed outdoor junk, cleaned out barns, poured new cement floors and planted flowers in front yards. Remodelling the house came last. As their farm operations expanded, they took advantage of any grants for farm improvements to install liquid manure pits and pipes, and build new stalls in dairy farms.

The next step, especially in farms located on clay soil, was to improve drainage. In many parts of Southern Ontario, farms contained wet, low-lying areas, only some of which were tiled. Beef farmers did not need tiling, but when these farms were converted to corn cultivation, this became a necessity. Tiling not only drained surface water, but retained heat, allowing farmers to start earlier in the spring. However, in areas that had never had drainage, Dutch farmers first had to persuade their non-Dutch neighbors to petition local politicians to organize the construction of larger municipal drains. Tiling, which required the use of specialized equipment, was expensive.[6] Cash-poor immigrant farmer had to pay others (usually Canadian farmers who specialized in tiling as a side-line), and were also expected to feed the workmen, creating extra work to women doing the cooking. Consequently tiling was often done in stages and could take up to a decade to complete. One farmer told me he put in drainage tiles whenever he needed to lower his taxes. In the meantime, Dutch farmers provided a lot of business for tilers. Other improvements during the first couple of years could include installing grain bins, setting up a feed mill, replacing the original low-grade cows with pedigree ones, adding a stable cleaner or erecting a new silo. Within a few years, Dutch immigrant farmers began acquiring additional land.

[6] Dutch farmers who were interviewed consistently reported tiling as one of their top expenses.

Yet, apart from buying one or two additional farms, or occasionally leasing land, most rural enterprises owned by Dutch immigrant farmers did not expand dramatically. They grew enough to sustain themselves, but continued to be average, though efficiently run, family farms.

Regardless of their energy and determination to stay in farming, once they had accomplished their goals, older Dutch farm immigrants were reluctant to expand further. They had already experienced the hardships of the Great Depression, the trauma of war and had struggled to buy a farm in Canada and keep it operating. In most cases, they had even managed to set their children up in farming. Unlike their offspring who grew up in a more stable environment, most of the Dutch immigrants who had grown up before the war did not want to take any more risks. But some of them were nevertheless willing to try something new or develop sidelines beyond farming.

Conversions and sidelines

Many first-generation Dutch-Canadians who had started out farming, veered off in other directions. One farmer, with a large family, who arrived in Canada in 1947, started a mixed farm with grain crops, cows and hogs in 1951. By the mid-fifties, he had not only set aside five acres to grow potatoes and turnips, but had also converted one of the barns to raise broilers. Another farmer, originally from Noord Brabant, started raising pigs on what had been a fruit farm near Niagara-on-the-lake in the mid fifties. Then, fifteen years later, he switched completely over to growing grapes. Other cases involved abandoning livestock and focusing exclusively on cash crops, starting up a white (milk-fed) veal operation, and converting a greenhouse operation from flowers to vegetables. Other Dutch farmers took on farm-related income-generating activities. Several of the farmers in my study had sold real estate on the side or became appraisers for crop insurance. One opened a small country store, and still others repaired farm machinery or did custom spraying. Many sidelines evolved out of innovations originally designed to improve their own farms. Thus, a farmer interested in lowering the cost of having a fodder crop milled or stored, might set up his own feed mill or a small elevator. It was a natural step to first grind corn for a few neighbors in order to fully utilize newly-purchased equipment, to then develop a small enterprise to serve a wider area. Secondary operations typically corresponded with previous farm activities, as when apple producers packed their own produce and then began packing apples purchased from other farmers. Sometimes the sideline might completely overshadow the farming, as in the case of a Dutch immigrant who set up a butcher shop.

Luit Miedema, who came from a farm background and wanted to become a farmer, started working for a Canadian farm sponsor near Simcoe in 1950. One day, the Miedemas ordered smoked, sliced horsemeat (*rookvlees*) from a traveling salesman. Luit's employer was shocked, and questioned whether it was "really a Christian thing to do,"[7] and gave his Dutch family piglets to fatten on the condition that they stop eating horsemeat. When the pigs were ready, he showed Luit how to butcher the meat. The Miedema family was allowed to keep half the meat for themselves and sell the remainder to other Dutch immigrants who belonged to their church. Four years later, Luit bought a corner of the farm where he worked and built a small slaughterhouse. At that time (prior to 1966) a special permit was not required. He eventually specialized in butchering and established several slaughterhouses in Norfolk and Oxford counties. Several Dutch immigrants who first developed livestock farms likewise set up abattoirs in other parts of rural Ontario.

In other cases, family members pooled their resources to finance a farm-related business. Harry Vanderzanden, who came to Canada in 1950, at age sixteen, farmed for over a decade before purchasing a feed mill, with a sister and a brother, also farmers. Their parents and an older brother had worked for George Atkins after landing in Canada (see chapter 4). His subsequent career as an industrial worker, farmer, and entrepreneur illustrates what Dutch farm immigrants who arrived as teenagers with primary education could accomplish, if they also had the ambition and family connections. Harry first got involved in farming when he was working for a trucking company. He rented thirty acres of land to grow oats and wheat for sale, and five acres of tomatoes, on contract. One often had to be creative to farm without money: Harry picked up a tractor from a dealership, claiming he wanted to try it out, but returned it when he had finished his ploughing. He borrowed a three-furrow plough from his father, and a disk and a plough from a fruit farmer he knew. In 1962, Harry set up a chicken operation in an old barn. At that time, he was living in a rented farmhouse while working full-time for a steel plant in Hamilton. A year later, he bought a chicken farm with three acres of land, which he sold to his father five years later. In 1968, the same year he quit his job, Harry bought a larger farm in Fenwick with a chicken and turkey quota. Harry Vanderzanden started expanding in 1975, when he bought a turkey operation in St. Anns (also in the Niagara Peninsula), where he put a hired man in charge. In 1980, Harry got involved in the feed business when another

[7] Interview (FS) with Luit Miedema, 12 June, 1989.

son started working full-time on the home farm. He knew the owner of a feed mill in Smithville and started working part-time selling feed to other chicken farmers. Four and a half years later, Harry bought that business, together with his relatives, when it went into receivership. His sister and brother were silent partners. In 1983, he also bought a grain elevator in Wellandport, and in 1988, added a fertilizer division.[8]

Other immigrant farmers invested in tile-laying equipment, which then became a sideline. For example, the Sanders, a large postwar family in Warwick Township (Lambton), started a tiling business to supplement their farm income. It was later turned into a full-time business by one of their sons. The son continued to share his tiling equipment with his brothers and his father as their business grew.[9] A sizable minority of rural enterprises owned by first generation Dutch-Canadian farmers expanded to the point where they were well above average in both assets and amount of land owned. However, the ability of Dutch immigrant farmers not only to establish viable farm enterprises, but also branch out into other ventures, sometimes created jealousy and resentment. Such resentment began even earlier while immigrants were still struggling to get established.

Resentment and Jealousy

Most of the Dutch immigrant farmers I interviewed said they got along well with their non-Dutch neighbours. However, they would sometimes mention that they thought some of those neighbors were also "a little bit jealous". When pressed for details, it became apparent that the resentment directed towards them was more than just a little. The following comments by an immigrant farmer illustrate how jealousy could escalate as the Dutch farmer started getting ahead:

> The established Canadian farmers initially did not show any resentment. When our barn burned down, they were ready to help us, especially someone who was our best neighbor. But as soon as I got my feet under my butt, that neighbor became very critical. It almost seemed like they changed overnight. For example, when we put up that blue silo, one of the first in this region, our boys were attending high school and they noticed how the stories started about how much money we must have.[10]

The theme of resentment came up in many interviews. One farmer told me that he thought Canadians were jealous because they could not

[8] Interviews (FS) with Harry Vanderzanden, 5 and 21 June, 1995.

[9] Another son started a machinery equipment business. Interview with Cornelius Sanders, Oct. 20, 1990.

[10] Quote from a farmer in Culross Township, Bruce County.

understand how the Dutch could accomplish so much so soon after arriving as "dirt poor immigrants." A family in Flos Township (Lake Simcoe) became aware of such resentment when their children heard stories at the separate (Catholic) school they attended about how the Dutch were taking over. They were referred to as "dirty Dutch kids" behind their backs. In some cases, resentment took the form of outright prejudice, as in the case of a farmer in Warwick Township (London region), who believed in a conspiracy by the Dutch to take over agriculture all over the world.

The Dutch farm immigrants who arrived in Ontario after the Second World War came from a country where farm families had a relatively high level of agricultural education compared to other parts of Europe, despite the economic difficulties and social disruption caused by five years of Nazi occupation. Their ability to learn and get ahead in rural Ontario demonstrates the utility of a general education, even at the lower level, as well as practical training. More educated, prosperous Canadian farmers who sponsored Dutch farm families were not surprised that their employees managed to set up their own farms. They recognized the skills the Dutch brought with them and their ability to learn and adapt. However, those Canadian farmers who operated marginal operations under difficult circumstances could not comprehend why their new Dutch neighbours did so well. The arrival of a large number of immigrants in rural Ontario after the Second World War, and their subsequent economic success in many regions, was bound to create tensions. Other non-Dutch farmers, and people beyond the farming community, admired the Dutch and thought they were all rich and successful. However the Dutch presence was not equally conspicuous in every instance; farm immigrants from the Netherlands were not evenly distributed in the province either before or after they started settling down. Consequently their impact on the natural and social landscape was more noticeable in some regions than in others.

VII. Dutch Farmers in Different Regions

When the Dutch started arriving in the late forties, they had sponsors all over the province. But, they became more dispersed as family members went off to work in construction sites, lumber camps, and mines. Within a decade, people started coming back together, as they gravitated to areas where affordable land was available. In some townships, settlement remained scattered. Other areas witnessed the clustering of Dutch farmers, who took over whole lines, changing the ethnic and religious composition of parts of Ontario. Most cases, with a modest level of spatial concentration, lay somewhere in between (see chapter 12). Where people ended up depended on many factors: availability of affordable land, the availability of off-farm work, and access to churches. In some areas, families belonging to the same church or from the same region converged, though never to the same extent as the prewar Dutch who settled the Holland Marsh.

This chapter will provide a series of snapshots of how groups of Dutch farm immigrants settled in various parts of Ontario, with an emphasis on the early phase of adaptation and integration. I selected distinct regions in South-western, Eastern and Northern Ontario in order to illustrate a diversity of experiences in three areas: 1) contact with other Dutch immigrants, 2) types of farming systems, and 3) availability of income sources beyond farming. These case studies also illustrate how social and religious divisions in Dutch society influenced where people settled in rural Ontario.

BOG FARMING IN THE GRAND BEND MARSH

Prior to the Second World War, Dutch farmers were involved in turning the wetlands at the southern end of Lake Simcoe into a major center for market gardening (see chapter 3). Postwar Dutch immigrants created a small-scale version of the same kind of muck farming in the London region, near Grand Bend located on the southeast fringe of Lake Huron. The marshes and shallow lakes of that region were drained and partially converted into arable land by the Canada Land Company, just after the First World War. There was no Dutch involvement at that time. After the Second World War, the market gardeners in that area faced fierce competition from the Holland Marsh. They encountered even more competition from postwar Dutch immigrants who developed a section of nearby bog land in Grand Bend.

Unlike the Holland Marsh, the farmers who cleared the Grand Bend Marsh were Roman Catholics. All but one originated in Noord Brabant. Jerry VanLeeuwen, who first worked for another Dutch farmer in the Marsh, was the only one who came from the province of Zuid Holland. His father had lost his small farm near Delft during the Depression, but Jerry gained greenhouse experience in the Westland region and took evening *tuinbouw* courses after the War. He saved up enough money to pay his passage to Canada in 1951 and, a year later, bought his own land in Grand Bend, from a landowner of German descent. He cleared four acres by the end of the following year and started working full-time on the farm in March of 1953. A younger brother who came to Canada a year later began working for him. Together they dug more drainage ditches and by the end of the year, had put up a provisional greenhouse structure for growing lettuce and celery for transplanting. The following year Jerry returned to Holland to get married. When the couple returned to Canada, they had to spend their first summer in a motel. It was to take another ten years before all of his land was cleared and tiled.[1] Like other farmers in the Grand Bend Marsh, Jerry used a Dutch system of drainage (a series of small ditches) running into the Goose Marsh Ditch, the main drainage canal. These farmers employed other Dutch immigrants and the children of those immigrants. They applied what they had learned at home to grow onions and turnips on virgin soil, but none had prior experience growing seed potatoes. But they learned quickly through study and experimentation. They were all subject to the vagaries of weather and plant diseases and the ensuing years of both prosperity and losses, and became embroiled in the controversies surrounding the bankruptcy of a co-op in Thedford that bought their produce.

The Dutch farmers in the Grand Bend Marsh applied their knowledge of agriculture, such as the importance of micronutrients, to produce a better quality product. The carrots they grew were sweeter and much in demand by local supermarkets, giving them a competitive advantage vis-à-vis the descendents of the non-Dutch farmers who had moved into the area prior to the Second World War. However, the Dutch were

---

[1] Interview (FS) with Jerry and Tilly VanLeeuwen, at their home in Grand Bend, 14 July, 1995; supplemented by his own written immigration account "The Story of Jerry VanLeeuwen", self-published, (circa 1995).

not the only immigrant farmers who arrived in the late forties and fifties; a group of immigrants from Poland bought and cleared land in the Smith Lake portion of reclaimed land near Grand Bend. These farmers were situated about halfway between the Dutch farmers and the older Canadian settlers to the south. Members of the two groups worshipped in the same church in Grand Bend, where the Dutch composed half of the congregation. The Dutch also co-operated with their Polish neighbors in setting up separate (Catholic) schools. However, whereas the Dutch farmers increasingly specialized in market gardening, Polish farmers became dealers. While the vast majority of potato farmers in the region today are Dutch-Canadians, they sell most of their crop to several Polish families who are both growers and dealers.

THE DRAYTON AREA

Drayton is a village in the township of Maryborough, in the northwestern part of Wellington County (Waterloo/Guelph region). Nearby villages include Moorefield and Rothsay. The soil consists of the harder clay of the till plain of Dundalk. When the Dutch appeared just after the war, most barns were not painted and land was cheap. These immigrant farmers transformed both the village and countryside. The influx of Dutch farmers to this region serves as a good example of the leadership role played by farm immigrants who were more educated or who had prior business experience. Drayton is also a good example of the relationship between Dutch immigrants who ended up on farms and those who moved to nearby villages.

The Flinkert family, who came from Zuidwolde, near Hoogeveen (Drenthe), was the first to buy land in the Drayton area. Wim Flinkert, an experienced farmer, started with a small dairy operation and eventually became one of the biggest milk producers in the province. He and his son were instrumental in bringing in numerous immigrants, mainly from the province of Drenthe, to the Drayton area. The farm immigrants who came to the area included a mix of small farmers, farmers' sons, and people who did not own any land back home. They were, for the most part, sponsored by fellow immigrants who were unable to provide them with work. Even the Flinkert family, who eventually employed many farmworkers, initially sent the families they brought to Canada to farms owned by Canadians not of Dutch descent. Furthermore, there were a few jobs in the region outside of agriculture, so immigrants had to earn money in other parts of the county. They worked in several fertilizer plants in Elmira, on the construction of a General Electric plant in the city of Guelph, and in the tanneries of Acton. Most moved to these places until they were ready to buy farms.

The postwar immigrants who settled in the Drayton area were Calvinists (Reformed), but religion had a different dynamic here than in the Holland Marsh. Although they initially worshipped under the auspices of the Christian Reformed Church, they belonged to various denominations (mainly *hervormd* and *gereformeerd*). When the Reformed Church of America (the closest equivalent to the *Hervormde Kerk*) began recruiting immigrants in Ontario, they appointed Berend Flinkert, the son of the Dutch farmer in Drayton, as a fieldman. In 1952, five families in nearby Harriston (Northwest Region) joined with the Flinkerts and others of *hervormd* background to set up a congregation separate from the one already established by Christian Reformed immigrants.[2] Each group eventually built its own church in Drayton. As soon as Berend Flinkert started working for the Reformed Church, he recruited families from his home region of Koekange (half way between Hoogeveen and Meppel), including Korrie Krabbe. She was a teacher and, her husband, also trained as a teacher, had worked as a government employee in charge of distributing cattle feed. They wanted to move to Canada to become farmers, so a former neighbour, then living in Drayton, passed their names on to Flinkert, who found a sponsor in Rothsay. Korrie was recruited to run a Sunday school for the Christian Reformed Church.[3] Another person with professional qualifications was Ralph Spikman, who became a key figure for the local Dutch farming community as manager of a co-operative feed mill. He started that co-op from nothing, as he had done in Holland.

Ralph's father, a miller, had made sure his son knew how to milk even though he had plans to become a farmer.[4] By the time Ralph emigrated, in 1950, at the age of 43, he had a position as director of a Dutch farmers' co-operative in the Wieringermeer polder. He left the Netherlands because he wanted to keep his family together, after his son, then eighteen years old, announced he was moving Canada. They ended up in Belwood on a farm owned by a Dutchman, but after just four weeks, Ralph bought his own farm from another Dutchman who

---

[2] Interview (FS) with Jan Reinders, on his farm near Moorefield, 13 January, 1991. His story was also included in Albert Van der Mey's *To All Our Children: the Story of Postwar Dutch Immigration to Canada* (Jordan Station: Paideia Press, 1983), 72-73; 232-33.

[3] They joined a United Church when they bought their own farm in Alma (in nearby Pilkington Township). Interview (FS) with Korrie Krabbe, at her farm home near Alma, 14 January, 1990; also archives of the Multicultural Historical Society of Ontario (MHSO), taped interview done by Mark Boekelman on 18 April, 1977 (DUT-0384-KRA).

[4] His story was written based on two sources of information: my interview with him, at a retirement home in Guelph, February 26, 1990, and a taped interview done by E. Hietkamp (MHSO) on May 8, 1978.

had returned to Holland. Ralph's son joined his parents when his sponsor no longer had work for him. They planted grain and twelve acres of cucumbers, and started working for a pickle factory. Ralph's job was to pick up cucumbers from all of the Dutch farmers. Six years later Ralph quit to spend more time on the farm. He also bought a truck to deliver his own cucumbers to the factory in Guelph. With the income from his first crop of cucumbers, Ralph bought some cows. He soon realized, however, that he was not cut out to be a farmer and decided to try something else. In 1952, Ralph Spikman found a feed mill in Drayton for sale for $4,000, and he jumped at the chance. Ralph's first customers were Dutch farmers, the first to start farming full-time. While Ralph had to pay cash to a larger co-op for feed, his own customers did not pay until they sold their animals or received their milk cheque. So, the manager of the co-op suggested that Ralph and his customers set up their own local independent co-op. Just like in Holland, Ralph became a manager on salary. Ralph Spikman turned the Drayton co-op into a viable operation, which soon employed twelve people. Ralph then turned his business talents to ventures of his own, and also became involved in public life as a member of the town council. Ralph also served for seventeen years as chairman of an annual provincial Dutch day for Calvinists.[5]

Ralph Spikman was not the only Dutch immigrant who lived in the village of Drayton. The history of the Hesselink family can be used to illustrate the intertwining fates of Dutch farmers who set up businesses in town and their rural customers. Originally from a farm in Winterswijk (Gelderland), John and Gerda Hesselink were living with their sponsor, a Canadian farmer in Mount Forest, when John fell from the hayloft and broke his back, leaving him paralyzed from the waist down. After ten months in Toronto hospitals, he and his wife decided to open a store in a rented house on the main street in Drayton, with the help of the Reformed Church. Within a year, they expanded from selling groceries and toilet paper, to hardware and specialty goods imported from Holland, reflecting the rising incomes of Dutch farmers. Within three decades, the business had become a small shopping centre managed by their son.[6] The arrival of people like Hesselink and Spikman enabled Drayton, once a small village of 200 inhabitants, to become a thriving town. Prior to 1950, women living on nearby farms preferred to do their

[5] The Dutch day rotated among several places, and was held in Moorefield, Stayner, Blyth, and Strathroy.

[6] See Albert Van der Mey, *To All Our Children* (Jordan Station: Paideia Press, 1983), 401-2.

shopping in Kitchener or Guelph, but over time, Drayton became a service center as more and more Dutch families settled in its rural surroundings. However, while Maryborough Township became a predominantly "Dutch" area, Dutch farms extended only as far as the top half of the neighboring township of Peel, a predominantly rural Mennonite area. This phenomenon of Dutch ethnic enclaves in the rural landscape was replicated in many parts of Southern Ontario, such as on the Niagara Peninsula, and to a lesser extent, in Eastern Ontario.

DUTCH FARMERS IN EASTERN ONTARIO

Like other parts of the province, Eastern Ontario witnessed the arrival of both Roman Catholic and Calvinist Dutch farm immigrants. People belonging to these two religious groups often landed and settled in different places, although they occasionally became neighbors.

Wolfe Island

When postwar Dutch immigrants discovered Wolfe Island in the early fifties, they did not recognize that Mr. Pike, a landowner and cattle dealer who sold some of them farms, was of Dutch descent; his original name (Snoek) had long ago been anglicized.[7] Wolfe Island was attractive to Dutch immigrants because its land, consisting of the clay flats found in other parts of Eastern Ontario, was among the most affordable in Southern Ontario. A Dutch journalist making a Canada-wide tour paid a visit in 1950 after he heard Wolfe Island referred to as "new Holland"[8] during a visit to the nearby city of Kingston. His escort around the Island was Peter Draaijer (Drayer in Canada), the first Dutch immigrant on Wolfe Island after the war. An "invasion" of newcomers from the Netherlands occurred soon after his arrival. Peter Drayer came to Wolfe Island in 1948, together with his wife and his father-in-law's family, the Hasselaar's (Hesler in Canada). His ancestors were plantation owners in the former Dutch East Indies, but Peter studied to be a marine navigator in Rotterdam. During the Second World War he went underground to escape the Nazis and hid on a chicken farm in Veenendaal (Gelderland) belonging to his future father-in-law. Just after the War, during one of his trips at sea, Peter was so impressed by the port city

---

[7] Interview with Len and Lientje DeBruyn, at their farm on Wolfe Island, 2 June, 1991.

[8] J.W. Hofwijk, *Canada: reis met onbekende bestemming* (Utrecht: Uitgeverij de Lanteern, 1952), 113. Sections of this book appeared as a series articles in the Dutch newspaper *De Maasbode* of Rotterdam.

of Vancouver, that he persuaded his wife to move to Canada. They bought a 50-acre farm on Wolfe Island, and Peter worked during the winter as a snowplough operator. As soon as the winter was over, he started working on a ship on the Great Lakes. In 1949, no longer interested in seasonal work, he started working in the local office of the Industrial Finance Corporation, which provided short-term credit for farm improvement. In the meantime, he bought his own farm, next door to that of his father-in-law. At the time of his meeting with the Dutch journalist, Peter and his wife had already bought some chickens and pigs and were slowly building up their farm.[9] The only drawback to living on the island was winter travel, especially crossing the channel between the island and the city of Kingston. Immigrants who arrived late in the season could only bring their belongings across with a horse and sleigh. In one case, a family had to leave their furniture on the mainland because of a sudden thaw.[10]

The Dutch on Wolfe Island are a good example of how postwar Dutch immigrants became an integral part of the social fabric of rural Ontario, especially if they found a religious denomination similar to that of their homeland. The Island had a Catholic Church, which soon became a center for social interaction between Canadian farm families and the Dutch Catholic immigrants who arrived after the War. With their large families, the Dutch soon formed a quarter of the parish.[11] The few Dutch Protestant immigrants on Wolfe Island joined the local United Church.[12] Members of both religious groups were accepted into a broader, close-knit island community, which came together at harvest time when the threshing machines arrived. All the men, including those who were immigrants, worked as a team, and the farmers' wives prepared a communal meal set out on a huge table. Dutch farmwomen were unfamiliar with this custom, but they were soon helping out and learning how to bake pies. However, the majority of the men and women who bought farms on the island had to work hard to make ends meet.

A typical example of a Dutch immigrant farmer on Wolfe Island is Piet van Hal. He and his wife, Mina, originally came from Langenboom (Noord Braband), where they had a small 1-hectare, mixed part-time farm. They emigrated in 1948, and worked for a farm sponsor before

---

9 Hofwijk, *Canada*, 116.

10 This section on Wolfe Island is based on information provided by a former resident, plus four people in Eastern Ontario who had a lot of contact with the Island or its Dutch inhabitants.

11 See a brief mention to these families in Van der Mey, 216.

12 Interview (FS) with Martin Vorstenbosch, at his farm near Lansdowne, 14 July, 1990 (formerly a resident of Wolfe Island).

landing jobs in a cheese factory. In 1951, they moved to Wolfe Island where they bought a 50-acre farm for $2,000, but they were not able to start farming full-time. For seven years, Piet worked in a tannery in Kingston just like many other part-time farmers. They would catch the ferry after an early morning milking and return home to work on the farm in the evening. Mina looked after the farm when Piet was gone, and raised their six children. Most of these immigrant farmers were able to eventually set up viable farming operations. According to a Canadian journalist who visited the Island in 1952, sixteen of the twenty-one Dutch immigrant families then on the island owned their own farms, and some were starting to expand.

The most prosperous farmer, "Papa" Cijsbertus Vollering (known as "Ben") had $20,000 worth of livestock, machinery, land, and buildings, with less than $4,000 owing.[13] He had also started off doing odd jobs, while living rent-free in a house on a vacant farm owned by the Catholic Church. The friendly owner of the general store supplied them with food on credit during their first winter. Ben continued to do odd jobs after they bought their own farm, after selling some of his wife's jewelry to raise money for a down payment. However, while the Vollerings established a viable farm within four years, others had to wait longer. Piet van Hal did not farm full-time until 1958. He expanded in 1969 and sold his farm in 1975, when none of his kids wanted to take over.[14] The De Bruyn family, who came to Canada in 1954, was the last to move to the Island in the seventies.[15]

Like other parts of Ontario, the number of farms on Wolfe Island has declined dramatically over time. There were 208 farms before the war, by 1951, there were 103, fifty remained in 1977, and, by 1990, there were only eighteen full-time farmers.[16] The fact that eleven out of those eighteen farmers were Dutch-Canadians (by now mainly second generation) indicates that those of Dutch background are more likely to stay in farming, even in less prosperous agricultural regions. Moreover, many of the Dutch farmers who were on Wolfe Island in the sixties left the island to farm elsewhere.

---

[13] Hilliard, "How the Dutch Made Good on Wolfe Island," in the *Star Weekly, 23* February, 1952, 2.

[14] The Van Hal story, obtained by Ad Wijdeven, a visiting Dutch writer, appears in *Oogsten op vreemde velden* (Uitgeverij Terra Zutphen, circa 1983), 48.

[15] They were still farming on Wolfe Island when I visited in 1991. He runs a dairy operation and his wife raises chinchillas. Interview (FS) with Len and Lientje de Bruyn, at their farm on Wolfe Island. 2 June, 1991. The Bruyns bought the Van Hal farm for $200,000 in 1975 (incorrectly reported as $40,000 in the Wijdeven book).

[16] Most of these figures are mentioned in the Wijdeven book.

The County of Dundas

The Dutch farm immigrants who left Wolfe Island moved further east to farm on the Glengarry till plain or the Winchester clay plain also in the former county of Dundas. They joined other immigrants who had bought farms there. As in other regions, Catholics and Calvinists formed overlapping clusters. The latter established a church at Dixons Corners, close to Iroquois. That community grew and expanded; several decades later, first generation families established a church close to Williamsburg. Both congregations continue to be predominantly rural. Dutch Catholic families, who had a choice of three churches, were more spread out. Nevertheless, they gravitated towards the Winchester township area where approximately forty families took over farms owned by French-Canadians. During the fifties and sixties, the Dutch immigrants still spoke Dutch as they lingered outside church after mass.[17] Dutch farmers soon outnumbered their non-Dutch neighbors on several concessions (a subdivision of land in township surveys).

The people who bought farms in the Chesterville area included a mix of big families and young couples. Most had lived in other parts of Ontario and even other Canadian provinces; one man first worked in the mines up North and several brothers were in Manitoba before moving to Ontario. A few others, including Gerry Brugmans and his wife, arrived directly from the Netherlands. They were married shortly before leaving the village of Erp (near Eindhoven, Noord Brabant) in 1954, to start a new life in Canada. His father, who later visited them in Canada, owned an average-sized Dutch mixed farm, with an emphasis on dairy. Gerry initially worked for a farm sponsor in Dixons Corner, followed by bush work in nearby Williamsburg. His next job was laying cement blocks for a construction firm based in Bells Corners, close to Ottawa. He worked together with several other Dutchmen all of whom had received prior training in various aspects of construction work. Gerry was able to earn more money starting in 1956 working on the construction of the St. Lawrence Seaway. That project provided work for countless immigrants, allowing them to save money to buy farms. The Brugmans put a down payment of $4,000 on a 125-acre dairy farm close to Chesterville in 1959. The family left the village of Elma (Williamsburg), where they had been living, to move to the farm, while Gerry worked at his job. At that time, several Dutch families were farming along the same concession road. The Brugmans, who started with

[17] All the English-speaking farmers would head straight home. Interview (FS) with Gerry Brugman, at his farm near Chesterville, 2 July, 1991.

thirteen cows and nine heifers, shipped their milk to a Nestle's plant. They gradually expanded, buying more land, and their fortunes improved even more. Gerry, the son who emigrated, surpassed his father: he eventually owned over 1,000 acres of land, combining cash cropping with a large enough dairy operation for several sons to live off of comfortably. While few Dutch immigrants in the county of Dundas became that big, or passed their farms down to the next generation, most farms became full-time operations.

THE DURHAM REGION IN GREY COUNTY (NORTHWEST REGION)

Not all immigrants bought farms on the same island or congregated in a village. In Durham, a group of farm families remained dispersed even though they came from the same Dutch area and had a common religious background. The town of Durham and its rural hinterland (not be confused with Durham county) is located in a southern corner of Grey County. Situated in the township of Glenelg, Durham is a service center for farmers living in Bentinck, Egremont, and Normanby townships. Geographically, this area includes the drumlin fields shared by Bruce, Huron, and Wellington Counties, as well as a section of the same Dundalk till plain as the Drayton area. There was no Dutch presence prior to the influx of immigrants, and the proportion of the population of Dutch descent today is between 1 and 3 percent, which is characteristic of the majority of townships in Southern Ontario (see chapter 11). Like other people who knew each other prior to coming to Canada, the Dutch immigrants remained in touch with one other. Yet, they were also absorbed into rural Canadian society. This pattern of dispersal and integration is typical of a lot of regions where only a handful of Dutch farmers became established.

Albert John Koeslag, a farmer with a medium-sized mixed operation in the old country was the first immigrant to arrive in the area in 1948. He and his family, including nine of their thirteen children, came from Laren, in the Achterhoek region of Gelderland. Their sponsor was a Toronto businessman who owned a farm close to Durham. Two more children soon joined them. Albert's oldest son, John Koeslag, was twenty-seven years old when they emigrated. He had an agricultural diploma from Wageningen, but was engaged to a woman not of farm background. His fiancée joined him two years later, and they were married in Durham in 1952. John took additional courses at the Ontario Agricultural College and then taught science in the high school in Durham (1955 to 1962). His brother Gerrit Willem (Bill), the first of the Koeslag children to get married in Canada (1950), married a woman from nearby Princeville.

She was the oldest daughter of another Dutch family (the Kloosterboers), who later bought a farm near Clifford. Soon other families from the Achterhoek area, many of them related through family ties (the Inthof, Meyerink, Woestenink, De Greef, and Schieven), showed up in the Durham region.[18] Canadian farmers sponsored some of these families, while others first spent a year elsewhere in Ontario. Because there were few non-farm jobs in Durham, they had to earn additional money by working in agriculture. Within six years all these families from the same Dutch region were operating their own farms within a twenty-five-mile radius of Durham, in such places as Princeville, Hanover, and Clifford.[19] The descendants of these families still live in the same area, although no one else married anyone of Dutch origin.

The case of Durham illustrates the rapid social acceptance of Dutch immigrants, especially if their neighbors had received assistance from the Dutch during the Second World War. Albert Koeslag and other members of his family were part of the Dutch resistance movement when they first came into contact with Canadians in April of 1945, during the liberation of Holland.[20] In Ontario Albert achieved wide-spread attention while still working as a farmhand. On May 8, 1949, the Lieutenant Governor of Ontario, the Hon. Ray Lawson, presented him with the King's Medal for Courage in Durham Park, for assisting sixty allied airmen shot down over Holland.[21] When their oldest son was married to his Dutch fiancée in 1952, a RCAF (Royal Canadian Air Force) officer

---

[18] Much of this information was obtained during a telephone interview with Gerry Koeslag (a son who did not become a farmer), St. Petersburg (near Kitchener), 18 February, 2002.

[19] Ironically, Albert Koeslag was the only one of the Dutch immigrants who arrived in this region who did not become a full-time farmer in Canada. Although he bought a piece of land with the help of his children, he and his wife went back home in 1957 due to illness, leaving their eleven grown children behind. The local press reported that the couple went back because they were not eligible for a pension in Canada, but their son John told me (in the telephone interview) that this was not the reason for their departure.

[20] Mrs. Koeslag's brother, an officer in the Dutch underground, was beaten to death by the Gestapo after he was caught spying on the movement of German army units. One of Albert's sons (another Albert) narrowly escaped deportation to Germany by jumping off a train.

[21] This and other events were reported in the local press. See Koeslag papers (1949-1975), including numerous newspaper clippings, in the Dutch-Canadian papers series in the MHSO, Acc. No. 21210, MU series 9418, MSR No. 6267. Information obtained from the newspaper clippings was supplemented by from a telephone interview with Mr. John Koeslag (Albert's oldest son), Quadville, Ont., on April 25, 1993. He also pointed our some minor factual errors in the Durham newspaper articles.

who was helped by his family attended the wedding.[22] Their integration into Canadian society can be demonstrated by their involvement in social functions and Canadian rural institutions, as reported in the local press. We learn about regular visits from Mr. and Mrs. Kessler, John Koeslag's Dutch in-laws, starting in 1957. During one of their trips in the sixties, this older couple provided the music for a meeting of the Grey County chapter of the Imperial Order of the Daughters of the Empire, held at John's home. When Mrs. Albert Koeslag returned to Canada after her husband's death in 1962, she became an active member of the local Knox United Church[23] and its women's organizations. Her son Bill (the farmer) became vice-president of the United Cooperatives of Clifford and the chairman of the Clifford Medical Building committee.

NORTHERN ONTARIO

Most of Northern Ontario is not suitable for agriculture with the exception of the "clay belts" (see chapter 2). It is these clay belt regions that saw the arrival Dutch farm immigrants, who took over farms whose former operators used to work in the bush or mines. Just as in Southern Ontario, Catholic and Protestant immigrants clustered in different areas.

The Region of Sudbury

The Regional Municipality of Sudbury with its outlying villages is known for its mines and slagheaps. The first sponsored Dutch immigrants to arrive were the Solominks. They arrived in 1948 to work in the bush for a farmer in Capreol.[24] However, few Canadian farmers in the north sponsored immigrants. The vast majority of immigrants, who had already worked on farms in Southern Ontario, came on their own after hearing about the high wages in mining and forestry. They usually did not stay longer than a few years. I have spoken with numerous Dutch-Canadian farmers who at one time or another worked in Sudbury to save money for a down payment on a farm in the Ottawa valley or other parts of Southern Ontario. Those who stayed to work for Falconbridge or the

---

[22] His wife's brother was also involved in the Dutch underground and died in a German concentration camp.

[23] The Koeslags, who had been *hervormd*, joined the United Church. The other Dutch farmers became members of this denomination or joined the Presbyterian Church. When Bill Koeslag married their daughter, he too, became Presbyterian.

[24] This family, who bought their own house in nearby Hanmer, started farming on a part-time basis. Interview (FS) with Adrian and Mary Smits, at their home in Sturgeon Falls, 16 April, 1991.

International Nickel Company (now Inco) did some farming on the side, but few became full-time farmers.

Prior to the excavation of numerous mineshafts in what is now Sudbury, there were mixed dairy farms in nearby Blezard Valley, Chelmsford, and Hanmer, where the stone-free light soil was ideal for potatoes. These farmers were predominantly French-Canadian. As a result of the devastating impact of the Depression, many of these farmers and their children started working full-time in the mining industry, abandoning their farms around the time the first Dutch arrived.[25] However, the growth of Sudbury generated a demand for fresh vegetables and milk, thus creating a market for Dutch farmers such as Adrian and Mary Smits, who farmed full time in the Sudbury area until they retired. Adrian's oldest brother was living on his own farm near Richmond, close to Ottawa, when Adrian and his wife emigrated in 1950. Adrian's father and another brother stayed on the farm in Ottawa, but Adrian and his wife went to Sudbury together with his brother-in-law to work in the Copper Cliffe mine smelter. Eventually, they all became full-time farmers. In 1952, they bought land in the nearby village of Hanmer and started a small potato farm. At that time, there were many part-time farmers, including newly arrived Dutch immigrants. In 1955, Adrian was still working in the smelter when they bought a dairy farm in Chelmsford. There were quite a few dairy farms at that time, owned by French-Canadians. A few years later, Adrian became a full-time farmer after buying more milk quotas.

Over the years, the Smits bought more farms until they had 250 acres to grow their own feed. Six of their children, like those of other Dutch farmers, attended the French-only separate schools and became fully bilingual in French and English. In 1970, they moved further east, close to Verner (half way between Sudbury and North Bay), where all of their neighbors were again French-Canadian. There, they bought two farms, one for themselves and another for their son, who married the daughter of a French-Canadian farmer. The land in Verner is slightly better than that north of Sudbury, but more importantly, the Smits found dairy farmers willing to support each other. Indeed, they were not the first full-time farmers to move out of the Chelmsford area. Any farmer close to Sudbury found that their kids preferred to work in the mines than do chores. By the end of the seventies, nearly all of the dairy farmers in Chelmsford had moved to the area around Verner or further east, within the narrow clay belt that runs to the Quebec border.

---

[25] Interview (FS) with Eddy Verhoeven, at their home in Blezard Valley, 11 April, 1991.

We can gain further insight into the experience of Dutch rural immigrant families in Sudbury, particularly those who became part-time farmers, by looking at the Verhoeven family. This family combined work in the mines with farming in Blezard Valley. Laurina Verhoeven was twenty-eight years old when she and her husband left Noord Brabant in 1949 to work for an English-speaking farmer in the province of Quebec, before moving to Southern Ontario where they lived as hired hands on a farm in Innerkip (Oxford County). Three years later they moved up to Sudbury to earn as much money as possible. She cleaned offices while her husband joined the many Dutchmen then working for Inco. He worked double shift for the next fourteen years, until he was injured and went on a disability pension. They bought their first house in Val Caron, one of the many settlements near Sudbury that sprang up around a mine shaft. Laurina stopped working when she was pregnant with her youngest son,[26] but they had enough savings to buy a lot in nearby Hanmer and began building a house close to the highway. One day, her husband read a newspaper ad for someone wishing to trade a house for a farm in Blezard Valley. When the Verhoevens started a part-time farm, there were a dozen other Dutch-Canadian families with farms in the area, including the Solominks. The Verhoevens had fifty head of beef cattle but, apart from one Italian family, they were the only ones who grew vegetables, specializing in turnips. Laurina used to wax the turnips at five in the morning, before her husband left for his job in the mines. They hired student labor in the summer to help cultivate cauliflowers and peas that were sold to local Loblaws and Dominion stores. As they expanded their production, they bought an additional twenty acres of land on three separate occasions. Laurina's four sons worked for mining companies. Two of them also started farming. Eddy, who was still working for Inco when I interviewed him in 1991, managed the home farm and the turnip business. They were the only landowners in the valley who had not sold the "sulphur rights," which allow farmers to claim compensation from Inco when crops are damaged by sulphur emissions. Whenever white spots appear on their more sensitive plants, Inco representatives come over to check and settle the damage. Eddy's older brother, who quit working in the mines, had already moved and become a full-time farmer in Warren (close to Verner).

---

[26] Interview (FS) with Laurina Verhoeven, at her home in Blezard Valley, 11 April, 1991.

The Lakehead Region

The name Lakehead refers to the north shore of Lake Superior (west of Fort William). Unlike the Dutch Catholic farmers in the Sudbury area, farmers here affiliated with the Christian Reformed Church whose members formed enclaves around Murillo, in the townships of Emo and Chapple, and in Fort Francis. These are all located on a clay belt that extends into the Rainy River District, where sixty out of seventy-five dairy farmers were of Dutch background in 1988.[27] While completely Canadianized (albeit within the context of a church with Dutch connections), the offspring of the original immigrants still marry other Dutch-Canadians. Unlike other regions in Northern Ontario, where the proportion of people of Dutch descent in rural areas hovers around 1 percent, the proportion in parts of the Thunder Bay region is closer to the 10 percent typical of "Dutch" rural enclaves in Southern Ontario. To understand this region we need to refer to Thunder Bay, its only significant urban center.

Located almost a 1,000 kilometers west of Sudbury by highway, Thunder Bay is easier to reach from Winnipeg, Manitoba, in Western Canada, than from anywhere in Southern Ontario. In the decade following the Second World War, this metropolitan area consisted of two towns: Port Arthur and Fort William. The most important industry was lumbering and pulp and paper, but in the rural area west of this urban region there had once been many small farms that supplied milk and other agricultural products. Their owners, who set up a dairy co-operative, combined small-scale farming and work in the forestry industry. The majority were of Finnish origin. The few Dutch farm immigrants did some market gardening or set up landscaping businesses in Fort William. These farmers had become completely integrated into Canadian society. For example, Klaas Kamstra, the son of a prewar farm immigrant, came to Canada in 1923, at the age of fourteen. The Kamstra family was well established in South Gillies, in the township of Oliver (west of Thunder Bay), by the time postwar immigrants arrived.[28] However,

[27] This figure was mentioned in an interview with a dairy farmer in the Wellandport area of Niagara, who had relatives in the Thunder Bay area. He had made trips there and used to meet regularly with Dutch dairy farmers during meetings of the provincial executive of what is now Dairy Ontario.

[28] Two Kamstra brothers owned one of the largest mixed farms, with a hatchery and fifteen milk cows. Neither of them had returned to visit their father's home in Friesland, nor did they associate with Dutch immigrants when interviewed in 1978. Both interviews, conducted by Henry Kamphof (with Klaas on 23 February and Cecil, 29 March) were recorded for the Multicultural Historical Society of Ontario. (DUT-2784-KAM & DUT-4987-KAM).

they had a lot of contact with postwar Dutch immigrants, being short of labor in an operation that required six full-time helpers. Their sponsorship of twelve families, several of whom stayed with them for three years, enabled them to expand.

The reason why so many Dutch Calvinist farmers ended up in the region after the war is the result of several events involving a few key people. Dick Maat, one of the early Dutch immigrants, lived in Thunder Bay at the end the war, where he ran a landscaping business. He was eager to help other Dutch farm immigrants. Around the same time, Abe Tuinman, the Dutch agricultural attaché who helped set up the Netherlands-Canada Settlement scheme (see chapter 4), promoted Dutch farm immigration to the Lakehead region.[29] Dutch farm immigrants were drawn to the region because of the low cost of land and possible milk subsidy for producers willing to expand into this part of the north. The third factor in the development of a predominantly Dutch dairy region, was the role of the Christian Reformed Church. Their fieldmen found sponsors, including local Canadian farmers, for people still in the Netherlands. Established members of this denomination, like Dick Maat, did volunteer work and billeted newcomers in their homes whenever necessary. Within several decades, these immigrants had built three churches. The Dutch farmers transformed a subsistence-oriented, small-scale form of dairy farming insufficient to meet local demand into a commercial and technologically advanced form of milk production capable of producing a surplus for export as far away as Montreal. By 1980, Dutch dairy farmers had displaced the Finnish-Canadians who had preceded them. These Finnish farmers, who became more involved in the lumbering industry, had already started leaving the area prior to 1968, the date of the collapse of their dairy co-op, and well before the postwar Dutch farmers became firmly entrenched in this sector.

The case studies presented so far exemplify the experiences of a diverse group of farm immigrants in different parts of the province. However, Drayton and the Thunder Bay area are only two examples of numerous enclaves of Dutch farmers belonging to several Calvinist denominations in Ontario. I could have chosen Norwich (London region), which is the site for an even more close-knit rural community, whose members belong to the Netherlands Reformed Congregation. Likewise, the cluster of Dutch Catholics in the Grand Bend bog, on Wolfe Island, or around Chesterville are only small-scale versions of what became de facto

---

[29] Interview with Fop Jensma at his home in Ottawa, 18 November, 1989 (a retired, former Dutch consul who worked in both Ottawa and Winnipeg).

Dutch Catholic parishes in the London region (see next chapter). Only there did Dutch immigrant families form communities comparable to their Reformed counterparts. To appreciate the impact of Dutch farm immigrants on such areas, we need to go beyond the stories of individuals, families, and population clusters to examine how Dutch-Canadian farmers formed broader networks, which spanned across many localities, and how they developed their own institutions.

VIII. Forming Communities and Institutions

Over time, Dutch immigrants became integrated into the economic life and, to some extent, the social fabric, of rural Ontario. Adults joined producers' organizations and attended county fairs, and young people entered 4-H clubs (see chapter 2). At the same time, they formed their own communities and institutions that endured well beyond the initial period of adaptation. To some extent, such communities replicated their counterparts in the Netherlands, since the Dutch pillar system at the time of emigration (see chapter 1) influenced patterns of social interaction and assembly, as shown in the last chapter. The intertwined social networks of these different communities linked Dutch farm families from different parts of the province.

THE FORMATION OF DUTCH REFORMED COMMUNITIES

Immigrants belonging to various Reformed denominations formed closely knit communities in Ontario. Two of these denominations were mentioned in the last chapter: the Christian Reformed Church (CRC) and the Reformed Church. Chapter 1 referred to other denominations in the Netherlands, some emerging from more recent divisions.[1] These were replicated in North America, with new English names, such as Canadian Reformed, Free Reformed, and Netherlands Reformed. Initially immigrants from all these denominations entered the Christian Reformed Church. Some people joined congregations founded during the interwar period, in Hamilton, Chatham, Sarnia and the Holland Marsh (see chapter 2). In the rest of the province, Calvinists organized their own prayer groups. Initially, they received visits from Dutch-American home missionaries covering huge territories.[2] The size of these early congregations fluctuated, since Dutch immigrants working on farms had one-year

[1] See Schryer, *The Netherlandic Presence*, chap. 9.

[2] The Rev. H. Moes covered the eastern half of Canada, while Adam Persenaire was in charge of the Niagara Peninsula. These ministers and many more appear in Van der Mey *To All Our Children*, 309-10.

contracts and tended to be itinerant. As more stable groups established themselves, their members moved from rented halls into their own buildings. Since there was a shortage of Dutch-speaking clergy, they summoned ministers directly from the Netherlands. The arrival of immigrant ministers allowed for the rapid expansion of a network of Reformed communities throughout rural Ontario.[3]

During these early years of adaptation, CRC congregations showed a spirit of co-operation as their members put aside doctrinal disagreements. Such divisions reappeared, especially in Southwestern Ontario, as rival denominations each set up their networks of churches. In the previous chapter, we saw how the Reformed Church established a separate church in Drayton. Farmers who had been *Christelijk gereformeerd* formed Free Reformed congregations in St. Thomas and Vineland. Eight families, one of whom had bought a farm in Puslinch, founded a church in Dundas because they found the Christian Reformed Church in Hamilton not strict enough. They persuaded a market gardener in Holland Marsh to become their minister.[4] Members of this religious group started looking for land situated close to this church. Similarly, farmers who had belonged to the Dutch equivalent of the Canadian Reformed church established congregations in Orangeville and other places. The arrival, in the mid-fifties, of immigrants who had been urban workers or run small businesses eroded the rural farm character of many such Reformed congregations,[5] and today, farmers form a majority in only a handful of those congregations.

Creating Parallel Institution

Within two decades Dutch Calvinists had become "institutionally complete".[6] Members of the Christian Reformed and related denominations established youth organizations and clubs, retirement homes, and a host of other associations, along the lines of the Dutch pillar system. They

[3] For example, families scattered over several counties in Central Southern Ontario who used to meet in Trenton, summoned their first minister in 1950. Four years later, those living further away east of Trenton, established a church in Belleville. Still later, families living to the west of Trenton started attending a new church in Brighton. Interview (FS) with Joe Dibbits, in Wooler, 3 July, 1991.

[4] Interview (FS) with Bill Brunsveld, on his farm near Crieff, 5 December, 1990.

[5] Most of the farmers who used to belong to the Free Reformed Church in Dundas went into construction work. By 1999, the Christian Reformed Church in St. Thomas had only two farmers.

[6] This term, first coined by Raymond Breton, refers to ethnic communities whose members become almost self-sufficient in most non-economic aspects of social life.

then set up their own schools, which can be seen in numerous towns and villages throughout Southern Ontario. However, while based on Dutch Calvinist principles, and founded by Dutch immigrants, parents are resentful when their neighbors call them "Dutch" schools, pointing out that an increasing number of students are neither Reformed nor of Dutch origin.[7] Yet farmers whose Dutch parents or grandparents were *gereformeerd* still dominate the committees and boards of such schools throughout rural Ontario. While fully integrated, Reformed Dutch-Canadians still tend to marry within their own religious communities well beyond the second generation, creating extensive kinship networks.

Kinship networks among Reformed rural Dutch

Figure 8, which shows members of an extended family, illustrates the extent to which kinship bonds link Reformed farm families across Ontario. The couple shown at the top was retired when they were interviewed in 1993. Their home farm, which remains in the family, is located in the Dunville area (Niagara region). All nine of their children, of whom five stayed in farming, married people of Dutch descent belonging to the same Reformed denomination. Those five farm couples live in three different counties or regional municipalities: Perth (London region), Niagara, and Haldimand-Norfolk (Niagara region). I have encountered other cases of extensive family links resulting from intermarriage between farm family members of Reformed churches throughout the province.

DUTCH-CANADIAN CATHOLICS IN RURAL ONTARIO

Dutch Roman Catholics, whose national identity includes loyalty to, and identification with, the House of Orange, had never heard of the Protestant Orange Lodges prior to their arrival in Canada (see chapter 2). Around 1950, members of these lodges were, in theory, forbidden to sell their farms to Catholics. It took a while for Protestant Ontario to accept the fact that there were "Catholic Orangemen," but religious divisions were not as strong as they were in the Netherlands; Dutch Catholic immigrants got along well with their non-Dutch Protestant neighbors most of whom did not mind selling them their farms. However, the Dutch were still seeking Catholic places of worship. They preferred to attend Mass celebrated in Latin, as it still was in much of the world. Young people still learning English were more at ease with members of the opposite

[7] This view came out time and time again in interviews with people whose children attend these schools. For a more detailed treatment of Christian Schools (and reference to a growing scholarly literature on the topic), see Schryer's, *The Netherlandic Presence*, 130.

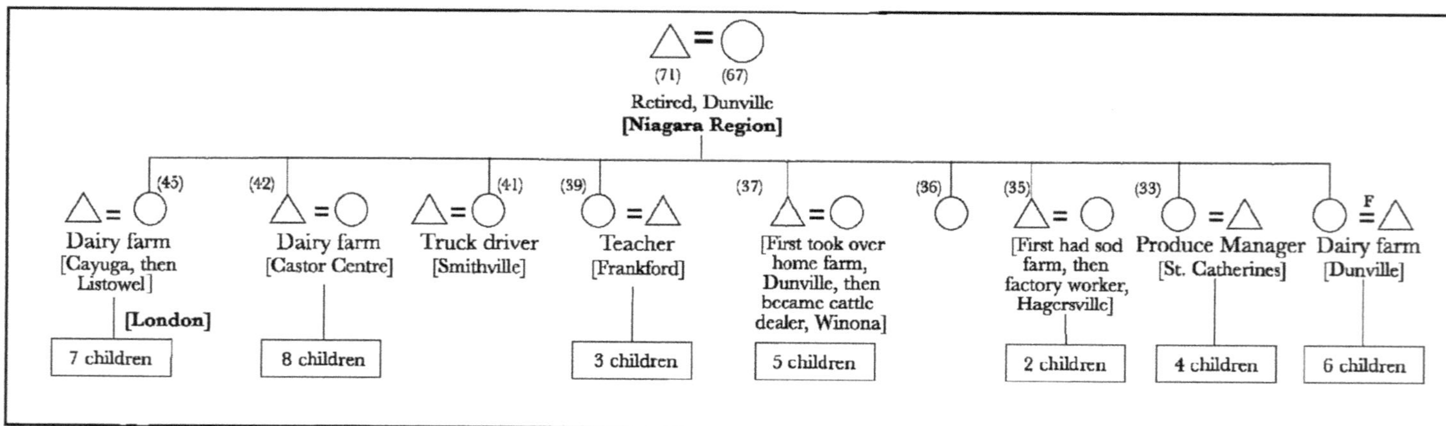

Figure 8. Kinship Network of Rural Dutch Calvinists.

sex in their own ethnic group and with the same values. They invariably married other Dutch Catholic immigrants. In contrast, younger siblings and their own children often married non-Dutch, albeit generally Catholic, partners. Dutch families were quickly absorbed into Canadian rural parishes, and by the first and second generation, marriage increasingly took place outside of religious and ethnic boundaries.[8] However, it would be misleading to make such generalizations for places where Dutch farm immigrants had a strong presence.

Roman Catholic churches in rural Ontario in the late forties were ethnically diverse. Apart from French-Canadians and prewar Flemish parishes, there were Irish-Canadian enclaves (see chapter 2) and parishes with a mixed membership of English, Flemish, German, and Irish descent. Some parishes saw the arrival of Dutch Catholic families, immediately followed by an influx of Portuguese or Italian immigrants. For example, approximately sixty Dutch immigrants in the countryside surrounding Strathroy (London region) began attending a Catholic church once catering to a largely Irish-Canadian congregation in the township of Caradoc. By 1953, Dutch farmers and farm workers constituted almost half of the congregation. With the arrival of an even larger number of Portuguese in the mid-sixties, the Dutch became a minority in an increasingly urban parish. There were soon two churches in Strathroy, one dominated by the Portuguese, although the Dutch still made up a third of the congregation of All Saints parish in 1990.[9] However, while parishes in larger towns saw ongoing changes in their ethnic composition, the Dutch remained a majority in other parts of the region of London. In such parishes, the trend of marrying spouses of Dutch background continued well into the second generation.

A pattern of endogamy among some Dutch-Canadian Catholic families does not necessarily reflect an onus on finding spouses of Dutch background. Although parents were usually pleased if their son or daughter married "someone who was Dutch," it was more important that they marry someone who was Catholic. The chance of meeting someone of Dutch Catholic background was higher in rural areas that had received an influx of Dutch Catholic immigrants. Many couples who had emigrated when they were teenagers, met at dances organized by the Regis club, a Catholic youth organization in the Diocese of London. The dense

[8] This was the case in almost all of the Dutch families I interviewed for both of my research projects.

[9] Interviews with John Stijbosch, in London, 18 April, 1989; and with Adrian de Bruijns, in Ilderton, 27 August; Katrien de Groot, Strathroy, 28 August; and Martin Strijbosch, Kerwood, 20 October, all in 1990.

networks among people in the same congregations, at the same separate (Catholic) schools, and who frequented the same dances, was not unlike that of Dutch Calvinists.

Dutch-Canadian Parishes

Dutch Roman Catholic immigrants usually found a church within commuting distance. However, they were usually unable to understand the sermons, nor confess in their own language, since there were only three Dutch priests in Southern Ontario in 1949. These priests belonged to an international religious order, the Sacred Heart Fathers (*congregatie van het Heilige Hart* or SJS).[10] The national council of bishops invited the Dutch branch of this order to send more priests to Canada when they became aware of the needs of Dutch Catholic immigrants. Father White, the director of immigration and rural settlement for the diocese of London persuaded his bishop to appoint a Sacred Heart priest, Father Jan van Wezel, to work full time with the Dutch. His order purchased an old homestead in the village of Delaware, which became the nucleus of a new parish, run by Dutch priests but meant for all Catholics who lived in the area. They, and other immigrant priests, subsequently worked with Dutch immigrants throughout the diocese and beyond. However, they had the greatest impact on the region of London where Dutch Catholics all but took over existing Canadian parishes in several locations. West of London, Dutch immigrant farm families came from the Dutch province of Noord Brabant, while east of London, around St. Marys and Stratford, a larger proportion of Catholics came from other regions of the Netherlands. A good example of such Dutch rural Catholics of diverse origins is a small cluster near Kinkora (in the county of Perth).

Kinkora is a very small hamlet of several houses surrounding an impressive looking church next to the Catholic school. It is the center of a parish once run by Irish-Canadians. In the late forties, at the time of the arrival of the first postwar Dutch immigrants, this school, run by nuns, was on the verge of closing due to low attendance.[11] A priest in Kinkora (not of Dutch descent) took the initiative to bring in Dutch families to revitalize the parish. He contacted Father Van Wezel from London and arranged for local farmers to sponsor Dutch families. The first Dutch family to come and live on a farm in this area after the war

---

[10] For a detailed description, see Schryer, *The Netherlandic Presence*, 112-116.

[11] The author visited this hamlet and its surroundings on several occasions. Much of the information on the Dutch in Kinkora is based on numerous conversations and five interviews with people who lived in, or still live in, this village and its surrounding area.

was the Vorstenbosch family. For the next two years, few other immigrants came to Kinkora, but in April of 1951, three more families did, as a result of Father Van Wezel's efforts: the Boersen (with 12 children), the De Groot and the Ganzevles. The Boersens and De Groots came from the Amersfoort region of Utrecht, and the Ganzevles were from Putten, in Gelderland. The Huitemas, originally from Friesland, the Willemsens, the Van Neses and other families came next, followed by several Dutch families who had first headed west, but then moved to Ontario. People kept arriving right up until the end of the fifties. Almost all of these newcomers had been farmers in the old country.[12] With increased enrolment of younger children from these Dutch families, the school in Kinkora was soon overcrowded. Local farmers, including a beef dealer who owned several farms, sponsored these Dutch families. Within a couple of years, most of them bought their own farms. Initially, they all had small farms (mainly hogs, and some dairy), but did not start expanding and growing cash crops until the late seventies. Some subsequently moved away from the immediate vicinity of Kinkora, but remained in contact. In 1989, people of Dutch descent constituted seventy percent of the membership of that parish,[13] and established a retirement home, Windmill Gardens, in nearby Stratford, to accommodate an increasing number of parents, then well beyond retirement age.

Setting up Separate schools and Patterns of Intermarriage

Rural Dutch Catholics put greater emphasis on the importance of having their children educated in Catholic schools than their non-Dutch counterparts. They were appalled at what they considered the lax standards of their non-Dutch Catholic neighbors, many of whom did not regularly attend church or drank too much alcohol. Dutch Catholic immigrants filled the pews of many rural churches, and took the initiative in petitioning for separate schools. Sometimes they joined forces with English-speaking Catholic families, at other times they clashed with non-Dutch Catholics unhappy with the idea of pulling their children out of public schools. Dutch farm immigrants spearheaded the formation of Catholic schools throughout the London region and beyond, including in Ingersoll, Uxbridge, Stayner, Forest, and Thamesville. In Watford,

---

[12] Most of the information in this paragraph and the next was obtained from an interview with Dora (Boersen) and Adrian Olsthoorn, at their retirement home in Ilderton, 29 April, 2002.

[13] I went over the parish list at the kitchen table with various members of the Willems family, (Jake, Lise and children) at their farm near Mitchell.

they did so in 1954 against the protest of local Orangemen. In Kinkora and Lucan, existing Catholic elementary schools were kept open with the arrival of Dutch children. A strong Dutch presence in such schools was apparent up until just two decades ago. In the Park Hill Catholic School, which draws students from the surrounding rural area, over 90 percent of the students were of Dutch origin in 1990.[14] The percentage of Dutch-Canadian students attending a separate school in Arkona numbered around 80.[15] In other schools, such as Holy Rosary School in Wyoming (London), Dutch children formed a majority until the end of the seventies.[16] The offspring of Dutch Catholic immigrants continued to intermarry in parishes where they constituted a majority, because the chances were higher in areas where a large percentage of students in primary, and later, secondary schools, belonged to the same ethnic and religious group. An example is Our Lady Help of Christians Parish in the village of Watford (London region). According to a fourth-generation young woman on a farm close to nearby Warwick, over 90 percent of the parish and student population attending St. Peter Canisius School (serving a parish of the same name) are children of postwar Dutch immigrants.[17] A similar predominantly Dutch-Canadian parish, St. Christopher, with a church in Forest, serves people in Bosanquet and part of Warwick township.[18]

The 1989 St. Christopher parish yearbook lists 382 families. A little over half of these families are Dutch, but the proportion is closer to 70 percent if we examine only those with rural addresses, particularly in the vicinity of Arkona.[19] The names of several large extended Dutch-Canadian families appear more than four times in the same yearbook,

---

[14] Interview (FS) with John Hendrikx, in his home Park Hill, 16 December, 1998.

[15] This estimate is based on a conversation (FS) with Mr. and Mrs. Arts, in Arkona, 21 November, 1988.

[16] I went over this yearbook with Mrs. Ann van Boxmeer, at her home in Wyoming, 4 June, 1991. Based on figures in the 1964-65 schoolbook, 50 percent of all students in grades one, two, and three, were children of Dutch immigrants. The percentage for grades five and six was slightly higher: Twenty-four out of forty-four students were Dutch.

[17] Conversation with Julie Straatman, with her brother Paul, 7 July, 1990.

[18] This once-small parish did not start to grow until after 1949, with the arrival of Dutch immigrants. Dutch Catholic families organized a Catholic primary school, St. Mary's, on the 6th line in Warwick township, in 1964. The founding of this school was recorded on four pages of handwritten minutes of meetings held between 24 November, 1953 and 27 December, 1964.

[19] I went over this yearbook with Margaret Hendrickx, at her home near Warwick, on 27 April, 1989. She is one of several postwar immigrants of Flemish descent, who became an integral part of this Dutch community, which consists primarily of descendents of postwar immigrants from the Netherlands.

while an additional number of farm families are also well represented in other parts of Southwestern Ontario. Although some of the older members of these families knew each other in the Netherlands, they all became part of a single rural community bound together through friendship and intermarriage. Figure 9 shows the extent of kinship ties among such Dutch-Catholic farm families. The four brothers (labeled A) near the top, whose parents emigrated to Canada in the late forties, all live in or near the village of Arkona. Their married offspring and spouses are included. The rest of the letters represent other surnames (maiden names, in the case of women). Present or original residence, other than Arkona, is included. Almost all of the couples shown are full-time or, in a few cases, part-time farmers. In several cases, brothers and sisters married siblings from another family, creating an even denser network of kinship ties.[20] No members of this family network have yet married outside their ethnic group, although other Dutch-Canadian families in St. Christopher parish have some in-laws not of Dutch descent.

This large extended family in Arkona formed links through marriage with other Dutch families in nearby villages. One of these villages, Park Hill, is part of a different parish that includes the hamlet of Bornish, once dominated by Catholics of Scottish background. Dutch-Canadian families in nearby Watford, Strathroy, as well as in Brussels, and even beyond the diocese of London, are similarly connected through ties of kinship. A smaller cluster of interrelated families is also found in Perth County, where Dutch-Canadian farm families continued to intermarry.

Worshipping together, and having children attend the same schools, are not the sole factors explaining endogamy. An important criterion for mate selection is farm background. As in the case of Reformed communities, family connections link Dutch Catholic farmers to small business operators, teachers, and other professionals in larger towns or in the city of London. However, unlike the case of Reformed families, such non-farm relatives almost always have spouses who are not of Dutch background. Young Dutch-Canadians with higher levels of education likewise have looser ties with their own ethnic group, unless they enter an occupation like farming. Marriage to non-Dutch spouses is also more prevalent when family members who were raised on the farm enter other occupations – such as welding, carpentry,or construction work – which require residence in, or constant commuting to, nearby cities. Figure 10

[20] Unmarried offspring living at home have been left out. The kinship diagrams were based on information on all Dutch-Canadian individuals listed in the same yearbook provided by Margaret Hendrickx from the previous footnote.

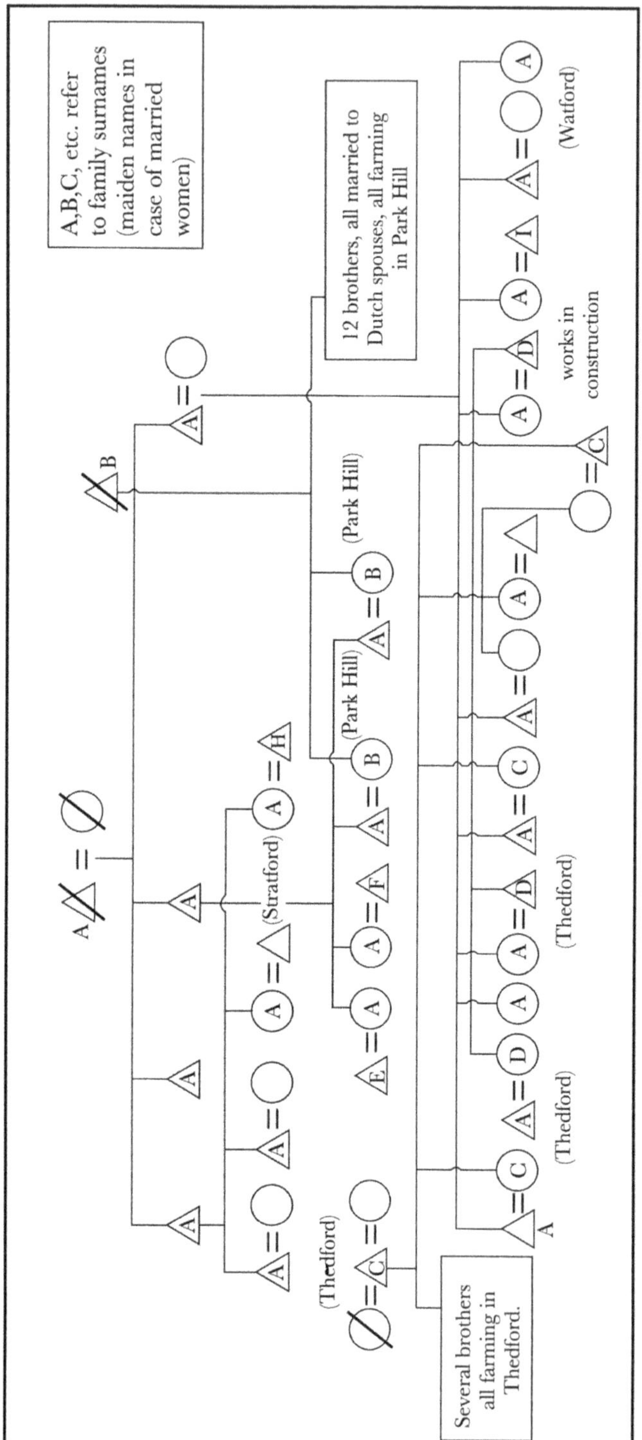

Figure 9. Kinship network of Dutch Catholic rural families.

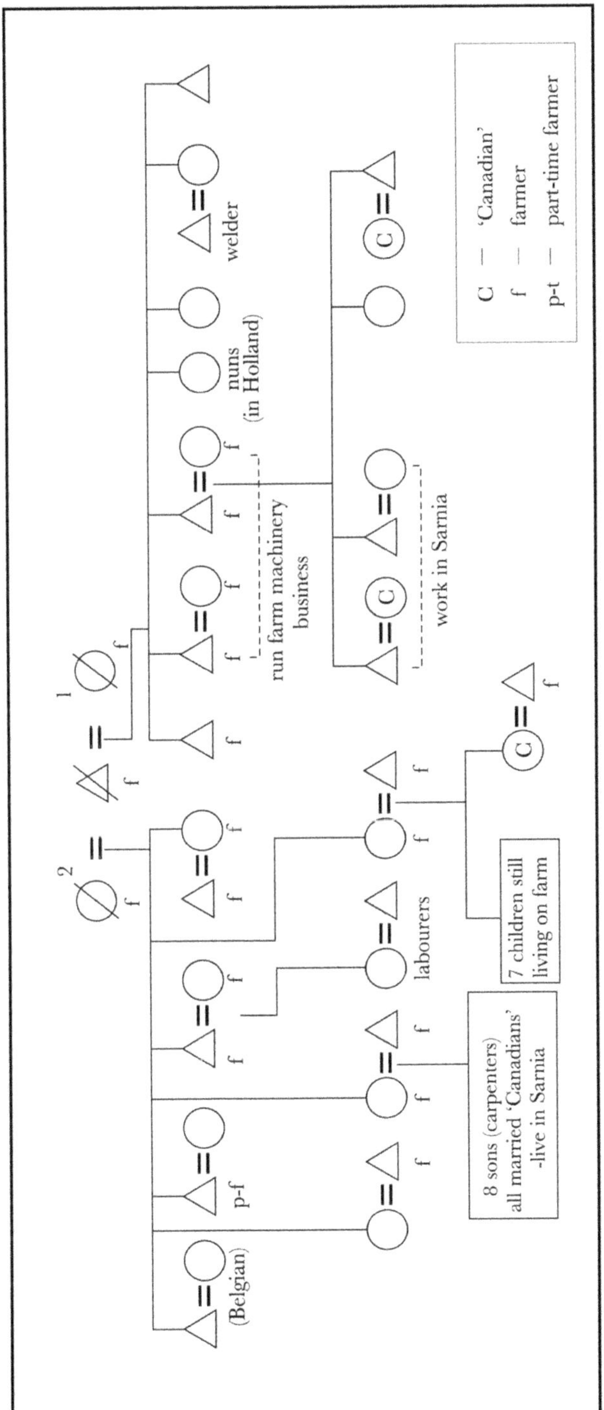

Figure 10. Kinship network of Dutch Catholic Family in Wyoming.

illustrates a typical pattern of marriage across ethnic lines for several Dutch Catholic families in Wyoming, near Sarnia.[21]

My final example is the parish of St. Ambroce in Brussels (Huron), where the Dutch Catholic families are a minority. Unlike Kinkora (further to the south), for many years, there were only two Catholic families in this parish. The two older daughters in one family had already started dating Dutch boys from one of the large Catholic families in the Kinkora area of Perth county, before the family moved to Brussels. After they moved to Brussels, their third child, a son, met and later married the oldest daughter of the only other Dutch farm family who attended St. Ambroce. Figure 11 shows how seven out of the nine children had Dutch spouses. In contrast, all but seven of the forty-seven members of the third generation, above the age of fifteen, married non-Dutch spouses. If we include three additional spouses who had one parent of Dutch descent, this number expands to ten, still less than 25 percent. This pattern of ethnic exogamy beyond the third generation is more typical of the rest of the province, where Dutch Catholic immigrants were absorbed into multi-ethnic parishes.

FAMILY GATHERINGS, REGIONAL REUNIONS, AND SOCIAL CLUBS

In addition to the formation of kinship ties based on marriage, people maintained contact with more distant relatives and neighbors from the old county. Extensive networks of members of Dutch rural families were replicated in Canada, even if they did not emigrate at the same time. For example, between 1949 and 1959, two branches of the Van Moorsel family, whose respective members originated in various villages located between Uden and Gemert (Noord Brabant) landed in Eastern Ontario, in Morrisburg and Williamsburg. A combination of family and regional identity was maintained even after one of these families moved to Southwestern Ontario, near Mitchell (Huron).[22] However, rarely did a group of closely related families leave for Canada around the same time

---

[21] This kinship diagram was constructed with the help of Mrs. Ann van Boxmeer, in Wyoming, 4 June, 1991. She provided kinship information about all of the Dutch families listed in the 1987 yearbook of the parishes of St. Philip (Petrolia), St. Ann (Oil Springs), and Holy Rosary (Wyoming).

[22] The Ontario branch of the family is given extensive coverage in Van der Mey, *To All Our Children*, 270-74. I heard about the Eastern Ontario branch during interviews in Eastern Ontario; and Elizabeth Allingham, who interviewed a Van Moorsel, was later able to ascertain that these two branches of the family are indeed related and aware of each other's presence.

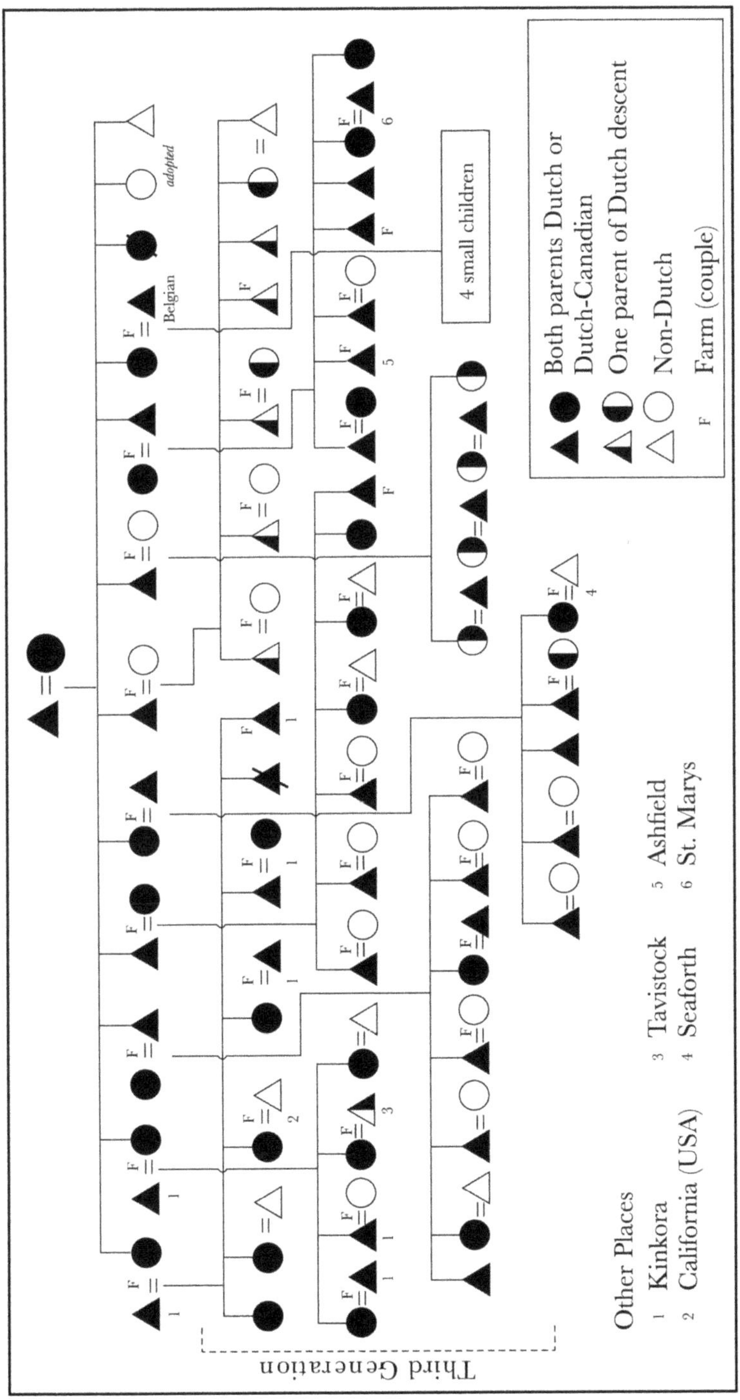

Figure 11. Patterns of Intermarriage in Brussels area.

and then end up buying land in the same township. The exception is the Van Donkersgoed "clan", six related families from Putten, in the Veluwe region of the Netherlands, who emigrated in 1948. After working for different farmers in Manitoba and Alberta, all these families moved to the Drayton area in Southern Ontario.[23]

Members of other large extended families scattered over Ontario and beyond, kept in touch in other ways. Two brothers of the Verhoeven family of Heusden (Noord Brabant) came to Canada in 1949. Three decades later, almost a hundred of them, scattered throughout Canada, held family reunions, one at a farm in Northern Ontario, in Blezard Valley, near Sudbury, and another in Ingersoll, in southern Ontario.[24] The Rodenburg family, which traces its roots back to Overschie, an old polder near Rotterdam, is even more dispersed. Its members, who live in different areas on both sides of the ocean, have kept in touch through a newsletter. I came across the first issue of their bilingual, *Rodenburger Courant*, during a visit to a farm in Kintore in 1989.[25] Other postwar Dutch immigrant families in different parts of rural Ontario have more informal networks activated through occasional visits.

Immigrants who came from close-knit rural communities in the Netherlands maintained a sense of identity based on a common place of origin, and maintained contact in Canada, even if they were unrelated. A good example is a series of hometown reunions held in the late eighties and early nineties. The first hometown reunion involved people from the two villages of Mariahout and Lieshout. Mariahout is unique, dating back to only 1933 with the expansion of Dutch agriculture onto former heath land. Its inhabitants not only had a strong sense of cohesion, but they were predisposed to heading out in search of new land. After the war, a third of its inhabitants emigrated to Canada, yet they maintained close contact with their hometown. When Mariahout celebrated its 50th anniversary, in 1983, their Dutch-Canadian friends were invited to attend the ceremonies.[26] In 1984, the Canadians invited their hosts to a Mariahout get-together in Canada, and they have continued to hold reunions in alternate years since that time. These *Mariahoutenaars* are almost all farmers, with a few skilled workers and businesspeople.

---

[23] See Van der Mey, *To All Our Children*, 284-85.

[24] Interview with Laurina Verhoeven, in Blezard Valley, 11 April, 1991.

[25] Interview with Adriaan Rodenburg, on his farm near Kintore, 30 November, 1989.

[26] At that reunion in the Netherlands, the Dutch-Canadians presented a totem pole to Mariahout, which now stands in its central square. Various interviews plus phone calls with Mrs. Freda Leenders, at her farm near Erin, 1992 and 1993.

One woman from Mariahout who lives on a farm in Erin, organized a similar reunion for her husband's hometown of Boekel.

Freda Leenders (neé VanderZande) was approached by a committee from Boekel and nearby Venhorst to organize a hometown reunion, when she and her husband made a return visit on their twenty-fifth wedding anniversary. With assistance of the Boekel committee, they compiled a list of names and started organizing. A year later, fifty people flew to Canada, accompanied by a Dutch film crew. Several hundred people converged on the Leender farm. Some of the film footage was subsequently transformed into a four-and-a-half hour video of conversations with former inhabitants of Boekel, interspersed with panoramic views of farmland and well-known Ontario landmarks.[27] These taped interviews are a good illustration of the geographical and occupational distribution of these largely agrarian immigrants. With the exception of two people – a woman who had been a social worker before emigrating and a retired priest now living in Ottawa – the people in the film were from a farm background and had either farmed or were still farming in Southern Ontario, including in the counties of Wellington (one), Perth (one), Halton (one), Lambton (six), Elgin (one), Haldimand-Norfolk (one), Niagara (two), Hastings (one) and Dundas-Stormont (three).[28] Similar hometown reunions were held for people from Uden and Bakel.[29]

Annual picnics of immigrants from the same region or province represent more stable, ongoing reunions that brought together people living in different parts of Ontario. Many farm immigrants would attend, and sometimes organized, such picnics. The driving force behind an annual get-together for people from the province of Friesland is Luit Miedema, a farmer-turned-butcher whom we met in chapter 6.[30] At the first gathering, 500 people showed up, and by the mid-eighties close to

[27] The conversion of a Dutch video to the North American system was done by Winnie Van de Berlo.

[28] Not all of them were full-time farmers or retired farmers in 1990. Two brothers in Niagara-on-the-Lake had worked at a large factory for over thirty years, while building up a vineyard and fruit farms in their spare time, with the help of family members; while a man in Chesterville had quit farming ten years earlier to become a small contractor. Additional people interviewed briefly during the reunion, included a bricklayer and owner of a welding shop.

[29] People from this region have shown me examples of newsletters or booklets about their hometowns; for example, *Heem Son en Breugel* (Driemaandelijkse uitgave van de heemkundekring Son en Breugel, nummer 3), 1986.

[30] He was born in his mother's village in Groningen and his mother tongue is actually the Dutch dialect of that province. However Luit Miedema grew up in his father's village of Buitenpost (Friesland), where he learned Frisian.

3,000 people were attending.[31] In addition to organizing the annual picnic, this butcher set up a folk theater group, which performed plays in Frisian. The actors were farmers who would perform three days at a time in many places in Southwestern Ontario, including Vineland, Chatham, Sarnia, Bowmanville, the Ottawa valley, and Picton.[32] The Frisian picnics and plays illustrate that the provincial reunions drew together people from different religious backgrounds. This is also true for similar annual picnics for people from Zeeland, who gathered at Springbank Park, near London.[33] A more popular get-together was the "Grunniger" picnic in Rockwood, organized by people from Groningen. Average attendance in the 1990s was 200 people, originating from all over Southern Ontario, including the Holland Marsh, Hamilton, London, Oakville, Woodstock, and Guelph.[34] Other gatherings have been held on a more sporadic basis. For about ten years, people from the region of Twenthe (in Overijssel) who settled on farms near St. Marys and Goderich, gathered at an annual summer reunion,[35] as did people from the Wieringermeer polder.[36] A similar annual picnic for the Andijkers used to be held in Aylmer.[37]

People from different backgrounds also established organizations founded on common heritage, which often grew out of informal gatherings with regional or religious ties. Most of the Dutch clubs in rural and small town Ontario were originally set up under the auspices of the Catholic Regis Organization.[38] Like ethnic social clubs in urban centers, they became secularized and attracted Protestants and Dutch-Canadians with no religious affiliations. Over time, non-farmer members outnumbered their rural founders. The only social club established by Dutch farm

---

[31] Including visitors from Friesland; about 10 percent of those who attend come from the United States. Interview, Luit Miedema, 12 June, 1989.

[32] Some of their old programs are included in the "Jarvis Frisian Club Papers" (1972-1976), located in Multicultural Historical Society of Ontario (Acc.#21210, Mu# 9411, MFN# 031, Subseries # 058-079, MSR# 1583). For additional information on the Frisians in Canada, including tournaments of a Frisian sport called *Kaatsen*, see a longer treatment in my book, *The Netherlandic Presence*, 142-144.

[33] Interview with Mr. De Voogd, sr., Chatham, 7 July, 1989; 2 December, 1992.

[34] Interview with Harry Klungel, in Rockwood Park, 17 June, 1989. See write-up in *Nederlandse Courant*, (15 July, 1989), 8.

[35] Conversation with a woman from Twenthe, at a Dutch exhibit in St. Marys Museum, 1 December, 1990; also interview with Peter Poel, at his farm near Thamesford, 16 March, 1989.

[36] See announcement in *Calvinist Contact* (1 July, 1989). This reunion used to be organized by Klaas Dekens in Drayton.

[37] Interview with Gerry vanderKley, Whitby, 7 May, 1991.

[38] Interview (FS) with Peter and Betty Poel, 16 March, 1989.

immigrants that still is predominantly rural is located in Arkona. Founded by the same group of men who once organized dances in the former schoolhouse close to Strathroy, it was set up as a company of shareholders, who also erected their own building. Today, few young people recognize the Dutch connotations of its name, Taxandria.[39] Nevertheless, the same interrelated Dutch Catholic farm families mentioned in the previous section still use the Taxandria club premises to hold wedding receptions, banquets, or to celebrate anniversaries. Dutch Calvinists did not need such ethnic clubs, since they could socialize and speak Dutch language within their own Reformed network of institutions.[40]

Secular ethnic clubs thrived as centers for social interaction, Dutch card games, beauty pageants, and parades well into the eighties. However, with an aging population of immigrants, many of these clubs declined and gradually disappeared. Concurrently, there was an upsurge in a new type of organization with links to the Netherlands: that of former Dutch marines (*Kontact Oud Mariniers*) and war veterans (*Bond van Wapenbroeders*).[41] Although officially branches of organizations in the Netherlands, their members and local executives in Canada are postwar immigrants, including Dutch-Canadian farmers of diverse regional backgrounds and religious affiliations, who served in the Dutch armed forces (during the war for independence of the former Dutch East Indies) prior to emigrating to Canada. The Association of Dutch Marines, in particular, frequently holds meetings and gatherings at members' farms.[42]

Social networks and formal gatherings based on distant family connections, common region, or hometown did not survive beyond the second generation of Dutch immigrants. The few remaining ethnic clubs and associations will disappear as the elderly Dutch die off. Even the strong kinship bonds created through intermarriage and common ancestry will fade away by the fourth generation in all but Reformed communities. Yet Dutch farm immigrants created more permanent institutions,

---

[39] John Strijbosch, one of the club's founders, named it Taxandria after a Celtic tribe that once inhabited the southern part of the Low Countries.

[40] However, Calvinists did set up at least one association defined along ethnic lines in Chatham. This club, which organized activities in nearby Wallaceburg and Dresden, showed films, had talent nights and organized tours. Dutch Canadian Cultural Club File, in Dutch-Canadian Papers, Multicultural Historial Society of Ontario, MHSO, SER. 058-129, MSR # 3421.

[41] For more detailed coverage, see Schryer, *The Netherlandic Presence*, 261-62.

[42] I attended a gathering at a farm in Embro on 10 June, 1990, where their Canadian president received a medal of honor from the Dutch consul, naming him member of the order of *ridder van oranje*, or Knights of Orange.

which not only have withstood the test of time, but have expanded well beyond their original ethnic and religious roots.

THE DUTCH CANADIAN CREDIT UNION MOVEMENT

In chapter 2 we saw how a fledgling credit union movement appeared prior to the Second World War. The immigrants who arrived after the war were ignorant of the existence of such credit unions, and that any group of people (parish, ethnic constituency, or company employees) could incorporate as a savings and loan association with just twelve signatures. Once they understood that this institution was the same as the co-operative banking system started by Friedrich Raiffeisen in Europe, or the Dutch *boerenleenbank*, they joined existing credit unions. Others built up their own co-operative financial institutions. The first Dutch-Canadian credit unions were tiny operations set up in someone's home or a church basement. The money collected could be as low as a twenty-five or fifty cents per person. Initially, these credit unions provided only consumer loans for used cars or fridges, while farmers might borrow enough to buy a calf. They all experienced difficulties. Some people did not want their neighbors knowing how much they earned or what they owed, and members unable to secure loans accused loan committees of personal bias. Even when problems were overcome, some credit unions did not grow beyond several hundred members and continued to operate as strictly savings and consumer loan operations. Others expanded, until they could offer more substantial loans, mortgages, and business credit, including loans to farmers.

The fifties was a decade of growth, with the establishment of fifteen Dutch-Canadian credit unions in Ontario. In 1965, they became united under an umbrella organization known as the Dutch-Canadian Alliance, in order to offer charter flights to their members.[43] Some credit unions were strictly urban and others completely rural, although frequently a more rural-oriented one would eventually include more city people. For example, a group of farmers in Smithville established the Dutch-Canadian (Lincoln and Welland) Credit Union, which later became DUCA Credit Union (Niagara) with its main office in St. Catharines.[44] In the seven-

[43] For a more detailed account of this alliance, and an overview of all the credit unions founded by postwar Dutch immigrants, see Schryer, *The Netherlandic Presence*, chap. 15.

[44] Interview (FS) with Jack Van Marrum (brother of one of these farmers), in Chatham, December, 1989. This credit union, which used to have branches in Beamsville, Smithville, and Grimsby changed its name in the early seventies and was absorbed by the much larger Niagara Credit Union originally founded by Mennonites.

ties, smaller credit unions started disappearing. The first were those associated with Reformed communities, in Thunder Bay, Brampton, Sarnia, and Barrie. The Christian Reformed (Clinton) Credit Union, started by farmers, was dissolved after sixteen years.[45] In 1979, there were ten left, with a wide range in assets and types of membership. The largely rural Quinte credit union, in Bloomfield, had 223 members and total assets of $193,876.[46] Some of these credit unions amalgamated with larger ones, while others disappeared. Between 1980 and 1990, additional credit unions folded, including the once successful Dutch-Canadian credit union in Chatham. With the exception of a credit union founded by Dutch Protestants in Strathroy, only the largest ones remain today: Toronto's DUCA (started by city immigrants), and St. Willibrord Credit Union, whose bond was originally Dutch and Flemish Catholic. Only the latter has a strong rural base.

St. Willibrord Community Credit Union

St. Willibrord Community Credit Union, which received a charter on January 25th, 1951, under the name Dutch Catholic Immigrants (London) Credit Union Ltd., provides financial services for farmers in many parts of Southwestern Ontario. Its name was changed to St. Willibrord (London) Credit Union on February 15th, 1953. Today it has ten branches in ten counties. The idea of setting up this credit union came from Jan van Wezel, the Holy Heart priest introduced earlier in this chapter. He persuaded a group of farm immigrants to sign up.[47] Most had been independent operators already familiar with the Dutch *boerenleenbank*, and some had been community leaders.[48] During its first decade, the main center of activity was the rural area west of London, although their central office was located in the city of London. Start-up capital came from deposits made by immigrants who had smuggled in "black money" (see chapter 1).

---

[45] All members were repaid their original investment and deposits, and all outstanding loans were paid back within six months, sometimes through new loans from the Clinton Community Credit Union. Interview with Kees Buurma, in his home near Homesville, 14 May, 1992.

[46] See Ger Graaskamp, *Dutch Canadian Mosaic*, Niagara Falls: self-published (circa 1980).

[47] Interview (FS) with Jack van Geel, in Lucan, 7 December, 1992.

[48] The names of these leaders, who had either bought their own farms or became farmers in Ontario, were mentioned repeatedly in my interviews. Many of them are also mentioned in a history of the credit union written on the occasion of its fiftieth anniversary. See Mark Kearney and Otte Rosenkrantz, *Roots and Branches: St. Willibrord Community Credit Union* (London: St. Willibrord, 2000).

St. Willibrord was an unusual credit union because it covered such an extensive area—the entire diocese of London. Because of problems associated with such a widespread membership, its first board of directors appointed collectors for different localities. Towns or villages that were service center for farmers, became sub-divisions (*afdeling* in Dutch): Wyoming (Sarnia), Woodstock, and Strathroy (1951), Chatham (1952), and Stratford (1953). While such subdivisions ("divisions" as of 1974) did not have legal standing at the time, they set a precedent of a decentralized administration that later characterized other credit unions in Ontario. Each *afdeling* had its own *bestuur* (governing council) and a *kassier* (treasurer) who had several collectors working under his supervision. Officially designated as advisory committees under the jurisdiction of a general board of directors, who met in London, their members saw themselves as running a local "board," which is one possible translation of the work *bestuur.* Older Dutch immigrants and their offspring used this term when speaking English and the expression "boards of subdivisions" crept into the official minutes of meetings held in the fifties.[49] The rural composition of these subdivisions, which later became branches, can be identified by examining several of them. Almost all of the original leaders of the St. Willibrord branches were Dutch farmers who had played similar leadership roles in the old country.

At the local level

The first meeting of Dutch farmers interested in forming a Canadian version of the *boerenleenbank* took place in Camlachie, half way between Sarnia and Forest. This meeting was held in the house of Huubert (Bert) Donkers who once owned a medium-sized farm in Boekel (Noord Brabant). The father of fourteen children, he had been involved in various organizations, including a Dutch farmer's union (*boerenbond*) and the *boerenleenbank.* Huubert was sixty-four years old when a farmer in the Sarnia region sponsored his oldest son.[50] A year later, the rest of the family moved to Ontario, and bought a farm. Huubert had more work on this farm than he did at home, but continued to play a leadership role by becoming involved in the credit union with some other immigrant farmers. They created their "board" in Wyoming (close to Sarnia) and elected Jack Boere as their secretary. Jack became one of the collectors and received money at his house in Forest. He also drove to Wyoming every Sunday where he accepted money after mass.[51]

[49] These minutes are housed at the head office of St. Willibrord in London.
[50] Interview (FS) with Adrian Donkers, at his home in Wyoming, 15 April, 1991.
[51] Interview (FS) with John Van den Broek, at his home in Wyoming, 17 April, 1991.

Another subdivision was Strathroy-Watford, locations referred to in connection with de facto Dutch parishes. Unlike the group of older farmers in the Sarnia region, its members were in their twenties.[52] John Strijbosch became the treasurer, while other members of their "board" were collectors in Kerwood, Thedford, Park Hill, and Grand Bend, all located in the townships of Bosanquet, Plympton, Warwick (in Lambton County), or Adelaide and West Williams (in Middlesex). Arkona, a central location for farmers throughout this region, later became the first site of a branch office outside of the city of London offering full-time services.[53] Initially a collection point for the Strathroy-Watford subdivision, it became the center of operations after John, who was to play a vital role in the development of St. Willibrord credit union, moved there in 1964.

John Strijbosch, who grew up on his family's farm in Asten, Noord Brabant had a good education. A graduate of agricultural and horticultural programs, he also received training in bookkeeping and co-op banking services.[54] At the end of the war, he was involved in the Dutch Young Farmers. When both his parents died, he assumed sole responsibility for his family and, in 1951, at age twenty-eight, served as chairman for his local *boerenleenbank*. That same year, John emigrated with all his brothers and sisters, ending up on a farm. He was doing odd jobs in Strathroy when Jan Van Wezel invited him to serve on the board of the credit union because of his past experience. His main source of income at the time was a job in a local factory. In 1953, John made a short trip to the Netherlands to get married. He continued as a cashier for the Strathroy subdivision, and two years later became a field rep, visiting branches in an advisory capacity and doing inspections on behalf of the London board. In 1956, John bought a farm in Kerwood (just outside of Strathroy) with his two younger brothers, still bachelors.[55] John's younger brother, who wanted to become a farmer, worked on

---

[52] Most of them grew up in Alphen, Baarle-Nassau and Chaam (in Noord Brabant) and three graduated from the same agricultural college in Breda. This information came from a variety of sources, including interviews (FS) with Piet Van Engelen, in Forest, 5 May, 1990; Chris van Loon, in Watford, 7 July, 1990; and Case Smeekens, near Thedford, 5 May, 1990.

[53] Peter de Groot, "St. Willibrord Credit Union in the East Lambton West Middlesex Area," an internal report written in 1981.

[54] See Kearney and Rosenkrantz, *Roots and Branches*, 50-51.

[55] I had long conversations with John Strijbosch on several occasions, at his home in London, and also interviewed his younger brother, Tony, at his farm in Kerwood, 28 April, 1989.

the farm, while the others earned other income. John conducted credit union business in a home office and started a bookkeeping business on the side, while also managing the farm. In 1964, he bought a house in Arkona, where he continued to operate his bookkeeping business with another postwar Dutch immigrant. That house became the new credit union office, resulting in the amalgamation of Wyoming and the Strathroy-Watford subdivisions. He now drove to Strathroy from Arkona instead of the other way around. John Boere, who still lived in Forest, continued to oversee collection points in Grand Bend and Forest, but was no longer in charge of his own subdivision. This change, the first of many in the evolution of the credit union, is indicative of a gradual shift in power and control from an older to a younger generation.

Farmers were involved in the formation of what were to become thriving branches of St. Willibrord credit union in other counties. The person responsible for the Chatham office (Kent) was Harry Wijsman, who emigrated in 1949. Born and raised on a farm in Katwijk, Zuid Holland, Harry worked for the Dutch flower auction in Rijnsberg, close to Noordwijk, his wife's hometown. He studied bookkeeping and received his *middelstand diploma* (general business training certificate) at the age of eighteen. In Canada, he worked for his brother-in-law, a sharecropper in Park Hill, before ending up in Blenheim where he took up house construction. That is how he encountered, and then joined, St. Willibrord. He did not become actively involved, however, until 1957, when the collector in Chatham moved away.[56] He discovered that the assets of the Dutch farmers he met in Harwich Township were constantly increasing in value even when they were not earning profits. So he sold the three houses he had built (as rental properties) and bought a farm. One room in the farmhouse was set aside for his credit union work. Harry eventually became a full-time employee of the credit union, managing the offices of both the Chatham (Blenheim) and Sarnia divisions.[57]

The sub-division in Stratford, in Perth County, was formed on March 15, 1953. The first meeting was held in the home of an "old timer," Peter Leyser, who owned a farm near Stratford. For the first three years, the office was located on a farm owned by its collector, in Easthope township. The membership covered a broad area, including the north-

---

[56] Interview (FS) with Harry Wijsman, at his home in Blenheim, 2 May, 1990.

[57] Farmers who "turn over a lot of cash", according to a recent manager, are the backbone of this branch. In 2000, it had the highest annual growth rate of St. Willibrord's branches. Conversation (FS) with John de Bruyn, at the credit union office in Blenheim, 2 May, 1990. See also Kearny and Rosendrantz, *Roots and Branches*, 124.

ern half of Huron County, so another collection point was opened in Seaforth. Their most enthusiastic supporter, who became committee chair, was Jack (Jake) de Groot, a farmer in Kinkora. Bill Van Westerop, a younger, more recent member, who was elected to be secretary in 1967, combined a full-time job at a large industrial firm with running a beef farm. Because of his city contacts, he tried to get more non-farm members to join the credit union.[58] This subdivision was slow to get started and was almost shut down in the late sixties. Jake was adamant that the Stratford office remain open and convinced José Cozijn to become its first female cashier. José, whose parents emigrated from Belgium, had married a Dutchman before moving to Canada with his parents. Her husband eventually set up a nursery and greenhouse operation. José, who had office experience, was working part-time and looking after two small children, when she accepted the cashier position, which she did out of her home for several years. When she took over, the majority of the several hundred members in her divisions were still farmers, mostly from the Kinkora area. Even in the nineties, with their own building and a full range of services for an increasingly diverse population of urban consumers and businesspeople, members of Dutch-Canadian farm families still dominated their annual meetings.

Expansion and integration

As early as 1970, St. Willibrord admitted Dutch people who were not Catholic. A transition from Dutch Catholic to ethnic Dutch took place with the incorporation of members of the Christian Reformed Church. This first occurred in Sarnia when a Reformed Dutch-Canadian became manager. The appointment of a Christian Reformed manager at the Watford branch, the centre of a predominantly Dutch-Catholic rural area, in 1985, shows how much the Dutch pillar mentality had declined. Membership diversified further with the removal of the ethnic bond. St. Willibrord became a community credit union in 1977, at least in the city of London, where it merged with the Co-op Services (London) Credit Union through a buy-sell agreement. From that point on, they played down their ethnic connections. The acceptance of non-Dutch members, as part of ongoing expansion, also occurred in rural branches. In 1980, the Chatham branch (whose office was then located in Blenheim) absorbed the Blenheim Credit Union. Originally run by Capuchin Flemish priests (and originally known as St. Mary's Parish Credit Union), it had long

[58] Interview (FS) with Bill Van Westerop, at his farm in Medina, 8 July, 1990.

been a secular and multi-ethnic institution, whose non-Dutch and non-Flemish members had been apprehensive about amalgamation with a "Dutch" credit union.[59] The Dutch connections were further de-emphasized to attract the non-Dutch-Canadian owners of downtown businesses interested in access to a broader range of financial services. After amalgamation, anyone within a ten mile radius of Blenheim was eligible to join St. Willibrord. The credit union became even more inclusive when small credit unions with no Dutch connections in both Watford and Stratford joined St. Willibrord in 1984 and 1988 respectively. The last breakthrough occurred when their bond was extended to all of Middlesex and Lambton counties, allowing the business sector of the village of Watford to join St. Willibrord when one of Canada's major banks closed its only branch there in 1994.

St. Willibrord, once a tiny operation run by a Dutch priest, developed into a large secular credit union, with a full range of services, including automatic tellers and on-line banking. By the turn of the century, its total membership reached 30,000. It has most recently expanded into the city of Kitchener, after absorbing a former industrial credit union. However, farm owners are still important and influential. Still the biggest lenders in the area, the retirement of many Dutch-Canadian farmers, starting in the eighties, brought an infusion of large deposits. This Dutch farm influence was still noticeable as recently as a decade ago. In a 1990 interview, one non-Dutch member of a branch council stated:

> St. Willibrord was and still is Dutch-oriented. You can certainly tell that at the annual meetings: more than 50 percent of those who attend are of Dutch origin. The Dutch are also the biggest depositors . . . The people who keep on defeating the motion to change the name are the older Dutch people . . . At the annual meetings I have also noticed that the Dutch farmers form into their own little groups and speak their native language.[60]

The farmers mentioned in this interview were from the Arkona and Watford branches and they were speaking the dialect of Noord Brabant. Such scenes will disappear in the not too distant future. However, the Dutch character of this credit union takes a less visible, yet more durable form. Very few current members (now called owners) would be aware that their system of choosing representatives resembles the way the members of the Dutch senate (*Eerste Kamer*) are elected by members of their

[59] Interview with John Cowan, Blenheim, 10 July, 1990.

[60] Taped interview with Dan McDonald, during a visit to the London Dundas Street branch, 29 June, 1990.

provincial parliaments.[61] The co-operative ideals of the Dutch Catholic farm immigrant founders of St. Willibrord contributed to an institution that has become an integral part of Southwestern Ontario.

SYDENHAM COMMUNITY CREDIT UNION

While some Dutch Protestants became members of St. Willibrord, another still active community credit union in the London region, was founded by, and continues to draw leaders and members from, the Reformed community. Sydenham Community Credit Union, formerly known as the Dutch Canadian (Strathroy) Credit Union, today has branches in Mount Brydges, Park Hill, Lambeth and Ailse Craig. Its origins are predominantly rural, and its historical development is similar to that of St. Willibrord. Both credit unions established or took over, existing co-operative funeral funds popular in the Netherlands. They both made the transition from Dutch to English for written correspondence. As in the case of St. Willibrord, city workers did more saving while farmers were the principal lenders. Finally, the Dutch-Canadian Credit Union in Strathroy also made the transition from a financial institution with an ethnic bond to one open to anyone within its region of operation. In 1978, they took over a small Canadian credit union and adopted its name Sydenham, but the majority of the board of directors continued to be Dutch-Canadians.

The history of Sydenham Credit Union goes back to its founding in April, 1957. While the charter does not mention a religious affiliation, it was originally classified as a "parish" type of credit union, and its meetings opened with prayers and readings from the bible.[62] Its board of directors met at the Christian Reformed Church. Of the twenty-eight founding members, eleven listed "farmer" as their occupation, not including several spouses of farmers, whose names also appear. About half of the rest had either a farm background or were later to attempt farming. Their first president, Luitzen Jacob Feenstra, ran a farm in Kerwood (near Strathroy). He had been a farmer in a Dutch polder region in Friesland. Likewise, all but one of the other members of the first board of directors had farmed in the Netherlands. Klaas Solomons, who joined the board a year later, came from a family of Dutch farmers in Groningen,

---

[61] This similarity in organizational structure was pointed out to me in a written communication from Dick de Man.

[62] Mentioned in "Fieldman's Credit Union Report" (1957) included with the minutes of meetings of the board of directors, which I consulted at their Strathroy office on 18 April, 1989.

who had cleared land in the Krim region in Drenthe. An ancestor of the same name was a founder of the *boerenleenbank*.[63] The credit committee consisted of well-established farmers, but most of those who borrowed money for farm operations worked part-time in town. Jelle de Shiffard, a member of the credit committee who used to visit the farms of loan applicants, recalls how those members would often have to sell a cow in the spring to raise cash. With a loan from the credit union, they were able to keep the cow until the fall and earn more.[64] Apart from farmers, plumbers, carpenters, and laborers, the credit union had small business owners as members and their first treasurer was an insurance salesman. However, farmers continued to have a strong voice, and Gerald Noordeman, a dairy farmer from Ailse Craig, took over as president of the board, a position he still held in 1990.

Apart from participating in the Dutch Canadian Alliance, the credit union in Strathroy maintained ongoing contacts with John Strijbosch of St. Willibrord. For example, John spoke at one of their meetings in 1959, and, in 1972, he attended the opening of their new office building in Strathroy. John and a group of Catholic farmers from Arkona had earlier collaborated with Protestant Dutch farmers in the joint purchase of fertilizers, and Gerald Noordeman became chairman of the short-lived Lambeth and Middlesex Farmers' Association, which had consisted of Dutch farmers from different backgrounds.[65] However when John Strijbosch suggested the credit union in Strathroy become part of St. Willibrord, they refused. They also raised legal objections to the opening of a branch office of St. Willibrord in Strathroy in 1974. Such tensions, and the continued existence of two credit unions with Dutch connections in the same town is a vestige of the Dutch pillar system. However, while Sydenham Credit Union is the only surviving credit union established by Dutch Protestant immigrants, Dutch farm immigrants belonging to the Reformed community were responsible for the creation of an equally durable, but very different, Canadian rural institution that involved farmers only.

---

[63] I came across an article on this other Klaas Solomons in a Dutch magazine clipping (no date or title), which I photocopied in Strathroy.

[64] Interview with Jan Tamming, in Trillium Village (a retirement home), Strathroy, April 21, 1989.

[65] Interview (FS) with Gerald Noordeman, 15 December, 1988. I came across at least one other reference to this short-lived association in another interview several years later.

THE CHRISTIAN FARMERS FEDERATION OF ONTARIO (CFFO)

Dutch Calvinist farm immigrants brought with them the same pillar mentality as their Roman Catholic counterparts. In Canada, they spent energy and money to build their own churches and to establish "bible schools", similar to those to which they were accustomed in the Netherlands. Farmer members of the Reformed community also created a Canadian equivalent of the Dutch Protestant Farmers' Association (*Christelijke Boeren en Tuinders Bond*).

The Formative Period

Dutch Calvinist farm immigrants were interested in pooling information and discussing common problems. In the winter of 1954, a group of farmers in Wyoming (London region) jointly purchased fertilizer. They paid Jan Bakker, who had been an agricultural consultant in Overijssel, to hold evening classes about Canadian agriculture. Bakker, then living in London, delivered similar courses in Strathroy, Woodstock, and Exeter.[66] Attendance in each location was around fifteen, and in one place twenty-eight farmers showed up. The next logical step was to form local associations. Representatives from Wyoming, Woodstock, Forest, and Strathroy met on March 6, 1954, to create the Christian Farmers Associations of Ontario. Additional chapters were formed in Exeter and the Lake Erie region (Jarvis and Aylmer).[67] They chose an executive with a *voorzitter* (chairman), secretary, and treasurer. Jan Bakker, whose job as sales promoter for Massey-Ferguson required him to travel throughout Southern Ontario, became their advisor. He offered more courses in the winter of 1955, and used his contacts at the Ontario Agricultural College to organize "Dutch" days to provide information to immigrant farmers. Apart from dealing with practical issues, their executive was interested in broader principles not directly related to farming. Their neo-Calvinist philosophy, as expressed in their constitution, put an emphasis on Christian principles in their daily living, including their farming. They saw their federation as an alternative to the Ontario Federation of Agriculture (OFA); they had refused to have anything to do with the OFA, against the advice of Tuinman, the Dutch agricultural attaché in Ottawa.

---

[66] Interview with Jan Bakker, Hamilton, 21 April, 1989.

[67] Minutes of meetings of the Christian Farmers Federation of Ontario (6 March, 1954), handwritten in a notebook kept in the office of the Federation in Guelph. Other historical information was also obtained from this notebook.

After an initial flurry of activity and letter writing, the Federation lost momentum. By April of 1955, their branches (*afdelingen*) in Exeter and Clinton were virtually defunct, and only four groups, including Wellandport, had paid their contributions. With little outside support, and insufficient members, the Federation dissolved in 1956, when only fourteen people attended a meeting in London. A year later, they lost their part-time advisor who moved to Toronto to become a high school teacher. Yet members of some associations continued to meet to discuss farming, buy fertilizer, and disseminate information about farming techniques. The idea of a federation was resurrected in March of 1961, when representatives from several groups met in Woodstock. They hired a part-time coordinator in 1964, although several branches operated completely independently. New groups were formed in Listowel and as far north as Owen Sound (Northwest region). During this interim period, the Federation stopped referring to the Reformed and Protestant character in their constitution and opted for a broader Christian label. English was adopted for their publications as a way of ensuring that younger farmers, raised in Canada, would become full participants. However, by insisting on exclusivity, they lost former members who had already joined the Ontario Federation of Agriculture (OFA).

Debates and Issues

The minutes of meetings (written in Dutch until 1969) reveal debates over a number of issues. There was little agreement on the value of the Milk Marketing Board. According to some members, it would lead to the elimination of smaller dairy farms. Although the executive voted to support the Milk Marketing Board in 1966, a year later, one farmer still expressed reservations, saying: "the farmer must remain the boss and not turn into a farmhand on his own farm."[68] In addition to disagreements over policy, there were also disputes over whether or not observing the Sabbath was compatible with having sugar beet scales open, or business travel. The Federation saw modest growth in membership through the sixties, and there were attempts to set up new associations in Bloomfield (Central South region) and Orangeville (Northwest region). Some groups, like the one in Blyth (in Huron County) were in the danger of folding, while others were reconfigured. The Belleville association, which did not have sufficient members, merged with the one in Trenton. Most of these local groups continued to send delegates to the annual meetings in the

[68] The Dutch sentence in the handwritten minutes of their meeting reads, "dat de farmer baas zal blijven en geen knecht worden op zijn eigen farm."

basement or hall of a Christian Reformed Church, initially in Woodstock and thereafter in Brampton. The executive (the provincial board) met at farmers' homes. During the sixties the centre of gravity shifted from the area west of London towards Drayton (Guelph region) and the Niagara Peninsula. However, even in those areas, many farmers affiliated with Reformed churches showed little interest. Nor did the organization attract farmers who were not Dutch. The election in December of 1970 of Martin Verkuyl, as chair of the board, marked the beginning of a new phase. He was a cash-crop farmer from Hickson in Oxford County (London region). Previous *voorzitters* had been older immigrants, often with prior involvement in public life in the Netherlands. Verkuyl, who came to Canada in 1947, represented a new generation of younger immigrant farmers.

New leaders

Martin Verkuyl was born in Haarlemmermeer, a prosperous cash-crop region in the Netherlands (see chapter 1), where his ancestors had farmed for generations. Before the war, he had received an above average education, graduating from the MULO (Dutch high school) and the *middelbare landbouwschool* before returning to his father's farm.[69] In Canada, he started off by working for an uncle in the Holland Marsh, where his Dutch bride soon joined him.[70] His subsequent farm career included working for Canadian farmers near Lindsay for a year and a half, before joining General Motors in Oshawa; buying and developing a mixed farm in Utica, near Port Perry; four years in Nova Scotia managing a large farm owned by a Dutch corporation; then returning to Ontario briefly to sell feed for that same company before going back to farming. When Martin bought a farm in Hickson to raise hogs and grow corn in 1965, he joined the Christian Farmers, but resigned from their board a year later when he became a sales representative for Vicon. After 1970, he had the time to become more actively involved. At the end of that same year he was elected president (the English term they had since adopted).

Martin Verkuyl felt that the CFFO should focus on influencing agricultural policy, but not be preoccupied with issues like milk pickup on Sundays. He argued that the membership had insufficient technical knowledge and proposed they disseminate information available from the

[69] Interview (FS) with Martin and Wilhelmina Verkuyl, at their farm home near Hickson, 23 November, 2001.

[70] His wedding in the Marsh was part of a double ceremony described in Albert Van der Meys's, *And the Swamp Flourished: The Bittersweet story of the Holland Marsh* (Surrey, B.C.: Vanderheide Publishing, 1994), 70.

University of Guelph's rural extension services. Once elected as president, he again argued that the organization would not expand until it adopted a broader outlook.[71] During his five-year term, they hired a full-time field rep: out of eight applicants, they chose Elbert Van Donkersgoed, a farmer's son with a BA from Calvin College, who had also taken courses in economics at the University of Waterloo. Van Donkersgoed satisfied all of their criteria: he had grown up on a farm; was young and educated; and was familiar with various Reformed organizations, including a Calvinist think-tank, the Institute of Christian Studies, in Toronto. Elbert recruited new members and, as research director, helped to articulate a coherent philosophy. The CFFO still believes in the family farm today. Their long-term vision is "a blend of entrepreneurial and co-operative agriculture." Because of his articulateness and command of English, Elbert became an effective spokesman for the organization. When a farmer inexperienced in dealing with government officials became president, he let Elbert do most of the talking at meetings in Guelph or Toronto. By 1980, the CFFO was gaining respect for offering input and advice to the provincial government, instead of complaining or demanding assistance. They became a driving force in developing proposals on manure management and land severances, which have since become government policy. Their membership grew rapidly from about 100 to 600.

By 1980, the CFFO membership had diversified even further when the first Catholic, but not Dutch Catholic, farmers started to give their support. More educated, younger farmers also joined. For example, Tom Oegema, who came to Canada with his parents at the age of eleven, received most of his education in Canada, including a BSc and an MSc at the University of Guelph, before becoming a full-time partner on his father's farm, a complete feed turkey operation, with his older brother. That was in 1971, just as his father was ready to retire.[72] Tom already knew about the Christian Farmers, but was not actively involved until later that decade. Next, he became a member of the provincial board where he served as president. Despite a vigorous leadership, the growth rate of the CFFO slowed in the eighties. One drawback was their annual fees, higher than that of other organizations. By 1988, the Federation still only had 625 members,[73] and it continued to be 95 percent Dutch-

[71] Mentioned in minutes of a meeting of seventeen persons from eight locals in the basement of the Christian Reformed Church in Woodstock, 3 April, 1965.

[72] Interview (FS) with Tom and Heiko Oegema, at their farm in Talbotville, 19 Feb., 2001.

[73] These members represent a minority of the approximately 2,000 Dutch-Canadian

Canadian and Reformed. Although most Calvinist farmers were aware of the CFFO, they did not necessarily agree with its policies, often perceived as "too radical," because they were critical of big business farming. Two founding members, whose agribusinesses were well beyond the scale of large incorporated family farms, had already quit in disagreement with the philosophy.

Expansion and Integration

In the mid-nineties, when new provincial legislation stipulated that all farmers had to support a farm organization through a check-off system (whereby all farmers allocate part of the fees they pay to a general farm organization), the membership of the CFFO increased dramatically. For the first time, farmers of Dutch Catholic background started joining what they or their parents had previously seen as a *gereformeerde* institution. They liked their philosophy or had heard Elbert Van Donkersgoed speak.[74] At the same time the CFFO attracted new members with no Dutch background, as well as Dutch-Canadian farmers who had previously not been involved.

Bill Ramp, a dairy farmer in Port Dover had not had any connections with the Reformed community in Canada before he joined the Christian Farmers. An immigrant who left the Netherlands in 1948, at the age of twenty, Bill joined the United Church. His earlier involvement in public life consisted of participation in a Farm Forum that produced a CBC radio program on farm issues. Much earlier, while still at an agricultural school, he had served as secretary of the *Christelijke Jonge Boeren en Tuindersbond*, the youth wing of the Dutch organization on which the Christian Farmers Federation of Ontario was modelled.[75] Half a century later, Bill joined the Haldimand-Norfolk Christian Farmers Association and was elected a director on the local board. While the philosophical foundation of the CFFO is still religious, it is very different from its post-war predecessor in composition and range of activities. Jenny den Hartog, who farms in Arthur, became its first woman president in 2002. At that time, their membership was 4,426 full-time commercial family farm entrepreneurs, plus part-time and hobby or lifestyle farmers, distributed among

Calvinists involved in full-time commercial farming in Ontario. I derived this very rough figure by checking on the average proportion of farmers in such congregations in the predominantly rural-agricultural Reformed congregations.

[74] Yet, the Federation did not make inroads in places like the Holland Marsh where one farmer told me that the Christian Farmers had little relevance to farmers specializing in market gardening.

[75] Interview (FS) with Bill Ramp, at their farm near Port Dover, 16 October, 2000.

twenty-three district associations, with anywhere from twenty-three to 450 farmers per district.[76]

Not all farm families were directly involved in the institutions established by Dutch farm immigrants. The number of first and second generation offspring with no direct ties was even greater. Nevertheless, St. Willibrord, Sydenham Community Credit Union, and Christian Farmers Federation of Ontario have become familiar names for many inhabitants of rural Ontario. Dutch immigrants, whose values and motivation were shaped in their country of origin, established all of these institutions. Yet they reached well beyond their original ethnic and religious roots. While still run largely run by Canadians of Dutch origin, they are no longer "Dutch" in name or membership. They survived and expanded because they became more inclusive and were able to integrate into Canadian society without discarding their philosophical principles. Originally created to serve the needs of immigrants, these institutions are prime examples of how Dutch farmers had a lasting impact on, and helped to shape the social fabric of, rural Ontario.

People in rural Ontario not of Dutch background may not always be aware of the religious differences among farmers of Dutch background. In some regions, they may assume that all Dutch farmers are Roman Catholics, in other regions that they are all "Dutch Reformed," while in still other parts of Ontario, people might not even know the religious affiliation, if any, of their Dutch neighbors. Canadian farmers not of Dutch background who meet their Dutch counterparts at meetings of farmers' organizations are also unlikely to know, or even care about, such internal differences. However, they are certain to be aware of the Dutch presence in agriculture and rural Ontario in general, especially when they encounter Dutch names or hear Dutch accents. Those who are older, and retired, might remember when a Dutch neighbor tried to persuade them to join a marketing board, or recall meeting Dutch immigrant farmers at political rallies. Such impressions helped to reinforce the positive stereotype of "the Dutch farmer" that was taking root in Ontario.

[76] An updated list of all of their associations, with names and addresses of their local executive, are available on their website, which is updated regularly.

IX. Getting Politically Involved

While building up their farm operations, some Dutch farm immigrants became involved in municipal government, ran as candidates in electoral politics, or took part in farmers' protests. Those born in the Netherlands, but who finished their education in Canada, assumed leadership roles on marketing boards and producers' organizations at the provincial, and in some cases, national level. Even immigrant farmers over forty years old, who had spoken Dutch for most of their lives, entered the political arena. This chapter will introduce most of the key political players, who were in large part responsible for how Dutch farmers were perceived by both other farmers and the public at large.

SETTING UP MARKETING BOARDS

Dutch immigrant farmers joined with other farmers to fight for marketing boards and supply management in the fifties and sixties. While alternative forms of marketing were already part of agricultural policy (see chapter 2), farmers had to be organized and agree on the kind of marketing board they wanted to establish. While non-Dutch farmers were more prominent in the Milk Marketing Board, at least during this early period, Dutch immigrants took the lead on "feather" (broiler chicken, egg production and turkey) boards.

Chickens

Prior to the Second World War, most farmers raised chickens, but there was no formal chicken industry. Chicken producers were not considered real farmers and banks would not lend them money. Starting in the late forties, with the development of a broiler industry (for six-to-eight week old chickens destined for slaughter), processors and feed companies promoted specialized chicken farming, which was often the only option for immigrants with little or no operating capital. On the Niagara Peninsula, Greek immigrants and Jews of Eastern European origin, as well as the Dutch immigrants, were chicken producers, while in other regions they were almost all Dutch. Feed mills supplied the feed, chicks, and technical advice. Those wanting to become egg producers signed multi-year contracts with feed companies. Farmers bound to three-way contracts with processors and feed dealers, felt they were unable to get ahead.

Egg and chicken producers were also faced with overproduction, which drove down prices, and were unable to cover costs. In 1953, the chicken farmers, mostly immigrants, formed the Ontario Broiler Growers Association. However, despite the existence of this voluntary association, chicken producers still felt vulnerable in an unregulated and unpredictable market.

In Huron County, Dutch broiler producers were particularly active in recruiting members and campaigning for their membership to approve the establishment of a marketing board. A good example of the leadership role taken by an immigrant is Cornelius Maaskant. A former greenhouse operator in Nieuwerkerk aan de Ijssel, in Zuid Holland, he brought his family to Canada in 1951. After a year of working for a farmer in Alliston, he moved to Sarnia to work in construction, along with several people from his hometown. At that time, Cornelius was still interested in becoming a greenhouse vegetable producer. However, he realized he would never be able to afford to buy land in the Holland Marsh. Instead, he bought a farm near Clinton (London region) that was no longer in operation, where the Maaskant family moved in 1956, and Cornelius converted the dairy barn for chickens, following the example of his wife's cousin, who farmed near Blythe. Only by growing chickens could he realize his dream to become a full-time farmer.[1] Since he could not yet afford to grow crops, he rented out most of his land. Cornelius then joined the newly established Broiler Growers Association. When the association established a price negotiation committee in 1963, Cornelius became its chairman. A year later, with the creation of the Ontario Chicken Producers' Marketing Board, he became a member of the regional chapter and was elected director of the provincial board.

It was not easy for Cornelius to persuade chicken farmers, especially those who were not immigrants or of Dutch background, to support supply management. In order to obtain approval for a marketing plan, which needed majority approval, he had to speak to almost all the broiler producers in his district (No. 2), which includes Huron and Perth counties. His school-age son helped him with his written English whenever he had to prepare speeches. Unfortunately, Cornelius did not complete his term as a board member because he died in an automobile accident in 1964. His son John, who took over the family farm with a brother, followed in his father's footsteps. Elected committee member for his district in 1974, at the age of twenty-five, John Maaskant went on to become a member of the provincial board, and chairman from 1992 to 1995.

[1] Interview (FS) with John Maaskant, at his farm near Clinton, 25 June, 1997.

In 1996, he represented the province at the national level, where he served as its vice-chairman.

Turkeys

The turkey business and specialized turkey farms appeared around the same time as the chicken broiler industry. Raising turkeys is labor-intensive, but producers are more likely to grow and grind their own corn, relying less on pellets and other forms of concentrates. The potential for growth and expansion, including on-farm processing of turkey products, is greater than for chicken growing. In Ontario, A. M. ("Mac") Cuddy, who today runs a $300-million business (including the world's largest supplier of day-old poults), started off on a 100-acre farm in Strathroy in 1950. A graduate from the Ontario Agricultural College and a war veteran, he bought his first incubator a few years after graduation and set up a slaughter house.[2] Many of his neighbors were Dutch immigrants, whom he employed in his turkey operation. He started with two helpers and, by 1958, a half a dozen Dutchmen were working for him.[3] In the sixties, other entrepreneurs entered the turkey business, such as Harry Beatty who founded Colespring Farms in Thamesville. He also expanded, and eventually became the largest turkey producer in Canada. Early on, he also hired Dutch immigrants with prior experience in the Netherlands to oversee the hatchery end of his business.[4]

Canadian farmers not of Dutch background were not the only ones specializing in growing turkeys. A typical example of a Dutch turkey farmer in the sixties is John Dortmans. A former Cuddy employee who set up his own turkey operation, John's farm could be written up as a model operation. Yet, he could not make money; John barely broke even after paying back the feed company. He consequently took up Mac Cuddy's offer to raise turkeys under contract, which he did for three years. But he wanted his independence, so he started his own dairy operation.[5] Other small-scale turkey farmers stayed in the business, but were worried about the fluctuating prices, which created boom or bust situations. In those days, it was not easy for Dutch immigrant farmers with little capital, and dependent on feed firms, to emulate the example of the Cuddys and Beattys. Just like chicken producers, turkey farmers

[2] www.cuddyfarms.com.

[3] By the mid-sixties, almost of all of Cuddy's workers were Portuguese immigrants.

[4] My information about Colespring Farms was obtained during interviews or conservations with four Dutch-Canadian farmers (at different times).

[5] Interview (FS) with John Dortmans, on his farm near Strathroy, 24 June, 1998.

banded together to form a larger voice in marketing. However, not all producers saw the merit in instituting a system of production quotas. The larger producers, particularly, were divided, while most small producers, including Dutch immigrants, were in favor of supply management. In Aylmer, St. Catharines, and Dunnville, Dutch producers were actively involved in organizing a marketing board for turkey producers, although it had no pricing power. It took a second vote, and low turkey prices, before a board that could set minimum prices and allocate quotas was established in 1970.

Two examples of people who became board directors illustrate how Dutch-Canadian farmers played a leadership role in the Turkey Marketing Board. Harry Vanderzanden, whom we encountered in chapter 6, at first produced chickens for a slaughterhouse while working for the Dofasco steel plant in Hamilton. In 1968, he was raising turkeys full time, and two years later, local turkey farmers asked him to represent them at meetings at the district level. In 1975, Harry became a director on the provincial board.[6] During that period, a third of the district committees and three out of seven directors on the provincial board were Dutchmen.

The Oegema family from Dedemsvaart (Overijssel) also became involved in raising turkeys in the sixties. They had come to Canada in 1958 when currency restrictions had just been lifted, allowing Sikko Oegema to bring enough money to make a down payment on a farm near Talbotville. He had cash-crop experience and wanted his Canadian farm to be an operation capable of providing income for the whole family, including two sons.[7] After investigating several forms of farming, they opted for turkeys. Their first contract, with Tillsonburg Grain and Feed, was for 10,000 birds. Their second flock was raised for Cuddy, but the contract almost fell through because turkey prices had fallen. The following year, the Oegemas decided to chance the open market on their own, although they continued to buy their poults from Cuddy. It was rough going, but they survived and were self-sufficient after installing special feed-mixing equipment. They grew, dried, and stored their own corn. In 1969, when the turkey quota was introduced, they had just had their highest level of production. Several members of the Oegema family also became involved in public service. Tom, the second son, who finished his education in Canada, was elected president of the Christian Farmers Federation

[6] Interview with Harry Vanderzanden (FS), at his farm near Wainfleet, 21 June, 1995.

[7] Interview (FS) with Heiko and Tom Oegema, in their farm office in Talbotville, 19 February, 2001.

(see last chapter). His older brother, Heiko, who had completed high school in the Netherlands prior to immigration, sat on the local committee of the Turkey Producers Marketing Board, and became a member of the provincial board in 1976. He served for twenty-one years and became a board member for the Canadian Turkey Marketing Agency at the national level. Such participation in public life is a family tradition; their father was similarly involved in organizations in the Netherlands, including one responsible for Dutch water control (*Waterschap*). Dutch farmers also supported, but played a less prominent role, in the Egg Marketing Board, established in 1972.

Dairy Ontario

The Milk Marketing Board (now known as Dairy Ontario), established on November 1, 1965, was a godsend to immigrants struggling to start dairy farms. Some had good contracts with dairies or creameries, especially if they had Guernsey cows, which produced milk with high butter content. But most producers received only half the value of the fluid or industrial milk they shipped. Although they could buy dairy farms with quotas assigned by a factory, there was no guarantee of how much money they would receive.[8] With the introduction of a system of quotas and prices set by a marketing agency, farmers no longer sold their milk directly to local processors. A bi-weekly milk check allowed numerous Dutch farm immigrants to stay in business.

The catalyst for supply management came from established non-Dutch dairy farmers dissatisfied with the fragmented nature of their industry – with its numerous local organizations and an inefficient marketing system. One of their main supporters was William (Bill) Stewart, Ontario Minister of Agriculture from 1961 to 1975, himself a dairy farmer. Another key player, and for a long time chairman of the board of directors of the Milk Marketing Board, was George McLaughlin, who owned a farm in Beaverton (Simcoe). While predominantly non-Dutch, dairy farmers of Dutch descent were involved at both the local and the provincial levels from the beginning.[9] The first Dutch-Canadian member of the milk board executive (acting as secretary-treasurer, starting in the

[8] There was also little quality control. One Dutch immigrant farmer recalls that a local dairy would accept the same cans they had been unable to accept the previous day, even though these can had frequently been left in the sun all day.

[9] Apart from interviews with numerous Dutch-Canadian farmers, I examined all the lists of local committee representatives, as well as members of the provincial board, housed in the archives of Dairy Ontario, whose head office is located in Mississauga.

seventies) was Francis Redelmeier, who graduated from the Ontario Agricultural College (OAC) in 1945. He was sixteen years old when his family emigrated. His father bought a farm just north of Toronto, a year after his arrival (see chapter 3).[10] However, not until 1975, did the first postwar Dutch immigrant, Peter Oosterhoff, join him on the board. With the addition of William (Bill) Schouten, they were three.

Bill Schouten of North Gower (south of Ottawa) emigrated in 1952. He bought his first farm in 1953.[11] In his hometown of Berlicum (Noord Brabant), he had been actively involved in farm organizations. A brother joined him a year later, and bought a farm in 1955. Of the eighty dairy farmers in their township, they were the only Dutchmen. Their first exposure to the difficulties facing dairy producers came when they discovered they were being overcharged for milk delivery. Within a decade of becoming a milk producer, Bill was going from farm to farm to organize a co-operative milk transport. People were hesitant because they remembered how an earlier co-op had gone bankrupt, but he convinced a third of them to sign up. By 1968, they were collecting their own milk. When Bill Schouten first came to Canada, he wanted to concentrate on farming, and was not interested in attending many meetings once the co-op was established. But, after a farm accident in 1968, when he lost his hand, he reconsidered his position. When his oldest son came home after studying at Kemptville (a nearby agricultural college), Bill became involved in public affairs. A year later, he was elected to the milk committee for the Carleton region, which already had one Dutch member. Between 1970 and 1980, four other Dutch immigrant farmers were elected to this committee. Bill became their local chair, which also put him on the provincial board. In addition to his role on the Marketing Board, Bill ran for local government and was elected town councillor. He was also chairman of the local chapter of the Ontario Federation of Agriculture (OFA) for a year and their director in Toronto for four years.

While farmers of Dutch descent never formed more than a fifth of the provincial board, they were conspicuous on many local committees from early on (see chapter 13). During the two decades that Dutch immi-

[10] Interview (FS) with Francis Redelmeier, in his farm office, in North York, near Richmond Hill, 30 January, 1990.

[11] Taped interview with Bill Schouten, 3 April, 1990, now archived in the Wellington Room of the University of Guelph Library. This interview was done by Edmund Briggs, the author of *The Challenge of Achievement: The Milk Marketing Board's first 25 years of operation—1965-1990*. Mississauga: Milk Marketing Board (1990). Frans Schryer interviewed Bill at his farm in North Gower (Richmond) on 2 July, 1991.

grants became full-time dairy farmers, at least one or two Dutchmen rotated on local milk committees throughout the province. In Thunder Bay and Rainy River, Dutch immigrant farmers completely dominated the local milk committees (see chapters 7 and 13). These same farmers also participated in other aspects of public life. Gerry Brugman of Chesterville, who was a member of the Dundas milk committee, sat on the board of the Chesterville co-op and was involved in the OFA.[12] In the London region, John Timmermans, who held all three executive positions on their milk committee throughout the seventies and early eighties, was a member of the Strathroy branch of St. Willibrord Credit Union, and president of its board of directors. By the nineties, the number of dairy farmers in most areas had shrunk to less than half of their 1966 numbers. Yet Dutch farmers started to play greater leadership roles, as shown by the increased occupancy of chair, vice-chair, or secretary-treasurer positions. Districts with only one Dutch representative in the earlier period, saw two or three in the eighties and nineties.[13] The ongoing influence of Dutch dairy farmers at the provincial level is exemplified by the election of Peter Oosterhoff as vice-chairman of the Milk Marketing Board in 1990.

Peter arrived in Canada with his parents in 1952. His father started up a dairy operation in Haldimand County (Niagara region) in the early sixties. They shipped their milk to a dairy in St. Catharines but it was sold in 1962. Within a year they had a contract with another dairy, but with only half of their old quota. His neighbors and relatives were interested in having "someonc take on the dairies." Peter's father encouraged him to attend a hearing in Vineland of the Hennesy commission, set up under Bill Stewart.[14] This was his first exposure to the possibility of province-wide milk pooling. In 1965, Peter and his brother bought out their parents, who were ready to retire. Peter welcomed the creation of the Milk Marketing Board, which enabled them to obtain new quotas. In 1971, he joined the Haldimand milk committee and, in 1974, was elected to the provincial board, serving two consecutive terms before becoming director and vice-chairman, a position he held until 1998.

---

[12] As reported in the *Chesterville Record* (spring, 1990); I read excerpts from that newspaper when visiting his farm in 1991.

[13] There was considerable inter-generational continuity since, in many locations, the same surnames reappear in the eighties and nineties, indicating offspring.

[14] Taped interview with Peter Oosterhoff, circa 1992 (located in Wellington Room, University of Guelph library).

Disagreement about supply management

Not all dairy farmers agreed with, or benefited from, supply management. A Dutch farmer in Grand Valley, who had a small dairy farm with thirty-five cows was satisfied with his contract with a local dairy prior to supply management. 75 percent of the milk he produced was for consumption and 25 percent for industrial purposes. He received money for all of the milk he produced. The Milk Marketing Board reduced his payments, even though he continued to produce the same amount. This young farmer felt the new system of supply management was implemented by a "bunch of bureaucrats", many of whom ran dairies. He especially resented the bigger producers in the Malton area (Toronto region) who were involved in setting up the Milk Marketing Board: "They were already thriving and later moved this way, with urban expansion, to set up dairy farms in and around Grand Valley and vicinity. Those guys were not interested in raising up the dairy producers at the bottom, but only in helping the ones who were already near the top."[15]

While admitting that other farmers benefited from the Milk Marketing Board, he told me he had emigrated to Canada to escape rules and regulations. With the implementation of supply management, he sold his milk quota, cows, and equipment and kept only beef cattle. He then bought land on the edge of town to build houses. While he withdrew from dairy farming and avoided any political involvement, other Dutch farmers who disagreed with decisions of the Milk Marketing Board tried to work within the system.

Eric was eight years old when his parents emigrated. He left home to study, and returned to his parent's farm in 1970.[16] He interrupted his studies to take advantage of a possible quota transfer, to get into dairy production around the time his father was ready to retire. Eric formed a partnership with several brothers interested in farming, and they expanded beyond the original twenty-five cows. By 1976, they were the largest milk producers in their county, with 150 cows. Eric, who served on the local milk committee, had views on supply management that were diametrically opposed to those of the provincial board, including their only Dutch-Canadian member. At one point, he had the opportunity to visit a large dairy operation with 1,800 cows in California where he observed milk pumped directly into tankers. So he made an appointment to see George McLaughlin, chair of the Milk Marketing

[15] Interview with a former dairy producer in the Grand Valley area, June 1998.

[16] The real name of this farmer, and any other identifying features, were disguised at his request.

Board, to convince him of the advantages of economies of scale. Eric made his pitch to their executive committee, which included a Dutch farmer, but they were unable to conceive of milking more than 500 or 600 cows, which was the upper limit on dairy farms in Ontario at that time. They claimed his plan would put most dairy farmers out of business and did not approve his plans for expansion. Consequently, Eric ceased milk production to specialize in other agricultural commodities, including seed crops. However, Eric continues to attend meetings of various organizations and is still outspoken on agricultural issues.

RABBLE ROUSERS

Dutch farmers and their offspring, some of whom became prominent agricultural producers, have been "trouble-makers" on both the right and the left end of the political spectrum. A good example of the latter is Dutch involvement in the Ontario Farmers Union.

Dutch Involvement in the Ontario Farmers Union

The Ontario Farmers Union, established in 1952 as an alternative to the OFA (see chapter 2), was particularly appealing to recent immigrants struggling to get established in farming. They thought the OFA was too complacent, and did not do enough to protect the interests of farmers. As the Union gathered momentum, it attracted numerous Dutch supporters, some of whom became actively involved at both the local and provincial levels. In Southwest Ontario, Dutch support was strongest in the London region, particularly in the Forest and Parkhill areas, but there were also many followers in Ashfield Township (in the Northwest region).[17] In Eastern Ontario, their strongest supporter was a young dairy farmer whose father was a pre-war Dutch immigrant. In Southwestern Ontario, an early and enthusiastic promoter was Andries (Andrew) Kuysten. An overview of his background and involvement in the Farmers Union can provide insight into the politics of protest, resistance, and cooptation.

Andrew Kuysten was born and raised on a farm in Sprang-Capelle in Noord Brabant where he attended a *landbouwschool*. He became a farmer but had to rent most of his land. They were a Protestant family (*hervormd*), but most of their neighbors were Catholic. Andrew's first experience in political resistance came during the War when he and a group of friends joined the underground resistance movement. In 1948,

[17] Interview (FS) with Jack Verhulst, in Guelph, 17 March, 1990.

he and his wife emigrated because he thought Canada offered greater opportunities. A beef farmer in Britannia Bay (now part of Ottawa) sponsored the Kuysten family. Andrew, his wife, and three daughters lived in a cottage in the middle of an apple orchard, and he essentially looked after the farm himself since their employer was often absent. After a couple of months, the embassy in Ottawa arranged for their transfer to a farm in Bloomfield, Prince Edward County. At that time, there was only one other Dutch immigrant family in the region.[18] By the time the Kuystens settled on their own farm near Springbank (London region), they had already moved four times, including a short stay in Norwich, where the family spent a season picking flue-cured tobacco. Andrew first rented land and then bought a dairy farm in 1958. He was a keen observer and befriended his neighbours. For the next ten years he wrote articles about farming and life in Canada, which were sent to the *Brabandse Landbouwblad*, a newspaper in the Netherlands. He also became an active participant in rural politics. By the mid-sixties, Andrew had joined the Farmers Union and became one of their recruiters.

Andrew Kuysten drew on a diverse circle of acquaintances. He was well integrated into a non-Dutch community as a member of a small Presbyterian congregation. When they arrived in the London region, they were members of a small Reformed congregation in Strathroy, but when it disbanded, the Kuystens started attending a church without Dutch connections in Springbank. While maintaining close contact with non-Dutch farmers, Andrew encountered other Dutch immigrant farmers at a feed mill operated by the United Cooperatives of Ontario. Most turned out to be Roman Catholics who wound up in the Parkhill area (see previous chapter). The main supporter of the Farmers Union in Parkhill, who became a close friend of Andrew Kuysten, was John Hendrikx. A well-educated farmer from Noord Brabant with Belgian connections (see chapter 5), he was a charter member and active participant in St. Willibrord Credit Union. Like Andrew, John served as chair of the local committee of the Farmers Union, and later as a board member, at a time when a third of its membership was Dutch. In an interview in 1988, he recalled that he and other Dutch immigrants were '*oproermakers*' (troublemakers).[19] However, a Dutch-Canadian presence in the Farmers Union did not materialize at the provincial level until a younger farmer with some Canadian education and good English skills took on leadership positions, including that of vice-president.

---

[18] Interview (FS) with Dinie (Kuysten) Twynstra 18 July, 1997.

[19] Interview (FS) with John Hendrikx, at his retirement home in Park Hill, 16 December, 1988.

Peter Twynstra came to Canada with his parents at the age of ten. A year later, Pete's father started working for Lawrence Kerr, a prominent farmer in Chatham (see chapter 4). Like the Kuystens, the Twynstras had been *hervormd* in the Netherlands, and briefly attended the short-lived Reformed congregation in Strathroy, then integrated into a non-Dutch rural community by joining a Presbyterian congregation in Ailse Craig. They, too, had a lot of contact with John Hendrikx and other Dutch Catholic farmers in the Parkhill area. While still a teenager, Pete had a reputation as a good equipment operator with a knack for fixing tractors. Everything he knew about farming and machinery he had learned on his own, having quit school before finishing grade nine to work full time on the family farm. In 1957, after a couple of years at home, his mother convinced the dean of the Western Ontario Agricultural School (now Ridgetown College) to accept Pete as a mature student. He became a model student, winning a trophy for engineering, when he graduated in 1960. He was immediately employed as a service supervisor for the International Harvester Company. To fully understand how a mechanically minded Dutch farmboy became a political activist requires some additional background information.

Pete's involvement in the Farmers Union occurred only after he met and married Dinie Kuysten, Andrew's oldest daughter. They were dating while she worked at a bank in London. In 1962, they moved to Ailse Craig where Pete took over his father's farm. Andrew Kuysten persuaded Pete to become a spokesman for the Farmers Union and many meetings were held at their respective houses. Up to fifty farmers at a time would come from as far away as Clinton and Aylmer.[20] The pinnacle of Pete's career as a political agitator occurred when he led protests against attempts by Bill Stewart to channel farm organizations and marketing boards into a single General Farmers Organization (GFO); in 1967, Pete and his friends disrupted a meeting in Toronto. A year later, while chair of the White Bean Producers Board, he initiated legal action when the recently established GFO wanted to pay $2,500 to the director of a central marketing board, with money collected from members of the various marketing boards. Pete saw that as a conflict of interest since the producer organizations were supposed to operate independently. So he took up the case with the civil liberties association and resigned from the board. They won their case, setting a precedent for other organizations who also demanded protection from government intervention in their financial affairs. Pete also joined a convoy of tractors, including

[20] Interviews (FS) with Peter Twynstra, at his home in Grand Bend, 18 July, 1997.

a larger contingent from Essex County (home of Eugene Whalen, then Federal minister of agriculture), on Parliament Hill, in Ottawa. Around that time, the leadership of the Farmers Union, including Pete, joined forces with farmers' organizations from other provinces to create a single National Farmers Union. He became their junior president.[21]

Dutch-Canadian farmers who became involved in the Farmers Union did not escape the attention of high-ranking government officials. Although Pete Twynstra was an opponent of Bill Stewart, the provincial minister of agriculture, Stewart appointed him as founding chairman of the new Ontario Farm Safety Association and member of the Farm Machinery Advisory Board. Stewart also solicited input from Hendrikx on such government initiatives as new farm loans. While John was serving as local chairman of the Farmers Union, Stewart would drop by, sometimes once a week.[22] At the same time the government was trying to co-opt these leaders, the Farmers Union started losing support in Ontario among both Dutch and non-Dutch supporters. The majority of farmers in Ontario, including those who were Dutch, could not relate to the Marxist discourse of the new national leadership. Most of their Dutch-Canadian members, including those who held executive positions, became disillusioned and Peter Twynstra channeled his energy into his own farm enterprise. However, he continued to be active in local politics, first as councillor of East Williams Township and then as township reeve and member of the Middlesex County Council.[23] Eventually, Peter became one of the most prominent farmers and processors of edible beans in Ontario (see chapter 13). Most people in the farming community, who are aware of his accomplishment in the world of agriculture, are unaware of his earlier involvement in the Farmers Union, nor would this register. Given the reputation of Dutch farmers as successful and entrepreneurial, one would not anticipate their involvement in radical agrarian politics. Yet, farmers of Dutch background continue to be "rabble rousers".

[21] See also Albert Van der Mey and Martin Mol, *The Dutch Touch in Ontario* (Toronto: Marten A. Mol and Associates, 1997), 143-44.

[22] Interview (FS) with John Hendrikx, Park Hill, 16 December, 1988. At the federal level, Eugene Whelan appointed Gosewinus (Gus) Sonneveld, a Dutchman involved with the Farmers Union, as a resource person for the Livestock Feed Board, a crown agency, 1968, to look at transportation policy. He became a full-time member of that board in 1972 and their vice-chair a year later. He held that position until 1989 when the agency was dissolved. Interview (FS) with Gus Sonneveld, at his farm near Blenheim, 10 July, 1997.

[23] See VanderMey and Mol, *The Dutch Touch*, 144.

Other Forms of Protest involving Dutch farmers

By the end of the seventies, the Farmers Union was an insignificant player in Ontario politics. Nevertheless, dissatisfaction with rising farm debts in the early eighties from high interest rates was expressed in other ways: tractor parades to Ottawa (1984) and penny auctions protesting foreclosures caused by high interest rates. Members of the Bruce County branch of the OFA, who helped organize such forms of political resistance, were particularly militant.[24] This part of the province even saw vigilante groups, whose aim was to dissuade bank representatives from evicting farmers who were going bankrupt. Such protests and direct action tactics, including carrying pretend guns, were supported by many of the same Dutch farmers who had earlier joined the Ontario Farmers Union.

The Catholic Rural Life Conference

Some Dutch-Canadian farmers again showed their displeasure with the state of agriculture in the eighties by becoming involved in the Catholic Rural Life Conference (CRLC). Originally set up by Bishop Sherlock in London at the time of the farm debt crisis in 1981, its main goal was to support policies that would help the family farm to survive. Their members joined the tractor protests in Ottawa. When the crisis was over, the organization continued to criticize the role of big business in agriculture and tried to raise farmers' awareness of social justice issues. Father Mooney, a priest who had earlier worked in Zurich, now in charge of the parish in Kinkora, drew in members of both Dutch and Irish background in Huron County. CRLC organizers were especially active in Bruce County where farmers were more vulnerable to the impact of broader economic forces. They recruited additional members in places further south, such as Tillsonburg, with the decline of the tobacco industry around 1987. While some Dutch-Canadian farmers, including some well-established ones, supported the CRLC because of their concern for social justice, others perceived the CRLC as too radical. For example, the Dutch-Canadian hog farmers in Kinkora wanted nothing to do with them when their parish priest tried to promote the organization. They were also upset about Father Mooney's comments about the dangers of monoculture, including specialized hog farming. In other parts of Southwestern Ontario, Dutch farmers who had little contact with the CRLC, or who disliked the leftist discourse of their leaders, accused

[24] Interview (FS) with Bob Bregman, at his farm home in Teeswater, 3 Feb., 1998.

Bishop Sherlock and some of their priests of meddling in affairs of which they knew little. They also portrayed the CRLC in a negative light. Several Dutch immigrant farmers I interviewed in the London area made comments such as: "Their supporters are poor farmers and they complain too much . . . they are failed farmers, whose farms are run down and going under, especially the Irish, but also Scots."[25]

Contrary to such negative ethnic stereotypes, Dutch-Canadian farmers made up between half and three quarters of the membership of the CRLC in the London area, including Ingersoll and the area around Arkona. In terms of leadership, several of their presidents or vice-presidents are of Dutch background. Even a person with an Irish-sounding name like Elsie Murphy, who became their secretary, is the daughter of parents who are Dutch and Flemish.[26] One could argue that farmers facing the pressures of increasing competition, and whose children are not likely to be able to continue farming, are more disposed to agree with the philosophy of the CRLC than those who are doing well. However, the leaders and many supporters of this organization do not operate marginal or "run-down" farms.

An example of a successful farmer who became a strong advocate of the CRLC is Tony Beernink, a man who was, and continues to be, actively involved in the Farmers Union. His experience as the son of an immigrant, and his own struggle to become a full-time farmer, formed his character as an idealist, who could also be pragmatic.

Tony came to Canada in 1949 at the age of sixteen. He was living with his parents when they settled in the township of Warwick, an area that has a concentration of Dutch farmers. His father, Theo Beernink, who used to farm in the Achterhoek region, where they worked horses, was not mechanically inclined. He was unable to make their farm succeed, so he became a part-time farmer and worked for wages in Sarnia, as did his three oldest sons. He also served on the credit and supervisory committees of St. Willibrord Credit Union. His son Tony was already on his own when, at the age of twenty-eight, he married the daughter of a postwar Flemish farm immigrant he had met through the Canadian-Dutch-Belgian Club in Sarnia. At that time, Tony was in the asbestos insulation industry, where he worked his way up to a supervisory position. That job required a lot of travel. In 1966, they bought

[25] These quotes, and the other details concerning Dutch involvement in, and opinions on, the Catholic Rural Life Conference were obtained from nearly twenty interviews, most of which deal with other topics.

[26] Telephone interview (FS) with Elsie Murphy, 28 May, 1997.

a farm but he continued to work off the farm. His wife, who loved farming, drove the tractor and managed the farm when he was away. However, before they could turn it into a viable operation, Tony had to fight his municipality for a drain so he could tile his land. In addition to feed crops and beef, he had chickens to teach his children how to earn money. His main specialty became cash crops, including soybeans and wheat. In 1971, while continuing to work in Sarnia, he bought an adjoining farm. His first setback came in 1978, after a period of heavy rainfall. With 1,400 bushels of corn that would spoil unless it quickly dried, Tony felt at the mercy of the elevator company in Forest, one and a half miles away. So he bought his own drier and storage facilities. An enterprising farmer, Tony sold chemicals for agriculture on behalf of the Ontario Farmers Union as a sideline and bought more land in the eighties. Only then, ten years before his retirement in 1998, did he stop working in Sarnia.

Apart from working hard to build up a farm enterprise, Tony was politically involved. From the time he bought his farm, he attended numerous meetings of the Farmers Union and set up marketing courses in Lambton County. He became a member of the soybean board in both Ontario and at the national level. He also served as a representative, and acted in various capacites, for other commodity organizations to which he belonged, including county director for the corn and wheat producers. It was all done on a volunteer basis. His philosophy was influenced by a group of American thinkers, who formed NORM (National Organization of Raw Materials).[27] Tony continued spreading his message after he joined the Catholic Rural Life Conference in the eighties, acting as a critic, and engaging in debates with agricultural economists at the University of Guelph. Retired since 1997, Tony regularly sends information to old contacts on issues related to agriculture, globalization and international trade.

CRLC's ideas were similar to those espoused by the policy analysts in the Christian Farmers Federation. However, the CRLC remained insignificant as a political force on the provincial level. It did not have sufficient members to be officially eligible for a share of the fees paid to the government by all farmers. It is not surprising then, that Dutch Catholic farmers who sympathized with, or belonged to, the CRLC, also chose to support the Christian Farmers.

---

[27] He first heard about them when they sent a speaker to address the Farmers Union. Interview with Tony Beernink, at his home in Forest, 28 May, 1998.

ELECTORAL POLITICS

In many regions, first generation Dutch immigrants who became farmers, or were of recent farm background, also became involved in electoral politics at the local level, sometimes within a decade of coming to Canada. Initially, the idea of a Dutch immigrant being elected as reeve or councillor was inconceivable. However, by the seventies and eighties, a surprisingly large number of Dutch-Canadians born in the Netherlands, had at one time on another served as mayors, part-time mayors, reeves, deputy-reeves, or township councillors, not to mention members of school boards. In Perth County alone, three out of eleven townships had Dutch-Canadian reeves. In the course of my research I encountered names of Dutch-Canadians who had been, or were still serving as, mayors and reeves in the Strathroy (twice), Bradford, Grand Valley, Chesterville, Osgood, Metcalfe (near Ottawa), Essex, Fruitland, Drayton, Londesborough, Fullarton, and Sydenham (north of Kingston), and as reeves or councillors in the townships of Bosanquet, West Williams, Metcalfe, West Nissouri, Adelaide, Westminster, North Dumphries, Adelaide, Harwich, and South Gosfield.

Two immigrant Dutch-Canadians held pubic posts at a higher level, as chairs of regional municipalities, including the Toronto region. Gerald Herrema, a dairy farmer from Leaskdale (near Uxbridge) was born in Tjumarrum, in Friesland. He came to Canada at the age of nine, in 1947. His parents joined the United Church, but they had a lot of Dutch friends, and Gerald kept up his Dutch and his Frisian. As a young man growing up in King City, where his father rented a farm, Gerald joined the junior farmers section of the OFA and became a liberal party supporter. Soon after moving to Leaskdale, where he bought a farm with his father, he ran for reeve and was then elected to the regional council. His next public post was mayor of the township of Uxbridge, becoming chairman of the newly created regional municipality of Durham (formerly Ontario County), a position he still held when I interviewed him in 1990.[28] However, no Dutch farm immigrants have ever been elected to the provincial or national legislature. An examination of the political career of the only Dutch-born farmer who ran for office at this level can provide further insights into the dynamics of electoral politics in rural areas.

[28] Interview (FS) with Gerry Herrema, at his farm near Leaskdale, 10 March, 1990.

Running for a Seat in Parliament (MP)

Gus Sonneveld, another Dutch farmer who was involved in the National Farmers Union, came to Canada at the age of eighteen, after completing all but three months of high school in the Netherlands, plus a *landbouwschool* diploma. He grew up in Pijnacker (Zuid Holland) where his mother managed the family's greenhouse operation after Gus's father died. Gus married the daughter of a Dutch (Frisian) immigrant a year after arriving in Canada and was running his own farm, on a share-arrangement, by 1950. They later bought a combined beef and cash crop operation near Chatham. Gus's interest in political life and social issues goes back to 1949. He was incensed to learn that the younger sister of his fiancée could not enter teachers college because her parents had not taken out Canadian citizenship. Gus complained to the local Member of Parliament (MP), and argued they should change this archaic law. His subsequent involvement in politics reflects his fondness for debate. He attended men's groups and ecumenical councils and also joined the Liberal party. Gus's enthusiasm for public life is part of a long-standing family tradition – his paternal grandfather had been an alderman in Pijnacker, and a founder of a Dutch branch of the Raffeissen bank.[29] Gus became increasingly active in the federal liberal party in the sixties, and in 1972, then Senator Keith Davey invited the young Dutch farmer to seek the liberal candidate nomination. After consulting with his wife, he ran. He won the nomination, only to be defeated during the elections. When the results were tallied, it was evident that in Dover Township, which is predominantly French-Canadian and Catholic, an overwhelming majority (98 percent) had voted for him even though he was Protestant. He also had support in a few other townships. In contrast, he won few votes in the more urbanized parts of this mainly White, Anglo-Saxon Protestant riding, and none whatsoever in the northern part of the county, the home base of the opposition candidate, then minister of revenue. One might conclude that, in the early 1970s, few people in rural and small-town Ontario favored an immigrant farmer, not even a Dutch farmer, as their representative in parliament. Gus had already pointed out to the then Prime Minister, Pierre Trudeau, that the riding was very traditional and only someone with a name like Jones or Smith would ever get elected. Trudeau, sceptical at first, had to admit he was wrong when confronted with the breakdown of the vote.[30]

---

[29] He had also helped to set up the Dutch auction system for a co-op. Interview with Gus Sonneveld, at his farm near Blenheim, 10 July, 1997.

[30] Interview with Gus Sonneveld, at his farm near Blenheim, 25 June, 1998.

In running for office, Gus also faced opposition from Dutch immigrants belonging to various Reformed denominations who supported the Conservative party. In their minds, the Liberal party in Canada was the same as the "Liberal" party in the Netherlands, the main opponent of the Anti-Revolutionary party closely allied with Orthodox Calvinism and thus perceived as anti-religious. In rural Canada, the political divisions associated with the Dutch pillar system also influenced how Dutch immigrants voted, as well as which organizations and credit unions they joined. However, ethnic and religious loyalties had less bearing on public posts in producers' associations or on marketing boards. When Gus finished his work for the federal feed agency, an appointed position, he became actively involved on the Ontario Wheat Board. He had already served as a local representative, when he ran for its provincial board and became second vice-chair in the eighties.

It is possible to generalize about the political orientation and public involvement of Dutch farm immigrants in Ontario, but only to a certain extent. Those who worked to establish marketing boards, especially in the case of broilers, were just as likely to be Calvinists, Roman Catholics, or members of mainstream Protestant congregations. In electoral politics, Dutch Protestants who were *gereformeerd* or *hervormd* in the Netherlands tended to support the Conservatives, or religiously oriented conservative parties such as the Christian Heritage Party. They generally shied away from radical protest movements, although some Christian Reformed Dutch-Canadian farmers supported the Ontario Farmers Union. Moreover, two leaders in the Farmers Union, Peter Twynstra and Andrew Kuysten, were of *hervormd* background. Another tendency is for Roman Catholic farmers of Dutch background to support the Liberal party. Yet, Gus Sonneveld and Gerald Herrema, who joined Liberal party, were also of *hervormd* background.

A common feature of the politics of Dutch farm immigrants, regardless of background or political orientation is that it never resulted in ethnic mobilization, not even during the brief period they were working as hired hands. The interests and inclinations of Dutch immigrant farmers and their offspring were too diverse to express with a single voice. Even those with the same religious affiliation diverged sharply. Nevertheless, Dutch farmers have left their mark on the political landscape of rural Ontario in other ways, as we have seen in this chapter. Those who became high-profile leaders or assumed important posts in the agricultural sector, especially in producers organizations, reinforced the positive stereotype of the Dutch as achievers and successful farmers. The image of the successful (and "rich") Dutch immigrant farmer had become well entrenched by the late eighties, just when a new group of farm immigrants was to arrive. However, the positive stereotype overlooks the fact that the majority of Dutch farm immigrants did *not* become rich and successful. It also misrepresents the diversity among Dutch farmers, and ignores the complexities of farming as a business and a way of life.

X. A Profile of Dutch-Canadian Farmers

Census data indicate that, on average, Dutch immigrant farmers have done well. They are, nevertheless, a diverse group. Some arrived as adults, with older children to help out; others had barely finished school. The time it took to buy a farm depended on education, socio-economic background, and just plain luck. Some immigrants became prosperous farmers who branched out into agribusiness ventures, while others did not progress beyond running a part-time operation. Some people became involved in farm organizations, while others channeled their energies into church life, service clubs, or politics. There were different religious beliefs and convictions. The experiences of men were not the same as those of women. They became an even more diverse group as their children started farming on their own.

Each person has an individual identity and each household has a unique history. Yet we can still identity commonalities. This chapter will present a profile of Dutch-Canadian farmers using these commonalities, as gleaned from a statistical technique called "dual scaling" (see appendix for a more detailed explanation). This technique, which generates graphical displays, allows a researcher to identify clusters of common traits. To breathe life into these statistical clusters, the profile will include biographical sketches of typical members from several statistical subgroups. The stories are real, but people's names are omitted (with one exception) because my own commentary and interpretations obscure individual stories.[1] To assure anonymity I do not refer to any specific township. My profile will demonstrate the heterogeneous nature of farmers of Dutch background, yet present enough commonalities to make generalizations without stereotyping.

[1] Some of these examples include farmers to whom I have referred in other parts of the book. However, most are not nor did I ask for permission to use their names, or show them copies of this chapter.

Table 1
Table of Items ('Variables') and their Values

Items (Variables)	Values [labels]	Definitions
1. Size of Agricultural Enterprise	1 [lof]	Small (part-time) farm
	2 [mdf]	Medium-sized farm
	3 [umf]	Big farm
	4 [hif]	Huge farm
2. Other Main Source of Income	1 [non]	No other source
	2 [wag]	Wages (manual labour)
	3 [bsp]	Business & Professional
	4 [ags]	Agribusiness
3. Cohort (generation)	1 [ad1]	First adult generation
	2 [ten]	Teenage immigrant
	3 [chi]	Child of immigrant
	4 [ad2]	Later adult immigrant
4. Partnership	1 [npr]	No partners
	2 [kpn]	Kin partner only
	3 [bpn]	Business partner & mixed
5. General Education	1 [elm]	Elementary (primary)
	2 [sec]	Secondary (high school)
	3 [hed]	Post secondary (college)
6. Agricultural Education	1 [no]	No
	2 [yes]	Yes
7. Dutch Farm Connections	1 [lkn]	Few if any connections
	2 [mkn]	Medium level of connections
	3 [hkn]	High level of connections
8. Public Posts (in Agriculture)	1 [lps]	Few if any posts held
	2 [mps]	Medium level of posts
	3 [hps]	High level of posts
9. Social Involvement (community/church)	1 [lin]	No or little involvement
	2 [min]	Medium level of involvement
	3 [hin]	High level of involvement
10. Spousal Support	1 [lsp]	Low level of support
	2 [msp]	Medium level of support
	3 [hsp]	High level of support

SELECTING RELEVANT VARIABLES

The first task in doing this type of statistical analysis is to select variables with a reasonable distribution. As a result of that selection, each farm household in my "survey" (see introduction) was assigned a series of numerical values, for a total of thirty-two possible attributes associated with ten variables. Table 1 lists labels and definitions for the attributes associated with each variable, including short abbreviations (e.g. [chi] for children), which appear in the diagrams. For some variables, an interval scale was divided into three or four discrete attributes. For other variables I amalgamated separate categories to create a composite variable, each with two or more 'values'. Establishing cut-off points involved making decisions based on my knowledge of the data. In the end, I had one dichotomous variable, with only two attributes (1 = yes, 2 = no), and the rest were assigned either three or four "values" (e.g., 1 = low, 2 = middle, 3 = high).

The first variable, "size of agricultural enterprise," refers to four types of enterprises: 1) small, part-time farms at the lower end of farm size labelled, "low farm" [lof]; 2) medium or average-sized farms labeled as lower-middle-size farms [lmf]; 3) larger, usually more diversified, farm enterprises (upper-middle-size farms or [umf]); and finally, 4) very large farms at the high end of the scale [hif].

The second variable is "other main source of additional income." I used this variable to distinguish among income-generating activities beyond cultivating the soil or raising animals, regardless of the size of the farm enterprise. These distinctions resulted in four possibilities: 1) no other source of income [non]; 2) wages [wag]; 3) other business and professional income including salaries [bsp]; and finally, 4) agribusiness [ags].[2]

The third variable, "cohort," consists of four attributes defined in terms of when and at what age farmers or their parents arrived in Canada. The first cohort includes those who were at least twenty years old who came to Canada prior to July 31, 1960. They constitute about two-thirds of those included in the "survey." I included two prewar farmers who emigrated as young men before the war, but developed their careers after 1945. All these adult immigrants were given the label [ad1]. The rationale for placing them into the same category as post-war adult immigrants is that none of them attended school in Canada, yet had had at least several years of full-time experience in Dutch agriculture. All of the members of this cohort were married prior to, or soon after, emigration, and spoke English with a noticeable accent. The second cohort [ten] comprises unmarried people who emigrated between the ages of twelve and nineteen, usually with their parents. These farmers had all first worked on a home farm in the Netherlands, while in Canada they worked for wages to help their parents. Unlike younger children, these teens had been socialized, and completed all or most of their education in the Netherlands; they faced psychological pressures leaving their social circle behind; and were maturing into adults at the same time their parents were struggling to adapt to life in a new country. They are bilin-

[2] This attribute refers to agribusinesses that require a high level of expertise and capital, as opposed to minor sidelines.

gual (Dutch and English) to a greater or lesser extent and might have a trace of an accent. I have information on twenty such people. The third cohort [chi] consists of twenty-three farmers who came to Canada as children (under age twelve), as well as those born and raised in Canada. They speak accent-free English even if their first language was some form of Dutch or Frisian, and started or finished grade school in Canada. These younger offspring of immigrants did not become farmers until the late sixties or seventies. The fourth cohort consists of nineteen immigrants who came to Canada after July 31, 1960. These latecomers [ad2] were roughly the same age, and started farming around the same time as the younger offspring of earlier immigrants, but they all grew up and received their education in the Netherlands.

The fourth variable, "partnership," refers to a farmer's relationships with people who had legal ownership rights at some stage in the history of the farm, not inluding spouses. Either there is no partner [npr]; or if there is, the partner is either related to the farmer through kinship [kpn] or not [bpn]. A kin partner could be a farmer's parent, sibling or mature son or daughter (see examples in chapters 5 and 6), who live on or near the farm, while non-kin business partners [bpn] are Canadians not of Dutch descent who have invested money or provided technical advice, but do not reside on the farm.

The attributes of the fifth variable refer to the highest level of general education achieved in either country: elementary [elm], secondary [sec] or higher education (college or university) [hed]. In order to distinguish between level and type (content) of education, I included the sixth variable, which asks whether or not the formal education of the farmer was specifically geared to farming. This is the only variable treated as a dichotomous variable [yes] versus [no]. "Dutch Farm Connections" (the seventh variable) is a composite variable that includes data on several measures regarding the total level of contact (low, medium, and high) that a farmer has had with other Dutch farm households through his or her lifetime.[3] These were labeled [lkn], [mkn] and [hkn]. The eight variable, with three "values", indicates the holding of public posts in agriculture, taking into account the number as well as the relative importance of posts held in both general farm and specific commodity organizations (including marketing boards).

[3] I first gave numberical values to a) total number of adult relatives (including in-laws) involved in farming in both the Netherlands and Canada; and, b) the proportion of farmers of Dutch descent included in the same census unit as the farmer interviewed (usually a township). These two numbers were added up and I then established what seemed like reasonable cut-off points between low, medium and high.

Three "values" for the ninth variable, "social involvement," refer to the level of participation in, and service to, churches, service clubs, or political parties. This was an attempt to distinguish between community participation and involvement in farm politics as separate attributes. The last variable, "spousal support," is a composite variable based on the amalgamation of what were initially measures of spousal contribution to farm enterprise through off-farm employment and both manual and non-manual labour.[4]

I eliminated several original variables that were ultimately irrelevant: Dutch regions of origin, agricultural regions in Ontario, and farm specialization were all excluded because too many farmers had lived in more than one region. Moreover, farmers often switched commodities, so type of crop grown is not a stable attribute.[5] Another variable, which was manageable, but turned out to be unreliable, is degree of ethnic identity. Originally I recorded two types of data for each farmer: observable indicators such as accent, membership in a Dutch club, or a windmill in front of the house; and a subjective feeling of how strongly someone identified as "Dutch". However, it soon became apparent that objective and subjective ethnicity were rarely the same, and people changed their ethnic identity over time.

Finally, while religion was significant in determining where people settled and with whom they interacted (as demonstrated in earlier chapters), this variable was also excluded. None of the denominations restricted where or what one could farm, nor did I have any reason to suspect that religion was associated with class or social involvement. Farmers from each of the religious denominations came from diverse class backgrounds, from *landarbeider* (farmhand) to prosperous farmer. Likewise, in Canada, Dutch Catholics and Reformed Dutch immigrants are well represented in every social stratum, from the most prosperous farmers to the least. Catholic and Calvinists farmers alike were involved in social and political life, albeit in different ways (see chapter 9). There were not enough farmers without any religious affiliation in my "survey" to draw any conclusions about them as a separate category.

---

[4] I initially assigned three separate values to each of these components, but soon found out I was unable to attain sufficient information for all farm households. Instead I made a global, and more subjective judgment, on the overall contribution of each spouse relative to other spouses around the same age, giving greater weight to the non-monetary component.

[5] We can only safely conclude that farmers in the horticulture and the greenhouse industry, especially in the Niagara region, came mainly from the Westland region in Zuid Holland (see chapter 13).

THE FINDINGS

When the data were entered into the dual scaling program, nine solutions were generated, which jointly accounted for 64.3 percent of the information. The rest of this chapter will interpret a series of figures corresponding to consecutive pairs of these solutions. Each figure includes a diagram or "mapping" of the data, plus a table with the relevant numerical output, including the R(jt) statistic for all variables, in order of importance. Dots representing households and triangles representing relevant attributes (the "values" of each variable) are shown only for the first two solutions. The rest of the diagrams show only triangles, while the general location of clusters of dots appear as ovals or circles. Such clusters represent the statistical sub-groups of farmers (note: reading the appendix at this point will help a reader not familiar with this way of reporting statistical data to better understand these diagrams).

The First Two Solutions

The diagram generated by the first two solutions (see figure 13)[6] shows the importance of the size of a farm enterprise and source of income for identifying contrasting groups of farmers. Two other relevant variables are public posts held in agriculture and connections with other Dutch farmers.

The first solution (1st horizontal axis)
The main contrasts are: (1) between huge farms [hif] and small, part-time farm operations [lof], and (2) between agribusiness [ags] and wages [wag]. These two sets of contrasting attributes, which represent the opposite extremes of size of farm enterprise and source of income, demonstrate the relevance of economic class for distinguishing between two clusters of farmers, shown as A and B. Each cluster represents farmers who share distinct combinations of attributes associated with poor rather than well-off farmers. The positions of the triangles representing upper-middle-class farmers ([umf] to the right) and lower middle-class farmers ([lmf] to the left), while not as far apart, are consistent with a class interpretation of dots and triangles in relation to the first axis. Overall the significant R(jt) statistics for farm size and other sources of income in the first solution were .858 and .844 respectively.

The position of triangles for other variables indicates what additional attributes are associated with farmers in clusters A and B and, to a lesser

[6] I have superimposed a diagram with normed rows onto one with normed columns, adjusting for differences in scaling, to facilitate the recognition of patterns.

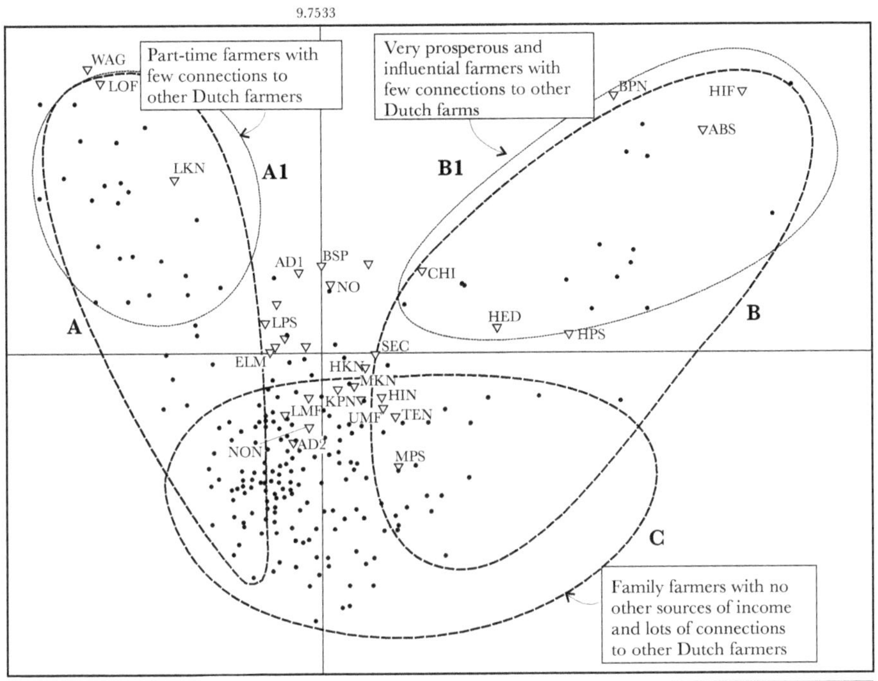

Item statistics (showing order of importance):			
R(jt) First Solution		*R(jt) Second Solution*	
Farm size	.858	Other sources of income	.823
Other sources of income	.844	Farm size	.790
Public posts	.600	Dutch connections	.454
Partners	.488	Partners	.404
Education (general)	.406	Public posts	.333
Dutch connections	.371	Spousal support	.327
Cohort	.344	Cohort	.319
Social involvement	.225	Education (agriculture)	.291
Spousal support	.160	Social involvement	.263
Education (agriculture)	.039	Education (general)	.033

Summary statistics:		
Cumulative delta (end of 2nd solution) = 21.3873		
	First Solution	*Second Solution*
Percentage homogeneity	25.5948	21.4573
Reliability coefficient alpha	0.67699	0.59270

Figure 13. First and Second Solutions.

extent, with the rest of the farmers whose dots are located on opposite sides of the vertical line in figure 13. Moving from right to left, we detect a contrast between farmers with the greatest influence in farm organizations (i.e., [hps], the holders of many public posts versus those who held few, if any, public posts [lps].) Holding a medium level of posts [mps], is also on the right, along with being an upper-middle-class farmer [umf]. The location of all of these attributes is what we would expect if level of post held positively correlated with level of economic power. It also makes sense that farmers with non-kin business partners ([bpn] at top right) should be included in the same cluster as those who owned agribusinesses.

The locations of other contrasting attributes are also consistent with a social class interpretation. Secondary education [sec], and more so, higher education [hed], appears to the right of the vertical dividing line, on the same side as more prosperous farmers. In contrast, the triangles for elementary education [elm], lower-middle-size farm [lmf], and small farm, are all located on the left. If we include the last two attributes, whose variable statistics are still above .300, we can detect other attributes associated with the economic class factors identified in the first solution. The attributes corresponding to the variable "cohort" show a contrast between the first generation of Dutch-Canadian farmers [ten] and [chi]; and both post-war and later adult immigrants [ad1] and [ad2]. These findings, which indicate that the offspring of immigrant farmers are better off and more educated, and also hold more agricultural posts than their parents, are consistent with what we know about ethnic stratification and intergenerational upward mobility.

While consistent with a social class interpretation, the low value of the statistics associated with the last variable, "Dutch connections," calls for caution when interpreting the diagram. While low connections [lkn] is associated with wages [wag] (since these two attributes appear in the same quadrant), the triangles representing medium and high connections [mkn & kn] lie closer to the center, although still on the same side as farmers who run agribusinesses and who own the largest farms. These minor associations can be misleading if we do not examine the larger picture, beginning with a different alignment of clusters that appear in the second solution.

The second solution (1st vertical axis)

The order in which variables occur in the second solution is similar to the first solution, with minor variations. For example, the R(jt) number for source of income is now higher than that for farm size. The order of the other significant variables remains the same. Moreover, the R(jt) numbers for farm size and other source of income are almost as high as those of the first solution. However, the contrasting attributes associated with these variables now align along the vertical axis, with dot configurations on opposite sides of the horizontal line of figure 13 forming three distinct clusters (A1 and B1, versus C). The main contrasts, as shown by distances between triangles, are between farmers with no additional sources of income in the bottom half [non]) and the other categories of farmers in the top half, who either have a job [wag], a career outside of farming [bsp], or who run an agribusiness [abs]. An almost equally strong contrast is between lower-middle and upper-middle class farmers [lmf & umf], shown in the bottom half against the owners of both small, part-time farms [lof] and very large farms [hif] in the top half of the diagram. In other words, the contrasting attributes associated with these variables bisect the class divisions that appeared in the horizontal axis. The second solution clearly shows that full-time farmers with no other sources of income have different attributes than farmers with other sources of income, regardless of whether those farmers with other sources of income are poor or wealthy. The second solution thus reveals an equally important determinant of what differentiates farmers: whether or not they make a living exclusively from farming. To discover other attributes associated with this social division, we turn our attention to the third variable, Dutch connections, which has an R(jt) number of .454. The triangles for middle and high connections, shown in the bottom half [mkn & hkn], although somewhat closer to the horizontal dividing line, are closely associated with two groups of farmers who have middle-sized farms [lmf & umf]. In contrast, the attribute "low Dutch connections" [lkn], found in the top half of the diagram, is closely associated with farmers who work for wages, and to a lesser extent with those who have other businesses or incomes as professionals [ags & bsp]. The tendency of people who are exclusively full-time farmers to have had higher levels of connections with other Dutch farmers (throughout their life) requires further examination.

Most commercial farms in Ontario are run as family enterprises even when they become relatively large-scale operations. We have already seen how Dutch immigrants had a competitive advantage over non-Dutch farmers in building up and maintaining viable farms, because of a combination of family support, a strong work ethic, and a value system that emphasized being a farmer. Hence it makes sense that Dutch farmers with a strong network of extended family members, friends, and neighbors who are farmers, and who could potentially provide support and advise, were more likely to succeed than those who did not have such connnections. Also it stands to reason that farmers with significant contact with other Dutch farmers with the same values are more likely to want to succeed in farming than part-time farmers with fewer role models. As a result, they may have had less pressure to compete with, or emulate, other Dutch farmers.

An examination of the only other variable whose R(jt) number is higher than .400 (partnership) can shed further light on the nature of the clusters identified by the second solution. It appears that middle-sized farmers without additional sources of income (cluster C) have kin partners [kpn], while farmers with big farms and agribusinesses (cluster B) are more likely to have business partners [bpn] than to rely on kin. This contrast indicates that the former, who are also more likely to have high or medium Dutch connections, utilize those kin connections to form partnerships to run those farms. Variables whose R(jt) numbers were between .3 and .4 do not supply new information, because they show the same association between class factors and traits such as level of post-holding and cohort as found in solution 1.

To get a better picture of which people are represented by the main clusters identified in diagram one, it is useful to describe several farmers representative of those clusters. I therefore selected one or more people represented by the dots most clearly situated at the outer edges of major clusters (where groups or sub-groups do not overlap).

Prosperous and influential agribusiness farmers (B1)

In my research, I encountered twelve prosperous farmers who also run agribusinesses. While these Dutch agribusiness farmers share a common socio-economic status, they are a diverse group in terms of age, and date of arrival in Canada. This diversity is reflected in the wider dispersal of point representing members of this cluster. Nevertheless, they were all actively involved with marketing boards or producers' associations, where they held important posts. They are thus more influential than any other group of farmers both in terms of economic power and political clout. I selected one man for a more detailed treatment to highlight this commonality. He happens to specialize in packing and shipping apples, but I could as easily have chosen a prominent farmer who runs a sod business or a turkey producer who owns feed mills.

The apple producer came to Canada with his parents and eleven siblings in 1951, when he was eight years old. His father bought a rundown farm in Southern Ontario a year after coming to Canada, which his boys ran while their father worked in construction. Their father became a full-time farmer six years after coming to Canada, but the whole family helped in growing vegetables and tending a small apple orchard that came with the farm. However, the father was more interested in being a dealer than a farmer. Within several years, he had built cold storage facilities on the farm and started packing his apples for delivery to nearby supermarkets. Eight years later, the fourth oldest son, then twenty-two, took over the family farm after marrying the daughter of another Dutch immigrant in 1965. He is the apple producer and packer whom I interviewed in 1998.

The apple producer started working in apple orchards at the age of ten, while still attending school and stayed home to farm, together with his brothers and his father, as soon as he had finished high school (grade 12). By the time of our interview, over thirty years later, he owned two other farms with apple orchards run by managers. He also supervised apple production on two other orchards in partnership with an entrepreneur (no Dutch connections), who runs several other enterprises, including a large hog operation. But the most important part of this Dutch producer's business is the packing facilities on the home farm. He employs twenty-five people full time, including seven delivery truck drivers, and hires Mexican seasonal laborers in the summer. He buys many more apples for packing and delivery than he grows on his own farms.

This apple producer was not interested in the broader political issues pursued by general farmers' organizations. He did, however, become actively involved in the Apple Marketing Commission. He was first elected to their board of directors in the sixties, not long after it replaced the Apple Association of Ontario. He subsequently served on that board for twenty-five years, representing apple packers. He was also on the executive of the Ontario Fruit and Vegetable Association, whose membership overlaps with that of the Apple Commission, and later became its vice-president, and eventually president. In that capacity, he became well acquainted with other prominent apple growers and packers in the province.

The farmers represented by B1 include several with agribusinesses much bigger than that of the apple producer. Almost all of them came to Canada as children or teenagers. Their level education ranged from primary school only to post-graduate university training. However, they all held important posts in one or more producers' associations or marketing boards. As members of an ethnic group, they are well represented in many such decision-making bodies (see chapter 13), which can explain why people prone to group stereotyping are convinced that most of the children of Dutch rural immigrants became rich corporate farmers.

The existence of such a positive group stereotype is not necessarily linked to a strong sense of group identity by members of that group. For example, not all of the farmers in this cluster of farmer identify strongly with their Dutch heritage, and those who do, see themselves as more Canadian than Dutch. Nevertheless, they are still considered "Dutch," especially in the field of agriculture. A few prominent Dutch farmers who came to Canada as teenagers or young adults have slight Dutch accents, identifying them as immigrant farmers, even if their surnames are not recognized as typically Dutch. Moreover, they are likely to be married to someone who is also of Dutch background. In contrast, several member of this cluster came to Canada as children, have absolutely no accent, no strong contacts with other Dutch-Canadians, and have married non-Dutch spouses. Yet, they have typical Dutch surnames that allow them to be identified. The fact that this minority of

Dutch-Canadian farmers has a high profile in both the world of agriculture/agribusiness and the world of farm politics further reinforces the reputation of "the Dutch" as good farmers.

Part-time farmers who hold off-farm jobs (A1)

The dual scaling exercise identified approximately the same number of farmers with small operations who also earn wages from off-farm jobs, as prosperous farmers with agribusiness who hold farm-related political posts. However, not all of the farmers in the former cluster had full-time jobs. For example, one man, who emigrated in the twenties, and later married a postwar immigrant, spent half of his life working as a sharecropper. He eventually bought half of the farm where he worked, which he later sold, and bought another farm where he stayed well beyond retirement. Throughout their married life, this couple took on an assortment of part-time jobs. Other farmers combined running a farm with a single full-time job, while still others alternated between full-time jobs and full-time farming. Apart from their lower levels of farm income, these farmers shared a lack of connections with other farmers of Dutch background.

The example I have chosen to represent this cluster is a man who emigrated in 1947. His parents had run a medium-sized farm in the Netherlands and he was relatively well educated, since he had obtained an agricultural diploma and was also a trained machinist. His first job was as a supervisor in the Noordoost Polder, but after getting married in 1945, he started renting his own farm. He emigrated because his wife's parents were already in Canada, and she wanted to maintain contact with her family. A well-off Canadian farmer in Southwestern Ontario sponsored them. The Dutchman became his cash-crop manager, a full-time job he held for the next twenty years. While working as manager, he was allowed to grow his own tomatoes on a share basis, on his own time. His boss supplied the fertilizer, the seeds, and twelve acres of land, but the Dutchman took care of the weeding, picking, and transportation of the tomatoes to two local canneries. He gradually expanded to twenty-five acres on a share basis. He quit farming when a local garage offered him a job, where he stayed another twelve years. Apart from a short stint running his own rented farm in the Netherlands, this farmer has always had a full-time job. In an interview with one of his sons, elsewhere in Ontario, I learned that his mother had not been interested in farming because she didn't want to risk the family losing a steady income from a paycheck to the uncertainty of running a full-time farm operation. This immigrant farmer, who eventually became a garage mechanic, had little contact with other Dutch immigrants apart from his wife's extended family, who were not farmers. This case illustrates an example of a man who did all of the farming, and also held down a job. His wife was a homemaker all her life. However, in other cases, the wives of part-time farmers worked outside the home, or cared for the farm while the husband worked in town. This cluster also includes all of the Dutch immigrant female farmers whose husbands were not involved in farming at all.

Another example from cluster A1 is a man from a farm background who emigrated as a bachelor in 1954. He had a few relatives who farmed in the Netherlands, but grew up in a mainly urban environment. Like other young men who attended school after the war, he was trained as a machinist but wanted to farm, the main reason he emigrated to Canada. He started out in a rubber factory in a small city in Southern Ontario, then worked at a slaughterhouse and finally found a job as a machinist for the Canadian National Railways. He still wanted to farm, but no bank would provide farm credit to a bachelor with a steady town job. So he continued to work as a machinist. In 1960, he met and married an English woman and, a year later, he was promoted to foreman. Because they wanted to raise their children in the countryside, they bought a 16-acre beef farm. He continued to work in a nearby city and tended the cattle in his spare time. In 1969, he borrowed money to buy a full-size farm, where he continued to fatten beef cattle and grow his own feed crop, with the help of his family. Four years later, he quit his job after experiencing back problems and farmed full-time for six years. He sold that farm after the barn burned down and resumed his machinist job. However, he continued to live in the countryside and bought a 30-acre farm because his children liked horses. When their son returned home from university, and said he wanted to become a farmer, this man bought yet another, bigger, farm so they could work together until his son could take over. When I interviewed this Dutch immigrant in July of 1990, he was still working as a machinist and farming with his son, then considering marriage. None of the other nearby farms are Dutch-owned, nor is he acquainted with many other Dutch-Canadian farmers.

Like so many other farmers not of Dutch descent, numerous postwar Dutch immigrants ended up running part-time farm operations while holding down a job in town, or else attempted farming on one or more occasions, only to return to non-farm jobs. Such part-time farmers are underrepresented in my study, since I focused on people who became full-time farmers, usually after a short period of working for wages. These part-time farmers, both men and women, did not have much of an impact on the world of agriculture. None of the people in this category of my "survey" held public posts in agricultural organizations or marketing boards, even though some were active in other aspects of public life. However, we cannot dismiss their numerical importance. Not only are Dutch-Canadians are overrepresented in the world of commercial agriculture, but they also constitute the second largest immigrant ethnic category in Canada who live in rural areas, including small towns (see introduction). The sheer number of such rural Dutch-Canadians, including part-time Dutch farmers, helped to reinforced the stereotype, held among the general public, that the Dutch "sure like farming." Almost everybody over the age of fifty in rural Ontario, including those who have at one time or another worked in a factory or at a construction site, is likely to have come in contact with, or heard about, Dutch coworkers living on small farms or saving to buy a farm. While not as visible or prominent as individual farmers, the sheer preponderance of part-time Dutch immigrant farm families, especially in the fifties and sixties, brought them to the attention of the public at large.

Family farmers with lots of Dutch connections (C)

The biggest cluster of postwar Dutch farm operations (figure 13) is the full-time family farmer with no sources of income beyond minor sidelines. Their farms are neither very large nor very small, and they are found in all parts of the province. This type of farmer usually specializes in just one or two agricultural commodities, although they are likely to have had a mixed farm operation, especially when starting. The main common denominator for these full-time family farmers is the strength of their connections with other Dutch farmers. My illustrative examples represent several invididuals portrayed as dots within the outlines of the cluster (C) shown in the bottom half of the second solution.

My two examples are both farmers in the London area. Both families arrived in Ontario in 1952, and are from the same region, in the western part of the Netherlands, although they did not know each other before coming to Canada. Both had, and continue to have, a lot of relatives who are farmers back home. The oldest member of the first family started off working for a farm sponsor but also did odd jobs, including in construction. A year later, the year he met his wife and got married, he was working in a rubber factory. He quit his job nine years later when he and his wife bought a farm, which later expanded to a more than average-sized operation. This couple combines dairying with a cash-crop operation. They also became well integrated into the local community, although none of their immediate neighbors are of Dutch background. However, through their church connections, they met a lot of other Dutch farmers in the larger London region. At the time of my interview, in 1992, their ten best friends were all farm couples also of

Dutch descent. Moreover, his brother also became a farmer in Ontario. That brother in turn has several children, who not only became farmers, but married spouses of Dutch background, as did several other large Dutch immigrant farm families in the region (see chapter 8).

The parents of the second Dutch-Canadian farm family also emigrated in 1952. They started farming full time in 1958, although their farm never became a big operation. Their initial Canadian sponsor had a mixed farm in Central Ontario, but they moved to the London region within a year. For the first few years she worked out of the home, first as a domestic worker and then in a boarding house kitchen. He worked in construction and as an industrial laborer. Together they saved enough to buy land. This couple started with some pigs, while he continued to hold down a job, but sold their land to buy a bigger farm with beef cattle. When they finally had a full-time farm, he switched to a purely cash-crop operation. Like the first couple, they met a lot of other Dutch-Canadian farmers in the region and developed an even more extensive network of contacts with Dutch immigrant farmers in other areas. They had eight children and, at the time of my interview, several of their children were planning to become farmers.

I could have picked many more examples to illustrate this group of farmers. For example, another farm family represented in this cluster bought a small, but more labor-intensive, truck farming operation in the Holland Marsh in the late sixties. Yet, they too are full-time farmers without other sources of income and also have extensive connections with other Dutch farmers, including some who live in the Listowel area. There are numerous farmers in Southern Ontario who ressemble members of this cluster of full-time family farmers, who have medium or high levels of Dutch farm connections, including relatives who at one time or another lent assistance or became partners in running one or several farms. Such connections, combined with their ability to mobilize the labor and resources of their own immediate family, gave many Dutch immigrant families a competitive advantage over their non-Dutch neighbors. In many parts of Southern Ontario, those neighbours could not help but notice, and often resented, the ability of Dutch immigrant farmers to stay in business when so many non-Dutch farmers went under. Their presence contributed to the positive stereotype of the Dutch as successful farmers.

The Third and Fourth Solutions

The secondary clusters identified by the third and fourth solutions overlap to a greater extent than those in figure 13. Figure 14 shows four sub-groups of farmers. Individual dots are not indicated but the numbers

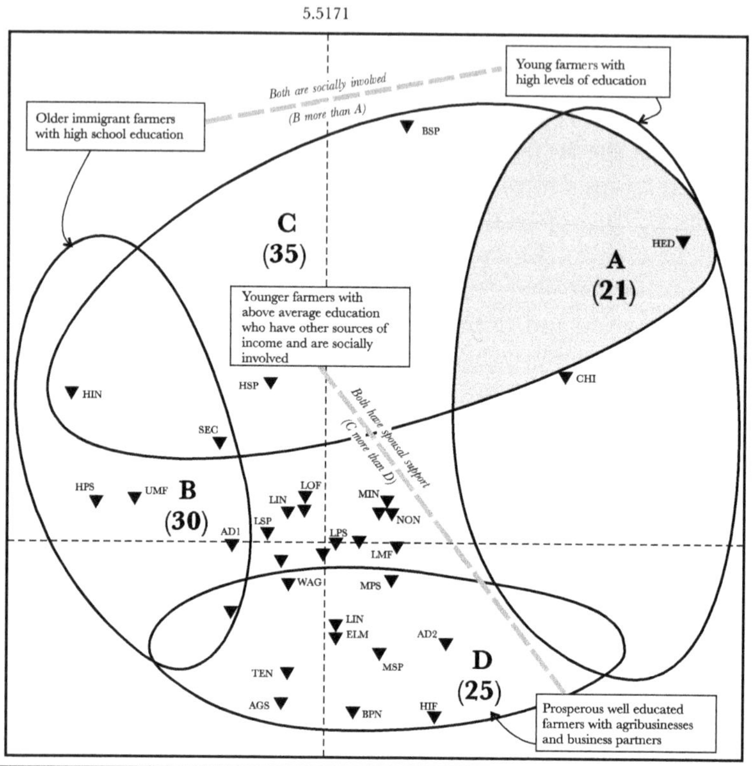

Item statistics (showing order of importance):

	R(jt) *3rd Solution*		*R(jt)* *4th Second*
Cohort	.628	Education (general)	.609
Farm size	.569	Other income sources	.583
Education (general)	.546	Spousal support	.577
Social involvement	.507	Cohort	.450
Public posts	.405	Social involvement	.416
Dutch connections	.323	Farm size	.266
Spousal support	.277	Dutch connections	.238
Education (agriculture)	.256	Partners	.205
Partners	.142	Education (agriculture)	.115
Other income sources	.125	Public posts	.101

Summary statistics:

Cumulative delta (end of 4th solution) = 36.5301		
	R(jt) *3rd Solution*	*R(jt)* *4th Solution*
Percentage homogeneity	17.2021	16.1121
Reliability coefficient alpha	0.46007	0.4314

Figure 14. Third and Fourth Solutions.

in parentheses indicate approximately how many farmers are in each group. The most relevant variables that show up in the graphical display are "cohort" (an indication of relative age and time of arrival in Canada), farm size, education, other sources of income, and spousal support. The interrelationships among these variables is far from straightforward.

The third solution (2nd horizontal axis)

The third solution indicates that farmers from different generations, regardless of immigrant status, have contrasting profiles. The corresponding R(jt) number for the variable whose values represent four cohorts is .628. In the graphical display, farmers on the right represent younger farmers, both those born and/or raised in Canada [chi], and to a lesser extent, those who arrived from the Netherlands after 1960 [ad2]. Members of both of these cohorts grew up approximately around the same time, at least a decade after the Second World War, although they went to school in different countries. The left half includes older postwar immigrants [ad1] and, to some extent, their teenage children [ten], who started farming on their own. The other variables associated with these contrasting clusters, whose corresponding statistic numbers are between .500 and .600, are farm size, education, and social involvement (in decreasing order of relevance). A weaker association (.405) involved public-post holding.

Looking at social involvement, together with education, it becomes apparent that Clusters A and B represent two sub-group of farmers whose members are better educated. The twenty one dots of cluster A represent younger, highly educated Canadian-born farmers of Dutch descent with a tendency to have medium levels of social involvement [min]. Yet the immigrant farmers with secondary education in cluster B have even higher levels of social involvement [hin]. A similar pattern shows up for political involvement: younger farmers on the right hand side were more likely to have held an average number of public posts [mps], while the older farmers in the left half held a high number of public posts [hps]. A higher college and university attendance for the younger sub-group is expected, given the increasing levels of education over time. Yet these better educated, younger farmers were not the most actively involved in community affairs or public post holding in agriculture, indicating a decline in both social and political involvement among more educated farmers over time.

While the third solution suggests links between level of education, chronological age, and social involvement, the relationships among their corresponding attributes are not linear. For example, the triangle for elementary education [elm] is located close to the central dividing line of the horizontal axis, indicating that farmers with primary education (who actually constitute the majority of farmers in my 'sample'), could not be distinguished in terms of contrasting attributes for other variables identified as significant in this solution. The position of the triangle for elementary education also indicates that the members of all the cohorts were roughly equally likely to have elementary education, although the younger farmers who came to Ontario after 1960 [ad2] were slightly more likely to only have elementary education than were their postwar counterparts [ad1], as shown the relative distances among these three triangles.

The fact that farm size is a significant variable in the third solution (.569) requires further analysis. The first solution showed that people with larger farms are likely to have more education. This finding is consistent with the strong contrast between two sub-groups of more highly educated farmers from opposite generations in the third solution; postwar immigrant farmers [ad1] with secondary education tended to have upper-middle-size farms [umf] while their Canadian-born offspring [chi] with higher education had even bigger farms [hif]. Yet, in figure 14, the attributes for having having a big farm [hif] and receiving only elementary education [elm] also occupy the same (lower right) quadrant, showing that younger farmers [chi] with elementary education nevertheless managed to establish very big farms, demonstrating that a low level of formal education [elm] did not prevent younger Dutch-Canadian farmers from advancing economically, as already shown in earlier chapters.

The Fourth Solution (2nd vertical axis)

The fourth solution highlights the connections among education, source of income, and spousal support. The two clusters associated with the fourth solution are farmers with above average levels of education [sec & hed] who derive additional income from business and professional activities [bsp] (C), and farmers with only grade school [elm] who have agribusinesses and business partners [abs & bpn] (D). Spouses in both groups provide support, but more so in the former [hsp] than the latter, with only average support [msp]. This finding is consistent with my interview data and observations. Wives of more educated farmers with non-farm careers, such as the case of the teacher mentioned in chapter 9, often assumed full responsibility for the farm, especially if their husbands were active in service clubs or church committees [hin]. It was especially important for young farmers who grew up in the sixties [chi], to have a spouse to help run the farm, since their children were less likely to look after the farm after high school than were their parents. Thus the spouse had to contribute more. The farmers in cluster D, who started farming in the sixties and seventies [ten & ad2] likewise started farming at a time when children went to school longer. However, they had a much lower level of social involvement [lin]. It stands to reason that they would need spousal support, though not quite as much [msp] as those of group C who had higher levels of spousal support [hsp].

Farmers with little formal education [elm] who nevertheless had very big farms and agribusinesses, would need at least a medium level of spousal support for such tasks as keeping records, even if they had business partners [bpn]. Such farmers were also more likely to have come to Canada as teenagers [ten].

Professionals who are farmers (where A and C overlap)
The four overlapping sub-groups shown in figure 14 represent additional patterns of shared attributes that can be used to distinguish among Dutch-Canadian farmers. It is possible to pick out individuals who are most representative of these sub-groups, as well as those who share the attributes associated with overlapping cluster. The area shaded in, where A and C overlap (see the right top quadrant), represents a cluster of younger members of Dutch-Canadian farm families who received a higher level of education, and who also ran farm operations while working as teachers, extension workers or other professionals [bsp]. All of these examples were men. These male professionals/farmers are exceptions, because most children of Dutch immigrants with post-secondary education not related to agriculture (both [chi] and [hed]) were unlikely to go into farming. Indeed, even members of Dutch-Canadians families with a degree in agriculture or related subject, were more likely to become milk testers or work for feed companies, than to work on a farm. The fourth solution (i.e. inspecting the relative distance of triangles when projected onto the vertical axis) also indicates that these younger, highly educated farmers with other occupations were more likely to receive a higher level of support from their spouses. Moreover, many are involved in social activities unrelated to farming, such as church leadership, running for political office, participation in service clubs, or a combination of such activities.

My first illustrative example for the shaded cluster is a first generation Dutch-Canadian man. At the time of the interview, he worked full time for the government, and was considering setting up his own consulting business. He also runs a larger broiler operation, together with his wife. The history of his family, in particular, shows how a strong desire to farm, and the ability to make money from farming, can be passed on from one generation to the next, even when the offspring of Dutch immigrants embark on other careers.

The man in question, the youngest of seven children, grew up on a mixed family farm in Ontario. As a young boy he enjoyed doing farm chores, especially driving a tractor at haying time, and he was later given the responsibility of caring for some pigs. Like several of his siblings, who were also born in Canada, this man continued his education beyond high school, studying animal science at the University of Guelph. When he graduated with a BSc, at the age of twenty one, he received four job offers. He could have gone into sales, but opted instead to work for the Farm Credit Corporation. While working, he continued to study, first obtaining certification as an appraiser, and eventually obtaining an MBA at the University of Western Ontario. In the meantime, he married the daughter of another Dutch-Canadian farm family. He was promoted to loans officer, and wanted to start a farm operation of his own. Two of his bothers already owned their own farms and the fourth oldest was about to take over the family farm. Another brother, who was a computer programmer, had also bought a 40-acre farm. The loans officer wanted his children to learn about working by living on a farm, but first had to persuade his wife. She was not of farm background and had likewise received a university education. In an interview with one of my research assistants, this man mentioned that his wife liked living on a farm and appreciated the additional income from their broiler operation, although they had not yet begun the badly needed renovations to their farmhouse. The loans officer changed careers when he became a project manager for the Ontario Ministry of Agriculture. He continued to help run the farm, which was a short commute from work. The farm was registered in his wife's name, and she did much of the day-to-day work as well as raising the children. They were both active in the Christian Reformed Church to which they belonged.

Another example of a professional who became a farmer is a high school teacher (also included in the shaded cluster). The son of a part-time farmer, he first studied urban planning, but then switched to teachers college because the job prospects were better. He raised pigs to earn the money he needed to pay for his education and by the time he was eighteen he owned three horses, which was his hobby. When he got married and started teaching full-time, he bought a 50-acre farm where he could raise his family, but also ran a cash-cropping operation, growing malting barley for a brewery. For four years he rented the equipment and had a local farmer do most of the work, which he oversaw. After selling his farm for a profit, he bought a 150-acre farm and invested in up-to-date equipment for a farrow-to-finish hog operation. At forty, he quit farming and moved to town, but soon tired of playing golf in his spare time. A few years later, when his father retired, he bought the small farm where he had grown up, and started raising pigs again.[7]

[7] He had mixed feelings towards the older Dutch farmers who had been his father's neighbors, expressing a combination of resentment, dislike for their narrow-mindedness, but also admiration.

In my research I encountered Dutch-Canadians with farm operations, whose primary occupation was real estate broker, bank manager, or loans officer for a credit union. These children of Dutch immigrants who became professionals and also ran working farms, represent a tiny minority within their respective occupations. Yet their presence, however insignificant, contributed to the further development of a positive stereotype of the dedicated Dutch farmer, among non-farm professionals, thereby reinforcing a broader awareness of the Dutch presence in agriculture.

The Fifth and Sixth Solutions

The diagram generated by the fifth and sixth solutions shows dots widely scattered throughout all four quadrants. Figure 15 shows four broad clusters that overlap with each other to an even greater extent than in figure 14.

The fifth solution (3rd horizontal axis)
The variable most relevant for the fifth solution is level of spousal support, a variable that also showed up in solution four. Another significant variable, with a higher R (jt) value, is whether or not farmers had attended agricultural school. However, this association is the inverse of the positive one between level of general education and spousal support (in the case of professionals who own farms), as illustrated in a previous example. In contrast, farmers who have attended an agricultural program [yes], regardless of country or level, have low levels of spousal support. They are included in cluster A on the left, shown in figure 15. In contrast, cluster B on the right, represents farmers who have not had formal training in agriculture [no], but who receive high levels of spousal support. The triangle for medium levels of spousal support lies in between, but still on the [no] side. This inverse association between a farmer's agricultural education and spousal support seems to hold true regardless of source of income or Dutch connection, level of education, or farm size. My conclusion is that farmers with agricultural education are more likely to regard farming as a career and therefore less likely to involve their spouses in either bookwork or barn work. Conversely, couples where the husband has no formal agricultural training, rely on each other more, and are more equally involved in running the farm.

Another significant variable, whose attributes are associated with level of spousal support, is "cohort." A low level of spousal support seems to be associated with being an immigrant farmer (both [ad1] and [ad2]), while a high level of such support is associated with both the older [ten] and younger [chi] offspring of postwar farm immigrants. Given the inverse relationship between spousal support and formal training in agriculture, this finding is consistent with the fact that the vast majority of more recent immigrants, as well as a good proportion of postwar immigrants, had attended a *landbouwschool*, whereas the teenage children of such postwar farmers, and to some extent their younger children, were less likely to have attended a Dutch, much less a Canadian, agricultural school. Formal training in farming did not become the norm until the second and third generation of postwar Dutch-Canadian farmers.

The sixth solution (3rd vertical axis)
The variable that most contributes to this solution is "cohort", with an R(jt) of .604. The graphical display shows a contrast between more recent farm immigrants [ad2] and members of all other cohorts, whose respective clusters are labelled C and D. This solution thus identifies the unique characteristics of immigrants who arrived after 1960. An examination of the other attributes that distinguish these clusters can provide insights into exactly how these more recent immigrant farmers differ in terms of social connections, partnerships, social involvement, farm size, and source of income, in that order. While the last two variables contribute the least to the solution, I will discuss them first since they provide information about the class and occupation of [ad2] farmers. The fact that the attributes [hif], [ags], and [wag] are close to each other in the top half of figure 15 indicates that later immigrants were overrepresented among both the biggest landowners who had agribusinesses, and wage earners. This finding is consistent with data reported in a recent Canadian national census indicating that recent Dutch immigrant farm operators are polarized into two distinct size categories.[8] The attribute [lmf], located near the center of the vertical axis, shows that [ad2] farmers were slightly more likely to belong to the lower-middle-size category of farmers, while upper middle size farms [umf] were more clearly associated with postwar immigrants [ad1] and their offspring.

The location of these class attributes can help us to interpret the other, more significant variables in the sixth solution. In the case of Dutch farm connections, [ad2] farmers are likely [with R(jt) of .586] to have medium connections [mkn], while high connections [hkn] are associated with the other three cohorts. "Low connections" does not differentiate

[8] According to the 1996 census, Dutch immigrants were polarized between very small and large-scale farm operators; this polarization was more noticeable for Dutch immigrant farmers than other immigrant farmers, or farmers born in Canada. See Charlene Lonmo, Farming is going Dutch in *Canadian Agriculture at a Glance* (Ottawa: Statistics Canada, 1999), 300.

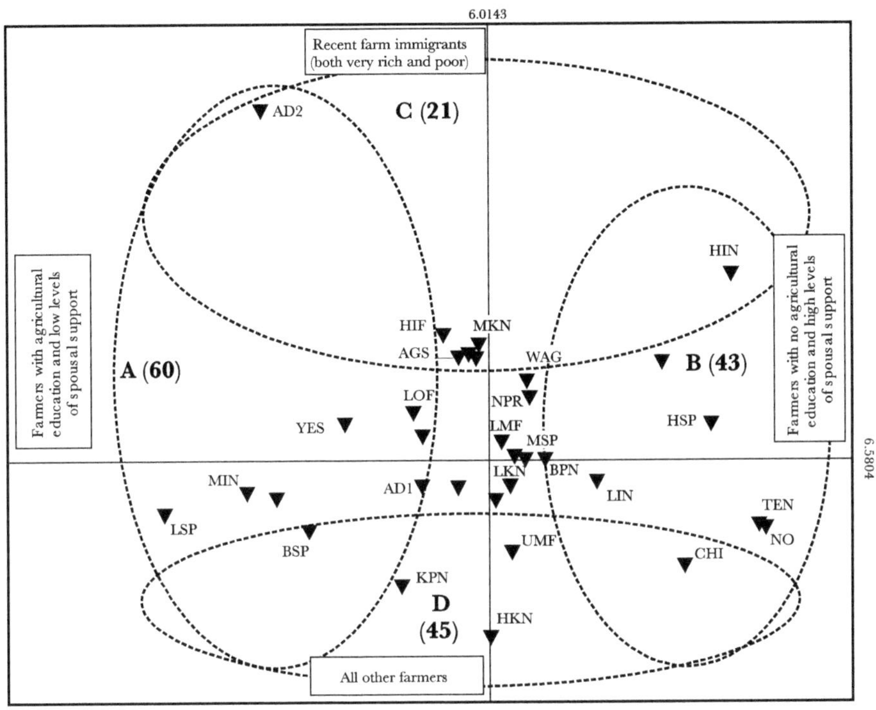

Item statistics (showing order of importance):			
	R (jt) *5th solution*		R(jt) *6th solution*
Spousal support	.610	Cohort	.604
Education (agriculture)	.609	Dutch connections	.586
Social involvement	.535	Partners	.389
Cohort	.493	Social involvement	.342
Public posts	.235	Farm size	.281
Education (general)	.200	Other sources of income	.275
Partners	.174	Public posts	.246
Other sources of income	.172	Education (general)	.244
Farm size	.086	Education (agriculture)	.215
Dutch connections	.040	Spousal support	.167

Summary statistics:		
Cumulative delta (end of 6th solution) = 49.1247		
	R(jt) *5th solution*	R(jt) *6th solution*
Percentage homogeneity	14.4768	13.2314
Reliability coefficient alpha	0.3440	0.19537

Figure 15. Fifth and Sixth Solutions.

between these two clusters. An examination of the upper left quadrant, which contains both [ad2] and [mkn] reveals another pattern: although the R(jt) numbers for farm size and source of income are low (between .2 and .3) for this solution, the triangles for [wag], [lof], [ags], and [hif] all lie close to [mkn]. This pattern contradicts the results of the second solution, where farmers with low connections were more likely to have other sources of income, i.e. [wag & ags] and vice versa. Why later farm immigrants who worked for wages plus those with agribusinesses have medium connections requires further examination.

The fact that postwar [ad1] farmers and their offspring are more likely to have high Dutch connections is easy to explain. The older postwar immigrants had more connections with Dutch farm relatives in the Netherlands at a time when the population density of the Dutch countryside was much greater. Once in Canada, they and their children continued to maintain close relations with farm relatives and other farm immigrants. In contrast, the relatively small group of more recent farm immigrants grew up at a time when families were becoming smaller, and there were fewer farmers in both Europe and North America. Moreover, few of them settled in parts of Ontario with concentrations of farmers of Dutch descent, hence meeting fewer farmers of Dutch background. Yet, the more recent immigrant farmers [ad2] nevertheless tended to have at least medium connections for the following reasons: while less likely to know other Dutch farmers in Canada, they had had strong connections with farmers in the Netherlands; furthermore, most had large families (contrary to the norm in the Netherland, starting in the sixties and seventies) and their older children for the most part also became farmers in Ontario within one or two decades of their parents' arrival.

Another distinguishing attribute of later immigrants is that they have few partnerships of any kind [npr], unlike all other cohorts whose members are likely to have kin partners. This is consistent with the fact that this cohort was better-educated and had access to more capital, making them less dependent on either kin or business partners. The only other variable in the sixth solution with a value above .300 was social involvement (.342). The location of triangles for attributes corresponding to this variable suggests that later immigrants have a high level of social involvement [hin]. However, this finding may be artificial, given the relatively small proportion of people with high levels of social involvement overall, and the smaller number of later immigrants (nineteen) in my sample.

Farmers with agricultural diplomas and low spousal support (A)

My illustrative examples for figure 15 all come from cluster A (in the left half), which includes include both post-war immigrants [ad1], who came in the fifties, and newcomers [ad2]. My first case is a man who had obtained an agriculture diploma in the Netherlands although he had also worked in construction. He and his wife came to in 1951, and they both worked off the farm for a couple of years. In 1956, they bought a dairy farm, and also raised hogs. He saw himself as a professional farmer and his *landbouwschool* certificate is displayed in a prominent place in their living room. This man involved his sons in their family business as they grew up, but his wife had almost no involvement. Like other full-time farmers who obtained formal training, he is proud that she "does not have to do any farm work."

A similar case was that of a young farmer who did not start farming in Canada until 1981. This newcomer, who was born in 1949, grew up on a Dutch farm, and attended an agricultural college for six years. However, there was no room for him on the family farm, so he went into road and dyke construction. He worked his way up to head supervisor, but continued to take a variety of business courses, even after marrying and starting a family. Ten years after their marriage, this couple, who wanted to farm, moved to Canada, together with his brothers and her sister. They bought a dilapidated dairy farm and started raising pigs full-time, but not until after he had completed more agricultural studies at Ridgetown college. As in the first case, there was no need or desire for this man's wife to be involved in the running of the farm.

The dual scaling exercise identified sixty cases of farmers with formal training in agriculture and a low level of spousal support, regardless of size of farm. The other attribute of this sub-group is a medium level of social involvement (mainly in church activities). One could argue that this sub-group of farmers more closely fits the conservative Dutch farmer stereotype, an image reinforced by the fact that all of the supporters of the pro-life movement in my sample belonged to this sub-group. Many members of this cluster have at one time or another supported the Christian Heritage Party or the former Reform Party, both of which emphasize traditional family values. Their support for such political parties, reinforced the stereotype of the Dutch as conservative and religious. Yet they represent a minority of farmers of Dutch background. In my study I also encountered Dutch farm families that do not aspire to traditional gender roles, or who have liberal views on social issues, even if they were active church-goers, with a high level of involvement in their congregations.

The Seventh, Eight, and Ninth Solutions

Normally the numerical output and diagrams generated by dual scaling are impossible to interpret beyond the sixth solution. Nevertheless, I detected some countervailing patterns usually ignored as "outliers" by other statistical techniques. While all the significant variables were similar to those of earlier solutions, their attributes were diametrically opposed to the earlier ones. For example, the contrasting values for Dutch connections in the seventh solution were the opposite of those for the sixth solution. Having kin partners is now associated with low connections, while farmers with no partners have high Dutch connections. Yet both groups of farmers were relatively well educated. The cluster of 22 farmers in group A thus represents farmers with a high school education with kin partners, even though they have fewer connections with other

Dutch farmers. In contrast, the farmers of group B have a university education but no business partners, yet high connections with Dutch farmers. These patterns of shared attributes are contrary to the finding of solutions 1 and 2, which showed that farmers with kin partners are more likely to have high connections with other Dutch farmers, while farmers with university education are more likely to have business (as opposed to) kin partners. However, the secondary, yet contrary, patterns revealed in solution 7 make sense under certain circumstances. My interpretation is that group A includes mostly late arrivals [ad2] who were not that well off [lmf] and needed to cooperate with adult family members [kpn] to set up and run farms. We already know that such farmers had fewer Dutch farm connections overall [lkn & mkn]. Yet they are more likely to have secondary education since that was the norm in both Canada and the Netherlands in the sixties and seventies.

The unusual combination of a higher education and high Dutch connections of group B requires a different explanation. This cluster represents farmers who came to Canada as older children [ten], or their younger siblings [chi] who decided to become full-time commercial farmers even after going to college (or university) in Canada. Such farmers were likely to have high Dutch connections if they bought, or inherited, their farms in parts of Ontario that already had a greater concentration of Dutch farm immigrants. Since dual scaling allows for the identification of individual dots, I was able to ascertain that several individuals who had this combination of characteristics were indeed located on the right hand side. For example, the single dot in the upper right quadrant is Peter Twynstra, whom we met in chapter 9.

The eighth solution distinguishes between a group of farmers with a low level of post-holding in the bottom half of diagram (D), and those with either medium or high levels of agricultural post-holding in the top half (group C). Group C includes farmers with kin partners [kpn] who own either small- or upper-middle-size farms [lof & umf], while the group (D) consist of farmers with business partners [bpn] who own lower middle size farms [lmf]. Adding the fourth variable (cohort), these clusters isolate two types of younger, more Canadianized farmers [ten & chi]. It seems that highly educated farmers [hed] who came to Canada as teenagers [ten], held a higher number of public posts [hps & mps] than farmers with secondary or primary education [sec & elm] who were the younger offspring of postwar immigrants [chi]. A further, but weaker, association shows that farmers in group D who were born or grew up in Canada [chi], were more likely to obtain extra income from nonfarm businesses or professional activities [bsp]. Group D also includes

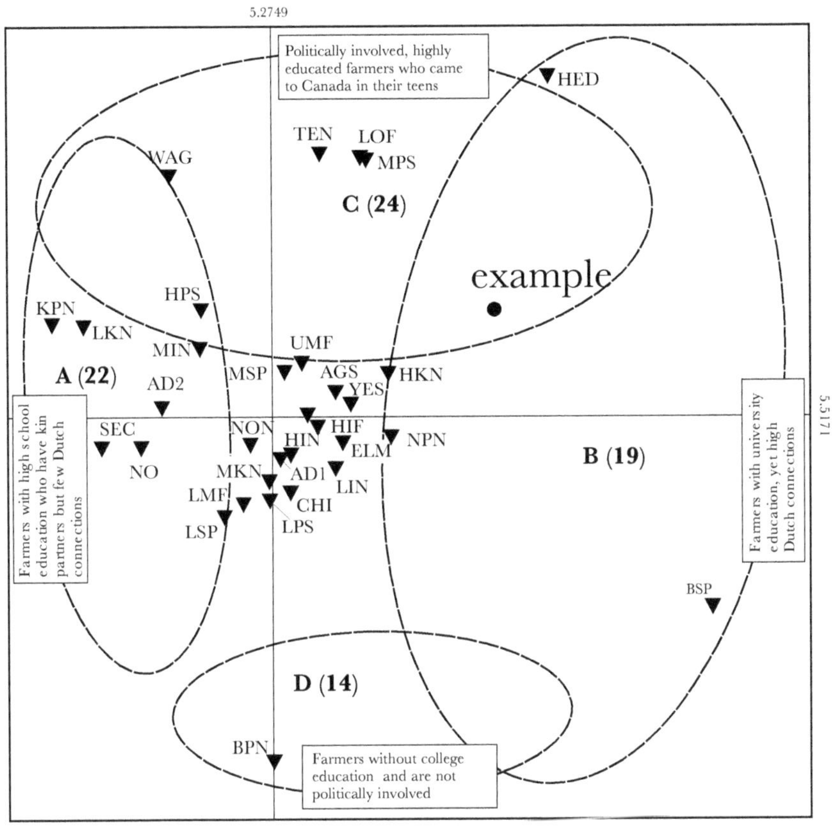

Item statistics (showing order of importance):

	R(jt) *7th solution*		*R(jt)* *8th solution*
Partners	.580	Public posts	.509
Education (general)	.503	Partners	.407
Other sources of income	.449	Farm size	.392
Dutch connections	.375	Cohort	.388
Education (agriculture)	.372	Education (general)	.372
Social involvement	.234	Other sources of income	.344
Public posts	.167	Dutch connections	.279
Cohort	.159	Social involvement	.254
Social involvement	.152	Spousal support	.183
Spousal support	.117	Education (agriculture)	.080

Summary statistics:

Cumulative delta (end of 8th solution) = 59.9167

	R(jt) *7th solution*	*R(jt)* *8th solution*
Percentage homogeneity	12.1376	11.6047
Reliability coefficient alpha	0.19537	0.16831

Figure 16. Seventh and Eight Solutions.

immigrant farmers raised in the Netherlands [ad1]. Cluster C thus represents two types of highly educated farmers [hed] who were politically involved in the world of agriculture. The first type includes college-educated men who became white-collar workers in the agricultural bureaucracy and then bought some small-scale farm operation [lof]. The second type are farmers with higher education who combined running a good-size farm operation [umf] with political involvement in the world of agriculture [hps & mps]. Group D, in contrast, consists of a combination of two distinct cohorts of Dutch-Canadian farmers, including those with business partners [bpn], who did not go to college (of any kind), but who nevertheless managed to develop a small business, such as a real estate business or a store [bsp], as well as running a small farm operation [lmf]. They held few, if any, posts in farm organizations [lps] because they were too busy combining farming with running a business that was frequently unrelated to agriculture.

The dual scaling program was aborted after the ninth solution. This last solution, which added an additional 4.41 percent to the total information explained, exposed further exceptions or minor sub-trends not previously shown. A mapping of the eight and ninth solutions together (not shown in a separate figure) showed two minor clusters, each with around ten farmers along the vertical axis. One of these clusters showed a few farmers with business partners deriving income from other business and professional activities and who have held an average number of posts. The other cluster consists of well-educated farmers who own small farms and hold a high level of public posts.

Farmers with a university degree who have high Dutch connections (B)

The last set of solutions of the dual Scaling exercise identified small subgroups of farmers who are anomalies. The most interesting one is cluster B, which represents highly educated farmers who have high connections with other Dutch farmers, and who also run agribusinesses or were professionals. An illustrative example of someone who belongs to this cluster is a Dutchman who works as an advisor and extension agent out of one of the regional offices of OMAFRA (Ontario Ministry of Agriculture, Food and Rural Communities). He came to Canada with his parents when he was ten to a part of Southern Ontario that already had a lot of Dutch immigrant farmers. The family also had prior connections with relatives who had already emigrated: an uncle who emigrated as a bachelor and his mother's sister who had originally invited them to come to Ontario. His father's sister and her husband also emigrated in 1965. They all belonged to the Reformed Church, which established rural congregations in several parts of the province (see chapter 8). His career trajectory provides insights into how he maintained and expanded his connections with rural Dutch-Canadian families long after graduating from university.

This OMAFRA official grew up on the farm his parents bought in the early sixties, a year after coming to Ontario. His father had a farm background, but had not been able to get into farming in the Netherlands, so he had started a milk delivery business. The proceeds from the sale of that business enabled him to start farming in Canada. However, only their older son was destined to take over this Canadian farm. The younger son was encouraged to finish his education and develop a professional career. This younger son started out in a one-room school house, then attended high school in a small city, and eventually received a degree in animal science from the University of Guelph. The same year he graduated, he started working for what was then called the Ontario Ministry of Food and Agriculture. He was still working for them when I interviewed him in his office in 1994. Not only did his work connect him with many Dutch farmers, but he and this wife also decided to buy a part-time farm. They did this partly so their children could grow up in the countryside, but also to earn extra income by running a commercial venture consisting of fifty sows for farrowing. This man's many connections with Dutch farm relatives and farm members of his congregation were further extended when he acted as an advisor for a community-based Dutch-language theater group to which his older brother belonged. As part of his official job, he also became a technical advisor to newly arrived Dutch farm immigrants, who started arriving in the area where he lived and worked, in the mid-eighties (see chapter 15). His knowledge of Dutch, a language he retained while growing up in Canada, enabled him to serve these clients.

Like other Dutch-Canadian farmers with higher levels of education, this man provides further exposure of Dutch-Canadian farmers to non-Dutch professionals. People like him might also give the impression that Dutch immigrant farmers and their immediate descendents form part of a single cohesive and united ethnic community. In fact, Dutch immigrant farmers formed several quite distinct, separate communities.

The profile of Dutch immigrant farmers and their immediate descendants presented in this chapter demonstrates the diversity of their life experiences. Yet, one cannot discount group stereotypes, which ignore such diversity. I have shown how the behavior of individual members of each cluster or sub-group of Dutch immigrant farmers, each in their own way, contributed to the emergence and subsequent reproduction of a single and simplistic image of Dutch farmers. Yet, even though the stereotypes associated with Dutch farmers belie the complexity and diversity of this rural ethnic group, a group stereotype does have a bearing on observable trends. The more people believe in a positive stereotype, and try to conform to it, the more closely the stereotype will correspond to social reality, at least to some extent. Few Dutch-Canadian farmers will ever conform to the image of "the" successful and enterprising, or "the" conservative, Dutch farmer. Yet group stereotypes are an integral part of social reality, which consists of a strange and often bizarre intertwining of belief and action, firm convictions and entrenched habits. Social life is characterized by both general tendencies and the deviations from such tendencies which may lead to new, often unintended, patterns of social interaction, and to new stereotypes.

XI. Three Generations of Farmers: Continuity and Discontinuity

The biographical sketches included in the previous chapter portray individuals and households typical of various sub-groups whose members share common attributes. However, even though these individuals belong to the same statistical cluster does not mean that they share a common identity or that they are more likely to interact with one another. People who belong to an actual social group, such as a family, might end up in very different statistical clusters. For example, one brother might be a highly educated and prosperous business farmer, while another brother, or his father, might be a part-time farmer who also works in a factory. Members of the same family may also have very different occupations, some with no connection to farming whatsoever. Nevertheless, some families do repeat patterns of farming. This chapter will trace several generations of farm families to illustrate both continuities and discontinuities, and concomitant inter-generational tensions. I will conclude with a counter-example of a family that should never have entered farming at all.

A CASE OF INTER-GENERATION CONTINUITY

The first case shows how a pattern of farming can be replicated over several generations. I will go back several generations, to a small village in the Dutch province of Friesland where most farmers had dairy operations. It was a prosperous region and very few families emigrated to North America, either before or after the Second World War. The exception was a man who owned a medium-sized farm that had originally belonged to his wife's family. This man (we will call him the "great grand-father") had six brothers and several sisters, none of whom became farmers. He in turn had twelve children and his oldest son, who attended a *lagere landbouwschool*, took over the original homestead. However, this son realized there was no more room for expansion and in 1950, at age forty-two, he emigrated to Canada, together with his wife and seven children (five boys and two girls). This man (the "grandfather") started off in Ontario as a hired hand, to gain Canadian experience, but bought his own farm two and a half years later. Originally a mixed farm, he gradually converted it into a more specialized dairy operation. The grandfather, who was the first Dutch immigrant farmer to buy his own farm in the region, took an active role in a local Dutch immigrant community. He helped to found a new congregation of the Christian Reformed church and played a leadership role in the establishment of their own school several years later. He was also the first farmer to buy a combine and do custom work for other Dutch-Canadian immigrants in the process of setting up their own farms. At age fifty-eight, he was ready to retire and hand down his farm to a member of the next generation, represented by another oldest son.

Another oldest son (the "grandson"), who was a teenager when his father (the grandfather) came to Canada, represents the next generation. He received an agricultural diploma after attending high school, and started off in Canada working off the farm and on it. His parents allowed him to keep some of his own wages and he used the money from the cream production of their family farm to buy his first pickup truck. In 1957, he married, and a year later, his parents helped him to buy his own farm. They worked separately, but shared machinery, and the grandson also continued to work for wages on his parents' farm. Ten years later, when his father was ready to retire, the grandson moved back to the family farm and sold his own. The father now worked for the son. This son was not as involved as his father in public affairs, but his son (the "great grandson") took after his grandfather and entered public life. The great-grandson (and last in succession) went further in school then any of his ancestors; he obtained a BSc from the University of Guelph. When I first interviewed his father, this young man was still single. He had never had a job, but always came home on weekends to help on the family farm. Three years later he entered into a partnership with his father after graduating and getting married. The first thing he did was to put computers in the barn. He now has children of his own.

Not everyone in this family made farming their full-time career. We have seen that the grandfather (the second generation) was the only one in his birth family who became a farmer. The father (third generation) had four brothers and two sisters. One of these brothers studied to be a mechanic after finishing grade eight, and then became a real estate agent. Another brother went into the computer business after finishing high school. Two other brothers farmed as partners on the home farm (after their father retired), but after three years they both started full-time jobs and the partnership was dissolved. One works for a rubber plant and the other started a furnace business. They both have hobby farms. None of his sisters married farmers, although one of them, whose husband is a master mechanic, lives next door to what was once part of the family farm. In the second to last generation, the grandson was the only boy, but his sister runs a kennel across the road from him. Her husband, a more recent Dutch immigrant, is a carpenter and a contractor.

In this case study, at least one member of each generation was a full-time farmer and their career trajectories were similar. In each generation, one or more members of the family ended up running a medium-sized to larger-than-medium sized farm, including the father, who ended up owning 500 acres. All of these farmers specialized in dairy farming and grew their own fodder. In Canada, this family was one of several Dutch immigrant families who took over existing farms. By the turn of the last century, the family was one of only two dairy farms left in their township, both owned by Dutch-Canadians. Most of the other farms had long since been converted into hobby farms, although several other Dutch-Canadian farmers (some part-time) still run chicken operations.

Other Cases of Continuity

I could have used many other examples of Dutch farm families where several generations of children carried on a tradition of family farming in Ontario. In other cases of inter-generational continuity, about half of the children ended up farming, although not always full time. In the case of one second-generation Dutch-Canadian farm couple with six children, in the Lake Erie region, three children ended up living on farms. The oldest daughter married the manager of a feed mill, but they bought a 180-acre farm, which is still operational. One son, the second oldest, started a dairy farm, like his father. He gave up milking cows to work as a driver for a local farm dealership, although he eventually went back to dairying after paying off his debts. He hopes to pay off the farm and still does cash cropping. The third son is a full-time dairy farmer. The fourth son, still single, works as a licensed mechanic and does not live on a farm. The next son still lives on the family farm and has just become a partner on his parents' dairy farm. Finally, their youngest daughter, who married someone in computers, no longer has farm connections.

Such patterns of continuity, with some diversity in occupation, are particularly noticeable among the first generation of Dutch farm immigrants. The father in the last paragraph had seven sisters and six brothers, of whom half were involved with some form of agriculture after they married. Of those, the biggest and more diversified farm operation features dairy and beef in several locations. Others operated nurseries, grew vegetables, raised chickens, and one had a greenhouse operation. The men not in farming include a taxidermist, and two brothers who run a manufacturing plant. Of his sisters who do not live on farms, one married a preacher and the other a baker.

In examining additional cases, it did not appear that women were as likely to marry farmers as not. Thus, in the case of one farm couple that had eight daughters and two sons, one son became a full-time dairy farmer, while the other ended up working in a gypsum mine. Of the daughters, half married farmers, and one other daughter worked in greenhouses before she married a produce manager. One of their neighbors, also a postwar Dutch immigrant, had seven children, all boys. In that family all but two of the offspring are farmers: two have pig farms, another is a cash cropper, and the two youngest took over the family farm. Such cases of continuity resulted in pockets of Dutch farmers throughout rural Ontario (see chapter 12).

A CASE OF DISCONTINUITY

Anecdotal evidence, and the non-random sample of the 161 farmers chosen for my study, suggests that rural immigrants from families with a tradition of running family farms were more likely to become full-time commercial farmers in Ontario. Those who came from a farm-laborers (*landarbeiders*) background, or small-scale, part-time farmers, were less likely to become farmers or stay in farming in Canada. However, there are exceptions. I will use the example of one of those former *landarbeiders* to illustrate a lack of continuity over several generations. I will call this person Piet Dykstra (a pseudonym).

Piet's parents owned a small farm in the Netherlands, but lost it during the Great Depression. Piet, who was one of 3 boys in a family with seven children, was attending grade school when this tragedy occurred. As soon as he finished grade six, at age nine, he started working as a hired hand for the same man who employed his father. His brothers worked for other farmers and they all milked, plowed, and helped with harvesting. When he was seventeen years old, Piet started going back to school to obtain an agricultural diploma. In 1953, still single, he emigrated to Canada, following in the footsteps of an older brother who had left the Netherlands several years earlier. He was still working as a farmhand in Canada when he met his future wife. She had come to Canada with her parents in 1949. Her family, also from a small-farm background in the Netherlands, had managed to start their own farm operation in 1953. However, she continued earning wages by picking tobacco.

They married in 1960, but Piet continued working as a herdsman for a Canadian farmer. The opportunity to farm for himself came five years later, when he received a junior farmer loan. He was eligible because he was under thirty. The farm they bought had 173 acres and came with Guernsey cows. Initially they teamed up with her brother and her brother-in law who were farming together at her parents' farm. Piet and his wife worked hard, and in 1967, they built a new silo. In 1978, they bought an additional forty-nine acres where they grew wheat. A new silo was erected in 1985, but Piet continued to use a chain system to get solid manure out of the barn. Piet and his wife retired from farming in 1990, after selling their dairy operation to a recently arrived Dutch immigrant, who immediately tore down most of the farm buildings. The family farming tradition, which had already been interrupted once in the case of Piet's parents, did not survive another generation. Their only son was not interested in farming, although he always had a job on the home farm. The only member of the family who has any connection to agriculture is a son-in-law who is a researcher for a government agency that promotes agriculture.

Such cases of discontinuity, which do not fit the stereotype of the Dutch family staying in farming, are common. In my research, I came across several examples of families involved in farming for less then ten years. However, because of the sheer number of Dutch farmers who bought farms originally operated by non-Dutch farmers, plus the fact that Dutch-Canadian families who did stay in farming for more than one generation are overrepresented in many areas, a generally positive image of Dutch farmers was reinforced. Yet, what actually happens on the ground is more complex and nuanced, even when Dutch immigrant farmers ended up with high-profile agribusiness farms.

A Case of Dramatic Upward Mobility

My last case study of several consecutive generations involves a rags-to-riches story, and illustrates the inter-generation tensions and personal problems that can be associated with dramatic success. In order to protect the identity of the narrator of this story, I have altered not only the names of people and places, but also the nature of the agribusiness they developed.[1] The narrator was sixteen years old when he came to Canada with the rest of his family. I will refer to him as "the entrepreneur." This man did not remember much about his grandparents on either side of the family but informed me they were not farmers. His father, who emigrated to Canada in 1952, used to work in the tulip fields of Noord Holland, as a laborer. They were a large family of ten

[1] The details regarding career trajectory and family dynamics are from one of my interviews subjects, who wished to remain anonymous, and are supplemented by additional information on this family obtained from other sources, including web sites.

kids, and the entrepreneur was the second oldest son. Like so many other Dutch immigrant families, they started off working together on sugar beet farms in Alberta. The oldest daughter went off to work in Ontario, where she met her future husband. That daughter (the entrepreneur's older sister) persuaded the rest of the family to move to Ontario. There, the whole family took jobs, some in construction, others in a cement factory, until they all started working for a nursery.

The entrepreneur's first steady job in Ontario was as a deliveryman in a nearby city, but he returned to the countryside when he learned that his father and two other brothers had started their own business on a rented farm. The father quit the next year when one of their first customers wrote them a bad check. He started working in a factory, where he stayed until retirement. However, the three brothers, at that time all married and with small children, continued the partnership. In 1967, they bought their own land. That farm operation expanded quickly and branched off into a major agribusiness with operations all over the province. Within two decades they owned thirteen farms (1,300 acres) and were employing 130 people. They leased an additional 1,200 acres and entered into joint ventures with a businessman (non-Dutch).

The family's troubles began as their children grew up and started their own families. They didn't get along and fought with their parents. The entrepreneur's eldest son, who went to university, disagreed with his father's business methods and there were personality clashes among the wives in the second generation. There were further personal frictions after the entrepreneur bought out his brothers twenty-five years after they started. One of the brothers, whose marriage dissolved, started working for the family business on salary, but squandered his money. Attempts at reconciliation failed, and the entrepreneur decided to sell off most of the assets, keeping only one part of the business. He is now semi-retired and one of his sons has set up his own company. The rest of the family is no longer involved in agriculture.

This last case contradicts the image of the cohesive Dutch farm family. Their problems were well hidden for a long time and few people outside a small circle of close friends and relatives were aware that this family empire rested on shaky foundations. Such tensions among family members, especially among members of different generations, while less apparent to outsiders, are not unique to situations of rapid upward mobility. Intra-family disputes can also be associated with feelings of resentment from the offspring of Dutch farmers who overworked their children in order to maintain a barely viable farm operation. I mainly heard such stories from younger people, who had already left a farm run by Dutch immigrants. In cases of difficulties among less well-known

Dutch farm families, it was harder to keep troubles a secret from their non-Dutch neighbors. In my travels throughout rural Ontario, I came across several non-Dutch people well acquainted with tensions among Dutch family members struggling to keep their heads above water. The most common impression that came up in casual conversations was that the Dutch were too strict with their children. Nevertheless, the overall impression, even among such critics, was that, "the Dutch sure know how to farm."

A COUNTER EXAMPLE

The non-random sample of farmers selected for my "survey" did not include farmers who went bankrupt, or postwar farm immigrants who never went into farming once they had fulfilled their one-year contracts. Likewise, I deliberately excluded Dutch non-farm immigrants who ended up buying hobby-farms for lifestyle or investment reasons, after many years of working in the city. However, in my last example, I will present an example of a Dutch-Canadian immigrant whose ten-year career as a full-time farmer completely contradicts the positive stereotype. I know about this man and his family because they used to live in the same village where I grew up, in what is now the Greater Toronto area. The Van der Sluis (a pseudonym), postwar immigrants who belong to the same generation as my parents, contradict the stereotype of "typical" Dutch farmers.

Teo Van der Sluis was the younger son of a wealthy Dutch businessman with no farm connections. Never really fitting in, he did not do well in school and married a woman unacceptable to his parents. He emigrated to Canada in the mid-fifties and started off working in a factory in a small town just outside of Toronto. His dream was to become a farmer, but he had no farm experience nor was the family able to save up enough money to buy even a small farm. His older children were already teenagers when an unexpected inheritance from his estranged parents allowed him to quit his job and put a down payment on a working dairy farm in the Northwestern region of Southern Ontario. The whole family moved to another part of Ontario in 1967, and he started milking cows and growing hay, while his wife, a nurse, took on a part-time job at a hospital to supplement their income.

Teo's working capital was enough to upgrade the barn and he had to invest in a brand new milk cooler in order to keep his milk quota. However, the family did not have enough money for a septic tank for an indoor toilet or to modernize their kitchen. Improvements to the house were put on hold for a couple of years, since the barn was a priority. The older children resented having to work on the farm and did not like living in the countryside. He was able to keep the farm enterprise going, but the family depended on his wife's earnings to pay their bills. While Teo learned the basics of repairing farm equipment, he was not a handyman and the family continued to struggle. Two years after becoming a farmer, he was still determined to make a go of it. However, the financial situation of this would-be-farmer deteriorated, and his oldest son refused to continue helping out. With mounting debts, Teo and his wife had no other choice but to sell the farm. They moved to a small city in Southwestern Ontario, where he worked as a gardener, looking after the grounds of a hospital and medical center. When my father visited shortly after this latest move, he asked Teo if he was still interested in farming. His answer was, "I don't have any more money, but if I did I would buy a gun and shoot every one of those cows."

This last example clearly demonstrates that not every Dutch person becomes a successful or even barely viable commercial farmer, especially if he or she is not of farm background. The man in question saw himself as someone destined to be a farmer. His distant ancestors used to run plantations in the former Dutch East Indies, and he aspired to carry on a tradition of working the land. However, he lacked the attributes shared by most of the male farm immigrants so far described in this book: previous experience working on a farm, formal or informal training in various aspects of agriculture (preferably combined with a trade), children and a wife who had grown up on a farm, and an extensive network of farm relatives. The reputation of his country of origin did not help him, except to feed his delusion until he finally realized that he was not destined to be a farmer. Such exceptions may have cast doubt in the minds of rural neighbors already been exposed to a positive stereotype of Dutch farmers; however, those same neighbors may have discovered that such failed farmers were "not really farmers," hence exceptions to the rule.

PART III – THE DUTCH PRESENCE IN ONTARIO AGRICULTURE

XII. Demographics and Spatial Distribution

People in Ontario tend to believe that the Dutch are more likely to live in the countryside. It is also said that the Dutch are trendsetters in agriculture and agribusiness. This dual image is part of the group stereotype of Dutch farmers mentioned in previous chapters. Researchers need to scrutinize such generalities by finding patterns, to construct a more nuanced and accurate picture. For example, the vast majority of people of Dutch descent actually live in urban areas and have no connection to agriculture. Yet, in relative terms, and as a percentage of population, the Dutch are in fact overrepresented in rural areas compared to other ethnic groups (see Introduction). However this overrepresentation represents a statistical average, which generalizes from a wide range of people in any rural region in Ontario of Dutch descent. The Dutch are more conspicuous in some counties or townships than others. Moreover, a Dutch presence in any rural region does not necessarily mean that all the Dutch in that region are farmers.

The demographic composition, and the proportion of farm versus non-farm rural Dutch immigrants and their offspring, has shifted over time. In the late forties, postwar Dutch immigrants were scattered all over the province as farm-laborers or tenants. They usually moved to a new farm within one or two years. Those who stayed in the countryside gradually converged in certain regions where they put down roots (see chapter 7). The spatial distribution of farm immigrants stabilized around 1970, at a time when anyone who wanted to buy a farm had already done so. Numerous immigrants who originally had intended to farm instead settled down in villages and towns where they worked in feed mills, small construction companies, co-ops, or canneries. The seventies and eighties saw an expansion in the amount of land owned by Dutch farmers as they bought additional farms from neighbors to accommodate offspring who also wanted to farm, resulting in an even greater concentration of Dutch farmers in some localities. The farm-to-non-farm ratios were further altered as Dutch farmers moved to town upon retirement.

In considering the spatial distribution of Dutch-Canadians in rural Ontario, we must keep in mind possible discrepancies between the number of people of Dutch descent in any given location and the *perception* of whether that area has a Dutch presence. The former can be extrapolated from census figures, while the latter requires a more qualitative approach, based on interviews. We need to consider the extent to which differences in religion and Dutch regional background are reflected in the spatial distribution of people. Finally, we must precisely locate farm properties at the concession and township levels, to better visualize patterns of farm settlement, as outlined in more general terms in chapters 7 and 8, and to appreciate how these patterns are currently changing as a result of the influx of new farm immigrants.

THE SPATIAL DISTRIBUTION OF RURAL DUTCH-CANADIANS IN 1986

The short population census of 1986[1] provides a useful snapshot of the spatial distribution of rural Dutch-Canadians: most postwar immigrants were still alive; those who came as teenagers in the fifties were independent farmers; and there was a good mix of immigrants, first, second and even third generation Dutch-Canadians. The fact that Dutch-Canadian distribution figures from a 1991 census are almost identical to those of 1986,[2] indicates that for at least the final two decades of the twentieth century, the spatial distribution of Dutch-Canadians remained relatively stable. The effect of the still ongoing arrival of the latest (smaller) cohort of Dutch farm immigrants, will not be known until the 2005 census, although their impact on Canadian farming was already becoming apparent in the 1995 census.[3]

The census of 1986, unlike the one in 1991, contains questions about ethnic origin and mother tongue. Assuming that the basic patterns of geographical distribution of rural Dutch-Canadians have not changed significantly during the writing of this book, the present tense will be used unless otherwise specified. I will refer to the census units and other administrative divisions of that time period. Unfortunately, the 1986 census does not include a question about religious affiliation. Nevertheless, I was able to roughly estimate of the proportion of neo-Calvinists (mem-

[1] The 1986 short population census counted one out of every twenty people. All of the numbers cited are statistical estimates (with a small margin of error). A system of random rounding off was used, which made it equally likely that any number between one and five would be assigned a value of either zero or five.

[2] Most of the minor changes in numbers between 1986 and 1991, mainly for census divisions with few Dutch-Canadians, were as likely to be the result of the random rounding off techniques used, as proportional increases or decreases in population.

[3] See Lonmo, Farming is going Dutch, 297-30.

bers of various Reformed denominations) in most areas by extrapolating from other sources.[4] The quantitative data was then interpreted with the help of information obtained through oral interviews, and observations made during fieldtrips.

Rural Dutch-Canadians

According to the 1986 Canadian population census, "rural" census divisions have populations of less than 10,000 people, including "townships", "rural areas", towns and villages with 4,999 inhabitants or less, and towns with between 5,000 and 9,999 inhabitants. By using this definition, almost half (47.12) percent, of people of Dutch descent in Ontario live in rural regions. However, we must in each case distinguish between percentages and absolute numbers. In terms of absolute numbers, major urban census divisions have more people of Dutch descent than their rural counterparts. Census divisions with more than 2,000 Dutch-Canadians are mainly found in the heavily urbanized Toronto region. Other cities with more than 2,000 Dutch-Canadians are Ottawa, London, Kitchener, Guelph, Welland, and Chatham. The urban census divisions with the largest number of Dutch-Canadians are London, with a total of 6,700, followed by Hamilton (4,325). By comparison, the actual number of people of Dutch descent in most rural townships is quite low. In villages and towns with under 9,999 inhabitants, the number of Dutch-Canadians rarely exceeds several hundred. However, if we distinguish between absolute numbers and percentage of total population, the Dutch proportion is totally reversed: the distribution of Dutch as a percentage of total population in large cities of over 200,000 is almost miniscule (usually under 1 percent). Even in London, this percentage is only 2.52 percent. In contrast, the percentage for rural census divisions (townships and towns) with a significant Dutch presence can reach as high as 29 percent!

The fact that Dutch-Canadians constitute a higher percentage of the population in predominantly rural-agricultural areas explains why they are more noticeable in the countryside, and why many people think most Dutch-Canadians are farmers or engaged in farm-related work, even though most Dutch-Canadians live in cities. Moreover, not all

---

[4] Data were taken from the 1987 *Yearbook of the Christian Reformed Church in North America* (Grand Rapids: CRC Publications, 1987); the 1987 *Yearbook of the Canadian and American Reformed Churches* (Winnipeg: Premier Printing, 1987); and the 1987 Annual Statistical Report of the Classis of Ontario for the Reformed Church in Canada. Adding up the totals for congregations in the same counties, Dutch-Canadian Calvinists represented anywhere from zero to 70 percent of the total Dutch-Canadian population.

Dutch-Canadians living in towns classified as "rural" population centers (especially towns with between 5,000 and 9,999 inhabitants) are involved in occupations that service the agricultural sector or agribusiness. Such towns have factories, construction companies, and other businesses that cater to a non-farm constituency. Yet, such towns are not without significance for Dutch farmers, as we shall see. I shall therefore include them in my discussion of the distribution of rural Dutch-Canadians. The two maps shown in figures 17 and 18, one for Southwestern Ontario, and the other for Central and Eastern South Ontario, indicate the distribution of Dutch-Canadians as a percentage of the population in rural administrative divisions. These maps include villages and small towns if they appeared as separate census units with a Dutch presence of at least 1 percent. Numbers and letters on these maps correspond to place names included in tables in this chapter. A map of Northern Ontario was not included, since that part of the province has few rural Dutch-Canadians.

Rural census divisions

In thirteen rural census divisions, Dutch-Canadians constitute over 15 percent of the population. Table 2 lists these census units in decreasing order of importance. The townships with the highest percentages of Dutch-Canadians are Adelaide, Warwick, and West Williams, where people of Dutch descent constitute more than a quarter of the total population. These three contiguous townships, located in the area between Sarnia and the city of London, represent a cluster of over 1,500 Dutch-Canadians, primarily descendants of postwar immigrants from the Dutch province of Noord Brabant after the War. Five other census divisions on the list are also located in the broader London region: the rural part of the township of Sarnia; the township (not town) of Chatham; the township of Hullett; Delaware; and the township (not town) of Goderich. The rest of these census divisions are found in other parts of Southern Ontario: One in the Waterloo/Guelph region (Maryborough); two in the Niagara region (rural parts of Grimsby and West Lincoln); two in the Lake Erie region (rural part of Norwich and Dunwich); and finally, the one township in Eastern Ontario where more than 15 percent of the population is of Dutch descent (Rear of Yonge and Escott).

When we add the rural census units where the Dutch comprise more than 10 percent of the population, the percentage goes up to 49. This longer list includes all of the townships where Dutch-Canadians have held elected posts (see chapter 9). There are 135 townships with a Dutch population of at least 5 percent. Thus, in about a fourth of the 569 townships in Ontario, one out of every 20 persons is of Dutch background. Yet, when we look at the location of Dutch-Canadians in Southern

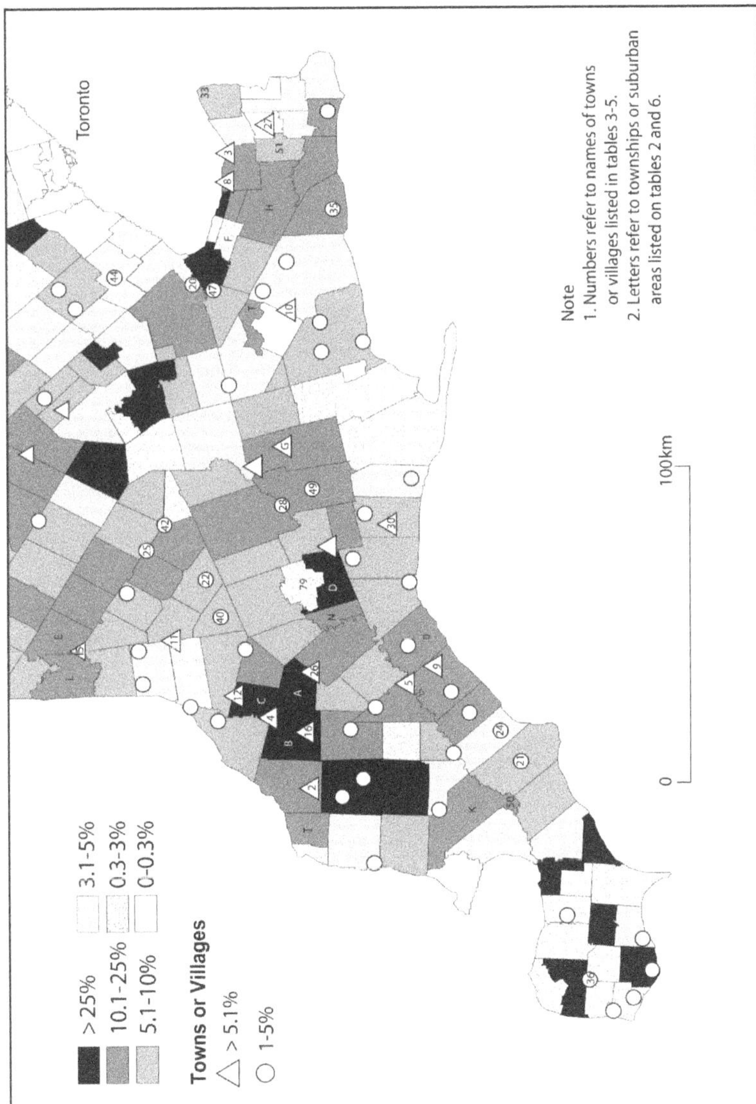

Figure 17. Distribution of Dutch-Canadians in Southwestern Ontario (percent of population).

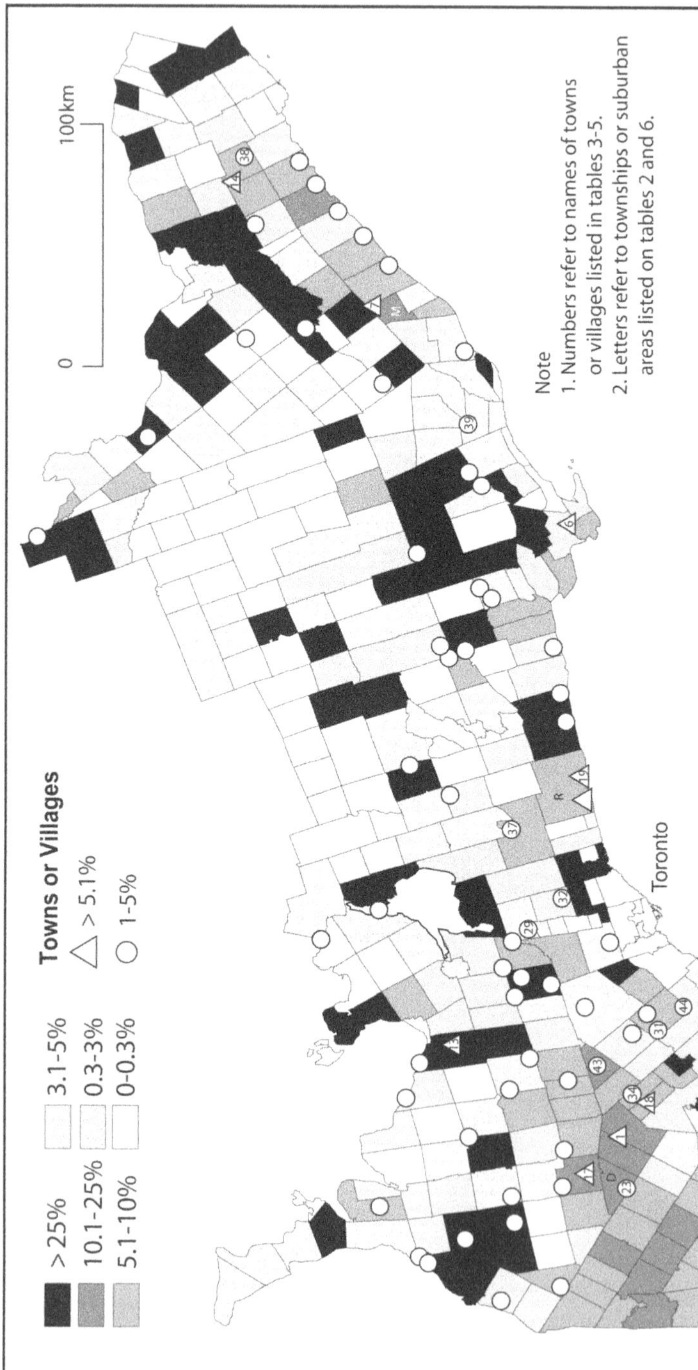

Figure 18. Distribution of Dutch-Canadians in Central and Eastern parts of Southern Ontario (percent of population).

Table 2
Rural Census Divisions with a Dutch Presence
[letters in brackets shown in figures 17 & 18]

Name of census division (with county, district or regional municipality)	Percentage of Dutch descent
Adelaide township (Middlesex) [A]	28.93
Warwick township (Lambton) [B]	28.81
West Williams township (Middlesex) [C]	27.72
Maryborough township (Wellington) [D]	23.15
Hullet township (Huron) [E]	23.06
Rural part of the town of Grimsby (Niagara) [F]	20.25
Rural part of Norwich township (Oxford) [G]	19.57
Rural part of West Lincoln township (Niagara) [H]	17.78
Dunwich (Elgin) [U]	17.42
Rural part of the township of Sarnia (Lambton) [I]	16.70
Rural part of the town of Chatham (Kent) [K]	16.39
Township (not the town) of Goderich (Huron) [L]	15.67
Rear of Yong and Escott (Leeds and Grenville) [M]	15.21

Ontario as a whole, we notice that they are almost completely absent on the Canadian Shield, where Dutch-Canadians constitute below 1 percent of the population in most districts[5] or counties.[6] These figures demonstrate that rural Dutch-Canadians are not well represented in areas whose economies revolve mainly around tourism, logging, and mining as opposed to agricultural enterprises. The pattern of spatial distribution of Dutch-Canadians, even within rural census divisions where they constitute a high percentage, shows further variation. For example, in Adelaide township, more Dutch-Canadians live in its southern half, between the town of Strathroy and the hamlet of Kerwood, than in its northern half. We also need to take into account the percentage of Dutch in towns and villages, as opposed to those living in farmhouses separated by open fields from their neatest neighbors.

Villages and towns

Settlements with the legal status of village may have anywhere from several hundred to over 2,000 people. Moreover, small towns (with fewer

[5] This also applies to the top portion of the counties of Renfrew and Victoria, and more than half of the townships in the districts of Nipissing, Parry Sound, Muskoka, and Haliburton.

[6] In none of the counties of Lanark, Frontenac, Lennox-Addington, Hastings, and Peterborough, and the district of Sudbury, do Dutch-Canadians comprise more than 5 percent of the population.

than 5,000 people) are often indistinguishable in size and appearance from officially designated "villages." Table 3 lists all such towns and villages where 3 percent or more of the population is of Dutch descent or where there are at least 100 Dutch-Canadians. Villages with lower numbers, but higher percentages, were included in this table if they appeared as separate census units. Beamsville has the largest number of Dutch-Canadian inhabitants in absolute numbers (455), while Drayton has the higher percentage (17.19). The towns or villages listed in table 3 service a Dutch-Canadian farm population representing more than 10 percent of the total population of their respective rural surroundings. Even when the percentage of that Dutch farm population dips below 10 percent, there are still more Dutch-Canadians (in absolute numbers) in rural hinterlands than in nearby towns and villages. The only exceptions are St. Marys, Exeter, and Ridgetown, yet each of these towns acts as a service center for at least 100 Dutch-Canadian farmers. Technically these population centers are considered "rural," implying farming, although some of their inhabitants may work full time in other occupations. Their inhabitants, including Dutch-Canadians, consist of a mix of farmers, retired farm couples, and young people who have "moved to town."

In these small towns, the visible Dutch presence is betrayed by Dutch-style lace curtains hanging in windows and the display of replica windmills in some yards. Buildings that indicate a Dutch-Canadian or Dutch-linked institution are also revealing. In downtown cores one might encounter a Dutch delicatessen, one or two Reformed churches, or a Catholic church where a large proportion of members have Dutch names. You might also see a Dutch retirement home or hear about a Dutch-Canadian social club or annual festival. A good example of an annual festival is found in the town of Clinton, which is surrounded by townships with a considerable number of Dutch farmers. All of these towns and villages with a Dutch presence are located in townships already identified as having a significant proportion of Dutch-Canadians. Indeed, in all but two cases, the Dutch population of the surrounding countryside outnumbers that of the villages in percentages *and* absolute numbers. This is also the case for villages where less than 3 percent of the population is of Dutch descent. For example, there are only twenty Dutch-Canadians in the town of Jarvis, constituting 1.59 percent of its total population. In contrast, its rural sector, a separate census division, has 1,080 Dutch-Canadians (8.67 percent).

Dutch-Canadians who live in or close to small towns and villages may almost all be Catholics or Calvinists. An example of a population center with mainly Dutch Calvinists is Drayton in the Waterloo-Guelph

Table 3
Small Towns and Villages with a Dutch Presence (under 5,000 inhabitants)
[numbers in brackets shown in figures 17 & 18]

Name (and county)	Number of Dutch Canadians	Percentage of people of Dutch descent
Drayton (Wellington) [1]	190	27.14
Wyoming (Lambton) [2]	300	17.19
Vineland (Niagara) [3]	385	17.12
Arkona (Lambton) [4]	65	15.12
Wardsville (Middlesex) [5]	45	12.50
Bloomfield (Prince Edward) [6]	80	12.03
Athens (Leeds and Grenville) [7]	115	11.39
Beamsville (Niagara) [8]	455	9.81
West Lorne (Elgin) [9]	120	9.02
Hagersville (Haldimand-Norfolk) [10]	180	7.78
Exeter (Huron) [11]	285	7.67
Park Hill (Middlesex) [12]	100	7.52
Stayner (Simcoe) [13]	125	6.67
Winchester (Dundas, Stormont & Glengarry) [14]	130	6.58
Clinton (Huron) [15]	185	6.09
Watford (Lambton) [16]	75	5.98
Harriston (Wellington) [17]	95	5.16
Elora (Wellington) [18]	135	5.13
Newcastle (Durham) [19]	85	5.06
Chesterville (Dundas) [38]	75	4.92
Flamborough (Hamilton-Wentworth) [20]	210	4.76
Blenheim [21] (Kent)	195	4.53
St. Marys (Perth) [22]	210	4.26
Strirling (Hastings) [39]	75	4.20
Listowel (Perth) [23]	200	4.07
Lucan (Middlesex) [40]	65	3.85
Ridgetown (Kent) [24]	110	3.60
Mitchell (Perth) [25]	100	3.44

region, discussed in chapter 7.[7] Another is Athens, a village that does not appear as a separate division in the 1986 census. Located in the Eastern Ontario county of Leeds and Grenville, it has a Christian Reformed church and a Christian school founded by Dutch immigrants, which serves the 300-plus largely rural Dutch-Canadian inhabitants of the township of Rear of Yonge and Escott, including the hamlet of

[7] According to my calculations, the percentage of Dutch people belonging to a Reformed church (compared to other churches) in Wellington county, to which Drayton belongs, is 60.4 percent, the third highest for the province, surpassed only by the district of Thunder Bay (75.3 percent) and Hastings county (74 percent).

Addison. An example of a completely Dutch Catholic hamlet in the region of London is Kinkora, in the township of Ellice (near Statford). Here, nearly all inhabitants of both the village and surrounding farms are Dutch Catholics, with Dutch names prominently displayed on mail-boxes and sometimes painted on barns (see chapter 8). Yet most of the 250 Dutch Catholic farmers in this township, whose Dutch population does not quite reach 10 percent, are concentrated in its northern half.

Sometimes even villages without a large Dutch population might have a visible Dutch presence. For example, Arkona, which is situated in the centre of a cluster of Dutch Catholic farmers in the townships of Warwick, West Williams and Bosanquet, has 430 people, of which only 15 percent are of Dutch descent. However this village had great symbolic importance as the original location of one of the main branches of St. Willibrord Credit Union founded by Dutch Catholic farmers (see chapter 8). Arkona also has a clubhouse, called Taxandria, established by Dutch immigrants, some of whom live close to places like Thedford, Forest, and Grand Bend. A short drive from Arkona, just south of highway 402, one reaches Watford. This village has only seventy-five Dutch-Canadians, but it is a service center for the farmers of Warwick Township, including the hamlet of the same name. In fact, this township, of which 28.81 percent of the population is of Dutch background, borders on the township of Adelaide, another area with a high concentration of Dutch farmers and which also had its own branch of St. Willibrord Credit Union. However, while the majority of Dutch farmers who live near Watford are Catholics, there are also Calvinists (mainly Christian Reformed) who own businesses in downtown Watford. They go to church in the village of Wyoming in the neighboring township of Plympton, a rare case where a small rural settlement has about the same number of Dutch Catholics and Dutch Protestants, including farmers. Wyoming not only has a Christian Reformed Church, but a Dutch store, and a Catholic school (originally located on the 6th line of Warwick Township) with a high Dutch enrolment.

A rural region that also has about an equal distribution of Dutch Calvinists and Dutch Catholics is the township of Wainfleet, which includes the hamlets of Wainfleet and Winger (not counted as separate census divisions). This township in turn, borders on the town of Dunnville, whose rural portion has more Dutch Calvinist farmers. Both of these census divisions are located in the Niagara region, although they belong to different regional municipalities. Their combined Dutch populations of 1,560 (mostly farm families) represent another Dutch farm cluster, although their respective percentages (13.09 and 12.51) were not high enough to be included in the townships listed in table 1. These lower percentages (of Dutch as a proportion of total population) are an indication of the more urbanized nature of this region. Both of these census divisions have more non-farm inhabitants than is the case for most townships and rural hinterlands of towns with populations between 5,000 and 9,999 inhabitants. Nevertheless, although Dutch-Canadians constitute less than 15 percent of the total population in this part of the Niagara region, oral testimony and other sources of information indicate that these rural Dutch represent over half of the farmers who cultivate the Haldimand clay soils characteristic of this region.

The town of Dunnville, which acts as service centre for farmers in this region, also has a Dutch presence, with 195 people of Dutch descent. It is included in table 4, which list all towns of between 5,000 and 9,999 inhabitants with a significant (above 3 percent) Dutch presence. Like smaller towns and villages, these towns are surrounded by rural census divisions whose Dutch-Canadian population is larger than that of the town itself. The only exception is Strathroy, with 845 Dutch inhabitants, surpassing the number of rural Dutch in its corresponding township. However, this proportion is misleading since Strathroy lies on the boundary between the townships of Adelaide and Caradoc whose combined rural Dutch-Canadian population well exceeds 1,000. Dutch farmers from these two townships, as well as the township of Metcalfe, all come to Strathroy to attend church or shop. Strathroy is another example of a town surrounded by both Dutch Catholic and Dutch Calvinist farmers. The latter mainly live in the rural part of Caradoc Township, which includes Mount Brydges (not listed as a separate census unit). Strathroy is also the site of the second branch of the Sydenham Credit Union originally founded by Dutch Calvinist farmers (see chapter 8).

Unlike very small towns and villages, the towns listed in table 4 have a large proportion of Dutch-Canadians who first came to Canada as non-farm, rather than farm, immigrants. While acting as service centers to farmers, these towns, technically designated as "rural," are likely to have Dutch bakeries or small businesses employing skilled workers (tool-and-dye makers, millers, electricians, and clerks) who grew up and received their training in Dutch towns and cities. The population centers listed in table 4, particularly Strathroy (London region) and Aylmer (Lake Erie region), generally have an equal mixture of Dutch Catholics and Calvinists, plus other people who joined other denominations or have no religious affiliation. However, even in this category of "rural" census units, some have a greater Catholic, as opposed to Calvinist presence. For example, Fergus, located in Wellington County (Waterloo-Guelph region), Essex (Essex-Kent region), and Bradford (Lake Simcoe region) have more Dutch Calvinists, while Ingersoll (London region) has more Catholics.

Table 4
Towns with between 5,000 & 9,999 inhabitants with a Dutch Presence
[numbers in brackets shown in figures 17 & 18]

Name (and county)	Number of Dutch Canadians	Percentage of people of Dutch descent
Strathroy (Middlesex) [26]	845	9.58
Pelham (Niagara) [27]	580	6.46
Ingersol (Oxford) [28]	350	4.22
Bradford (Simcoe) [29]	330	3.78
Aylmer (Elgin) [30]	275	5.32
Acton (Halton) [31]	275	3.75
Whitchurch-Stoufville (York) [32]	245	3.47
Niagara-on-the Lake (Niagara) [33]	215	4.12
Fergus (Wellington) [34]	205	3.33
Dunville (Haldimand-Norfolk) [35]	195	3.81
Essex (Essex) [36]	190	3.18
Port Perry (Durham) [37]	170	3.36

The semi-urban 'rural' fringe

Some urban centers with populations above 10,000 that have a significant Dutch presence (anywhere from 1.6 to 9.5 percent) are still surrounded by, or border on, rural agricultural townships with Dutch farmers. Table 5 shows such townships, or the rural sections of bigger towns and cities in this category (excluding the rural sections of Grimsby, Chatham, and Sarnia that already appeared in table 1). The only large cities with over 200,000 inhabitants that have such semi-rural areas with a significant Dutch presence, are London and Hamilton. Those near Hamilton not yet mentioned are the township of Flamborough and the rural part of Ancaster. In all of these census divisions, fewer Dutch-Canadians live in the countryside than their corresponding urban centers. The only other case where the ratio is reversed is the town of Milton, located within the Toronto region. Its rural section is included in table 5 even though the 260 people of Dutch descent represent only 3.46 percent of the population. In the fifties and sixties, the numbers and percentages would have been much higher, but by 1986, all Dutch dairy and most of the cash-crop farmers of Dutch background in this area were long gone, as a result of skyrocketing land prices associated with urban speculators. However, there are still quite a few chicken farmers left. This area also had, and still has, a larger number of Catholic Dutch immigrants who ran their own Dutch Canadian Country Club,[8] whose most active mem-

[8] See Schryer, *The Netherlandic Presence in Ontario*, 151, 258.

Table 5
Rural Census Divisions with a significant Dutch Presence close to Urban Centres of more than 10,000 inhabitants
[numbers in brackets shown in figures 17 & 18]

Name (township/rural) & nearby city	Number of Dutch Canadians	Percentage of people of Dutch descent
East Galafraxa (Orangeville) [43]	170	17.65
Southwest Oxford (Tilsonburg) [49]	1145	13.76
Rural part of Zorra (Woodstock) [54]	830	13.29
Downie Twp. (Stratford) [42]	250	10.99
Township of Norwich (Tilsonburg) [49]	225	10.84
Ancaster (rural) [47]	430	9.73
Milton (rural) [44]	260	3.46

ber was a Dutch immigrant farmer specializing in cabbages. However, care is needed in interpreting census figures for urban-rural ratios in the grey zone between town and countryside, where many people with urban occupations are rapidly moving into once rural areas.

The townships or rural zones of towns and small cities with significant Dutch populations, located on the edges of expanding metropolitan areas, are rapidly being transformed into suburbs. These census divisions have a mixture of urban and rural Dutch immigrants and their descendents. However, many of these semi-urban areas have only recently shifted from being predominantly rural-agricultural to becoming an integral part of urban conglomerations. Table 6 lists such intermediate rural-urban zones that in 1986 were quickly becoming satellite cities of larger metropolitan centers. Two smaller semi-urban census units listed as townships and where the Dutch constitute more than 7 percent of the total population, are included in this ambiguous category.

The anomalies of Westminster and Onondaga

Westminster, which was not included in table 2, has the highest proportion of people of Dutch descent in the province (32.99 percent). It therefore did not come as a surprise to learn that two of the six candidates from this area running for London city council in 1993 were Dutch-Canadians.[9] Yet, this area is not perceived to have a lot of Dutch people, not even by those of Dutch descent who are living there! This anomaly required investigation. Westminster, which includes the village of Lambeth, had close to 6,000 inhabitants in 1986 compared to an

[9] See "Two Dutch-Canadians Vie for Seat," *The Windmill Herald* (Surrey, B.C.), 23 March, 1993, 14.

Table 6
Ambiguous ('semi-urban') census divisions
[letters in brackets shown in figures 17 & 18]

Name [census category]	Total Population	Total Dutch	Percentage Dutch	Proportion (DMT)
Westminster [O] [township]	5,865	1935	32.99	7.49
Delaware [N] [township]	1,865	235	12.60	59.57
Onondaga [T] [township]	1,357	175	12.90	28.37
Flamborough [P] [rural only]	21,493	2390	11.12	30.54
Nanticote [S] [rural]	12,457	1080	8.67	29.63
Newcastle [R] [rural]	17,272	1285	7.44	40.47

average of 2,000 for townships in the rest of Middlesex County. At that time, Westminster was well on its way to becoming part of a suburban fringe. Indeed, it was upgraded to a town (with a rural component) soon after the census was taken, and later absorbed into the city of London, in 1992. Unlike most other townships, the close to 2000 Dutch-Canadians in this former township in 1986 were much younger (second and third generation), as shown by the fact that less than 10 percent of them reported Dutch as their mother tongue (DMT, see table 6). This number stands in stark contrast to the figures for more established Dutch-Canadian localities, where the proportion of people reporting Dutch as mother tongue was around 30 or 40 percent.

At first glance, it may appear that the demographic composition of Dutch-Canadians living in Westminster simply reflects an influx of young Dutch-Canadians into towns and villages from nearby rural areas in the seventies, similar to places such as Drayton, in the fifties and sixties. However, by the seventies and early eighties, the dynamics of rural-urban migration involving Dutch-Canadians were quite different: younger, and, for the most part, better-educated younger people in rural townships either stayed on the farm or moved directly to larger towns and cities,[10] while older, retired Dutch-Canadian farmers began moving to

[10] Arkona, which also has a much younger Dutch-Canadian population, showing a percentage of DMT (Dutch Mother Tongue) as 15.12 percent, is somewhat exceptional in that regard, because several Dutch-Canadian or Dutch-linked institutions and businesses provide employment opportunities for better-educated young people.

nearby villages and small towns. This explains why the DMT figures for villages like Drayton in 1986 were much higher than those of their rural surroundings.[11] In contrast, the fringes of larger cities, including London, were more likely to see an influx of younger people from both rural and urban areas. Indeed, on the basis of several phone calls and interviews in the early nineties, I was able to ascertain that the young Dutch-Canadians in Westminster (mainly in Lambeth) comprise both younger professional couples originating in the city of London, who bought new houses in that area,[12] and the offspring of rural Dutch-Canadians moving closer to the city. The only farmers are a handful of apple producers.

A similar anomaly was detected in the census division of Onondaga (in the Lake Erie region). Nearly 13 percent of the population of this township is of Dutch descent, but, like Westminster, this township also lacks a reputation for having a Dutch presence. However, the demographics are different. More Dutch-Canadians in Onondaga, which has no villages or hamlets, live in farmhouses.[13] Moreover, the Dutch people who live in Onondaga, are older since nearly a third reported Dutch as their mother tongue. I learned from other sources that the children of Dutch farm families in this area tend to stay close to home growing up, with ready access to jobs and shopping facilities in several nearby urban centers. One would therefore assume that these Dutch-Canadians, who make up 12.9 of the township, would be more "visible."

The discrepancies between the census data and the perception of where Dutch people reside in both Onondaga and Westminster, can be explained through a combination of the diverse composition of its Dutch-Canadian population and the fact that members of various Dutch sub-groups interact with each other in places other than the census division where they reside. In the case of Onondaga, Dutch immigrant farmers from a variety of backgrounds, and who had connections with other places, happened to buy land in the same area.[14] Dutch families belonging to different churches had little need to get together, and did not know each other unless they lived close together; Calvinist immigrants attended one of several Reformed Churches in nearby Brantford, while Dutch Catholics

---

[11] 47.37 percent for the village, compared to 35.29 for Maryborough township.

[12] Lambeth is a preferred place of residence for young couples working in London, but wishing to live within a short commute. Telephone interview (FS) with John Feron, 28 February, 1991.

[13] The Dutch who live closer to, or right within, the suburbs of Brantford appear under the census division for the bordering township (not city) of Brantford, where they constitute only 4 percent of the total population.

[14] Telephone interview (FS) with Mrs. Guichelaar of Brantford, 7 February, 1992.

had contacts with Dutch priests in the villages of Caledonia and Hagersville, just to the south, in Haldimand County. Not only was there little intergroup contact, but outsiders (both non-Dutch and Dutch-Canadian) would be less likely to identify as "Dutch" those places with no buildings, such as churches or ethnic stores, characterized as Dutch. In Westminster, including Lambeth, established institutions were also located outside of this census division; the main social center where Dutch Catholic immigrants gathered used to be Delaware, just outside of Westminster. Moreover, no branches of St. Willibrord Credit Union, founded by these immigrants, are located in Westminster, and Dutch-Canadians that belong to Reformed denominations established their churches in the city of London. Dutch-Canadians in Westminster who wanted to socialize with members of their own ethnic group, regardless of religious affiliation, could join a club built by Dutch immigrants on the outskirts of another part of London. And Dutch delicatessen goods are easily found a short distance away, at one of two Dutch stores also located in the city of London. Hence the younger, more acculturated people of Dutch background (both rural and urban) who moved here from elsewhere, are less likely to have met, much less be aware of the heavy concentration of people of Dutch background where they live.

Places not Mentioned in the Census

With the exception of Onandaga and Westminster, the presence of Dutch-Canadians in all of the census divisions mentioned so far, corresponds fairly well with where I was informed one would find "lots of Dutch farmers."[15] Villages or hamlets that are not listed as separate census units, and hence do not appear on the two maps, would be subsumed under townships or the rural component of more urbanized townships. The places that emerged in conversations, or that I visited, include several that saw the arrival of mainly Dutch Catholics: Brussels (township of Morris in the Huron part of the London region),[16] Kennilworth (township of Arthur in the Waterloo-Guelph region),[17] Kintore and Thamesford (Zorra Township in the London region),[18] Phelpston and Elmvale (Flos Township in the Lake Simcoe region), Stayner (Nottawasaga Township,

[15] My fieldnote files (including those entered into the NUDIST program) include 187 place names, with anywhere from one to twenty-eight references to the same place.

[16] Many of the Dutch farmers in the Brussels area live on farms located in Grey township (Perth county).

[17] This hamlet, located on Highway 6, half way between Arthur and Mount Forest (in Wellington County), once had a Dutch priest.

[18] Located on Highway 19, just north of Thamesford, in Oxford County.

also in Simcoe), Winchester (in the township of the same name), Lansdowne (Rear of Leeds and Landsdowne, both in Eastern Ontario), Kincardine (London region), Kingsbridge (in the township of Ashfield, on the northern tip of the London region) and Wallacetown (in Dunwich township, in the Lake Erie region).

Villages and hamlets with a Reformed Dutch presence not yet mentioned include Blyth (Morris Township, in the London region); Troy (part of the town of Flamborough, near Hamilton); and Puslinch (in township of same name, at the southern edge of the Waterloo-Guelph region). In the Northwest region one finds Lucknow (Kinloss, township in Bruce County); and the townships of East Garafraxa and Mono (in Dufferin County), both serviced by Orangeville. There are also numerous locations in the Niagara region: Fenwick (rural part of Pelham), Attercliffe, Wellandport, and St. Anns (all located in the township of West Lincoln), Smithville (in the rural part of Grimsby), Vineland (part of the town of Lincoln), Boston and Waterford (rural part of Nanticoke), and York (part of the town of Haldimand). Most of these places also have one or more churches associated with various Reformed denominations. Ansnorsveldt and Springfield, both in the Holland Marsh region deserve special mention, not only because of their Dutch presence today, but because they were founded by pre-war Dutch immigrants (see chapter 3). They are located respectively within the boundaries of East Gwillimbury (York Regional Municipality) and West Gwillimbury (Simcoe County), both included in my larger Lake Simcoe region. Additional places that have a religiously mixed Dutch rural population include Williamsburg and Lancaster (in townships with the same names), and Iroquois (in Matilda), all located in Eastern Ontario; the two townships of Tecumseth and Essa in the Simcoe region, which include the town of Alliston, the hamlet of Thornton, and the village of Beeton; the village of Ayr (North Dumphries) located in the southwestern corner of the Waterloo-Guelph region; Embro (London region); and finally, in the Northwest region, Grand Valley (straddling the townships of East Luther and Amaranth).

Northern Ontario

With the exception of a few enclaves, Dutch agrarian immigrants did not settle outside of the Ottawa and St. Lawrence valleys. Most of Northern and Central Ontario, with few people of Dutch descent, present quite a different form of spatial distribution from Southern Ontario. The ratio of urban to rural Dutch, is for the most part, inversely proportional to that of Southern Ontario. We can look at the districts of Algoma and Nipissing as examples. The cities of North Bay and Sault

Ste. Marie have 195 and 345 Dutch-Canadians respectively, with only one or two families in their corresponding rural regions. This pattern also appears in towns further north, such as Timmins and Dryden.

Only in the northern clay belt regions, where agriculture or cattle raising is possible, do the urban-rural ratios typical of Southern Ontario surface. For example, there are more Canadians of Dutch descent in rural portions of the flatter, arable land around Verner (on the north side of Lake Nipissing), than in the nearby towns of Cache Bay and Sturgeon Falls. Similar agricultural pockets with rural Dutch can be detected in the districts of Cochrane and, to a greater extent, in the Thunder Bay and Rainy River areas. In the case of the latter, Dutch-Canadians constitute 20, 10.62, and 7.23 percent of the townships of Morley, Emo, and Chapple respectively, located east of the Lake of the Woods region. In contrast, the percentages for the towns of Fort Frances and Rainy River are 10.36 and 2.03 respectively. In the district of Thunder Bay, the Dutch are mainly clustered in the townships of Paipounge (10.99 percent), Oliver (4.86), and South Gillies (6.86). As in the south, there are more Dutch-Canadians in the city of Thunder Bay in absolute numbers, but they represent only .71 percent of the total inhabitants. In contrast, the 130 Dutch-Canadians who live in the rural component of that city represent 3.74 percent of the total rural population. The tendency of people belonging to the same religion to live in the same area also shows up in the North. While Dutch-Canadians in the Thunder Bay and Rainy River areas are mainly Calvinists, Dutch farmers in the district of Nipissing and in Sudbury are Roman Catholics (see also chapter 7).

THE SPATIAL DISTRIBUTION OF DUTCH FARMS

Census data, designed to avoid identifying individuals, do not allow us to pinpoint the exact location of specific farms, their owners, or the extent to which landownership is concentrated at the micro level. However, most rural people have a good idea of who owns which farm in their neighborhoods and anyone can consult the land registry office to obtain this information. The following section will therefore examine several maps created with the help of people interviewed, illustrating different patterns in the spatial distribution of farms owned by Dutch-Canadians.

Except for the village of Ansnorsveldt in the Holland Marsh, where a group of prewar settlers were allocated contiguous strips of land in a single block, Dutch immigrants rarely bought farms in one location simultaneously. A more typical scenario for postwar Dutch immigrants is one or two families or friends buying farms fairly close together or in the general vicinity of a church. In parts of the province where people of Dutch background constitute a small percentage of the rural population, their farms are still scattered, with no more than one or two farms on each line. Typically farmers in such regions come from a variety of regional or religious backgrounds. An example is Teeswater, in the township of Culross (Northwest region), where people of Dutch descent constitute just below 3 percent of the population. Figure 19 shows a small part of that township, where Highway 4 intersects with Bruce Country Road 6.

In this part of Culross, no farms are owned by Dutch people east of Highway 4, an area inhabited by Canadians of Catholic-German descent. On the west side, only seven out of sixty-six farms within four blocks of land are owned by Dutch immigrants, including one family who came to Canada after 1990. Going west along Bruce County road, you would not know from mailboxes alone that the wife of one of the non-Dutch farmers was originally from the Netherlands (not shown on the map). There are few Dutch farms, and they are not located adjacent to one another. The Dutch-Canadian families who live on these farms represent different religious affiliations, including one who does not attend church. To locate an area nearby with Dutch farms closer to one another, one would have to go to Kinloss, one township further west, in the vicinity of Lucknow, which borders on the township of Ashfield, in Huron County. In both townships, the Dutch represent around 7 or 8 percent of the total population. Dutch Catholic farmers settled close to the villages of Kintail and Kingsbridge, while their Protestant counterparts (mainly Christian Reformed) can be found along the tenth line in Ashfield and in the vicinity of Lucknow. A more concentrated cluster of about fifteen Dutch-Canadian farm families is located further south, in the Grand Bend Marsh (see also chapter 7).

A much denser clustering of Dutch-owned farms is located even further south, in the London region. The map shown in figure 20 shows several adjacent blocks of land (close to 5,000 acres) in the township of Warwick (Lambton County), where the percentage of rural Dutch goes well beyond a quarter of the population. As in so many parts of rural Ontario, many of the original 100-acre farms – originally owned by farmers of English-Protestant background – have consolidated into larger units and most extend "from road to road", meaning that where there were once three or six one-hundred acres farms within a square section of land located within four concession roads, one finds only one. One can observe that, out of the forty-seven hundred-acre farm lots shown on this map, all but ten are owned by Dutch-Canadian families. Figure 20 includes some additional information on individual farms or their owners.

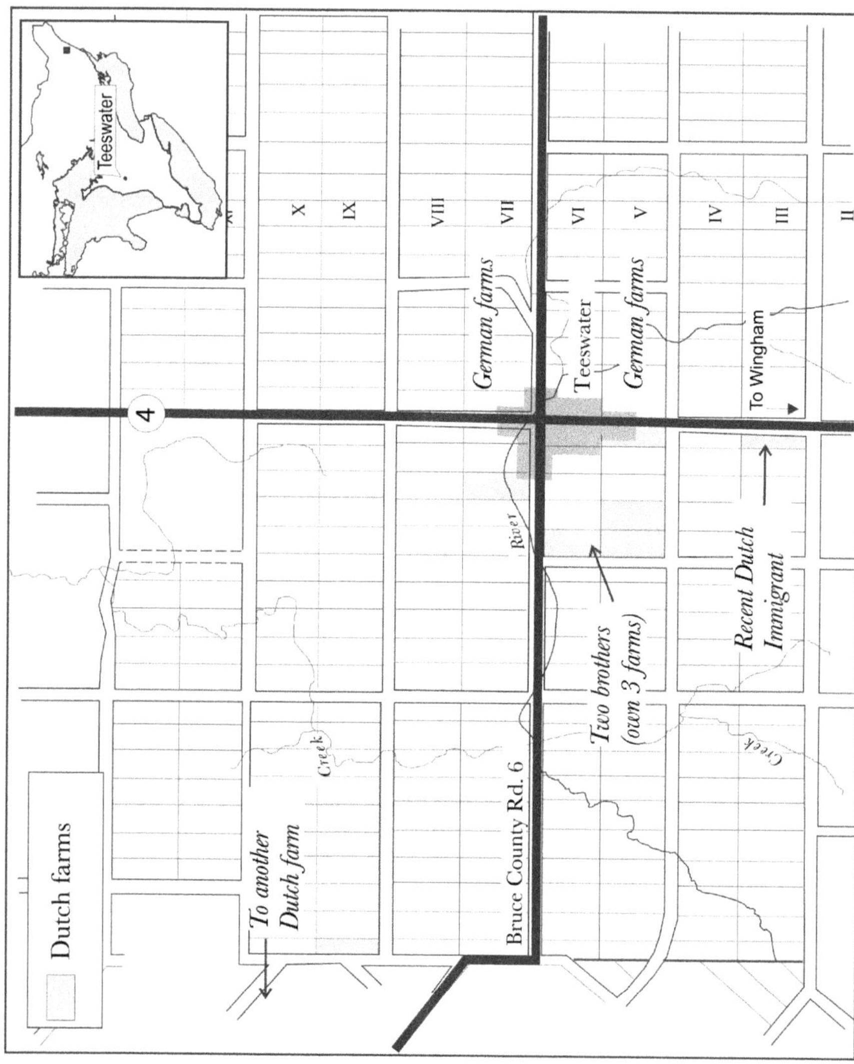

Figure 19. Dutch-Canadian Farms in the Vicinity of Teeswater (circa 1999).

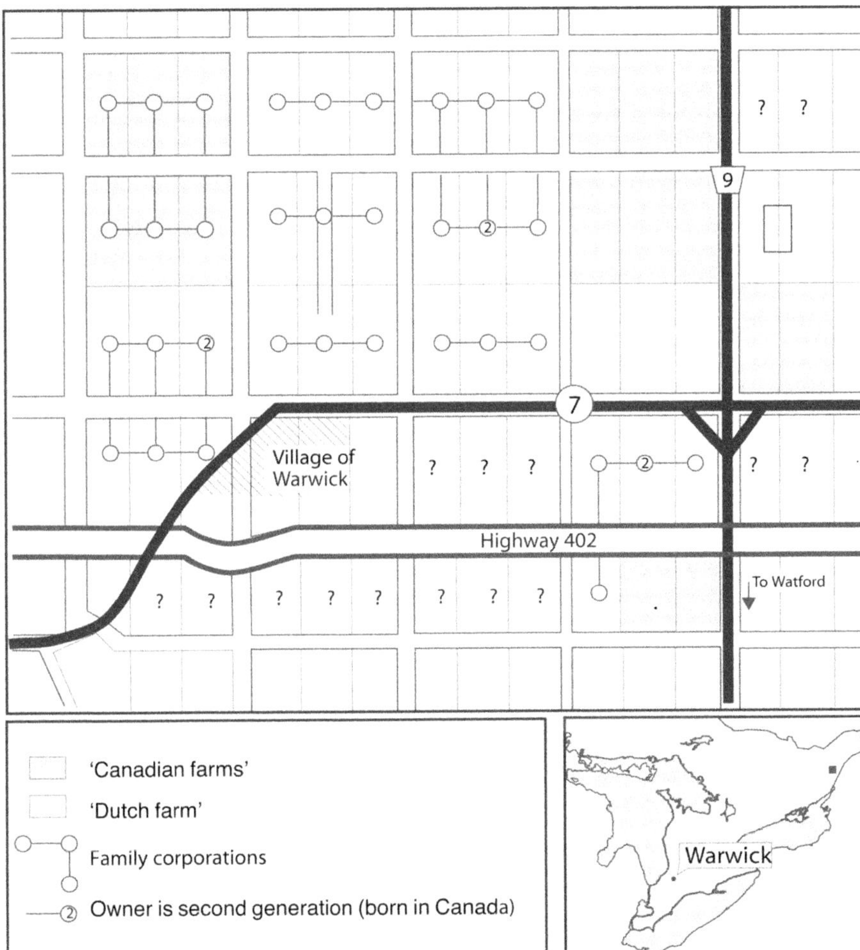

Figure 20. Dutch-Canadian farms in Warwick area (circa 1998).

This level of concentration of Dutch farm ownership is repeated for most of the second line of Adelaide township (Middlesex County), on the other side of Highway 402. Similar, albeit smaller, pockets of Dutch farms are found in some of the other areas already identified as having a significant Dutch Catholic presence, including several tracts of land or near the hamlet of Kinkora, in Perth County (also in the London region), and just east of Chesterville (Winchester Township), in the county of Dundas in Eastern Ontario (see chapters 7 and 8). Here, at a farm owned by a Dutch-born dairy farmer, I was told that most silos, as far as one could see, belonged to other Dutch farmers. The farmer quoted a local expression, "It smells of the Dutch around here."

In areas with a concentration of Dutch farmers who belong to Reformed denominations, a similar pattern of ownership of entire blocks of farms is evident. Such blocks are common in sections of the townships of Maryborough and Peel, especially all along the 8th line (south of Moorefield). These townships, located in the most northern part of the Waterloo-Guelph region, constitute the "hinterland" of the town of Drayton (Wellington County). In fact, the Drayton area is part of a string of Dutch Protestant (Reformed) rural communities that straddle the northeast corner of the London region and the southern edge of the Northwest region. One of these communities is centered in Listowel, located on the southernmost edge of the township of Wallace. Its Christian Reformed church is an offshoot of the one founded in Drayton (see chapter 7) but farmers who attend that church also live in Morningside and Elma, in the northern half of Perth County. Although each of these townships has a Dutch rural population that constitutes only about 5 percent (compared to 23.15 percent for Maryborough), Dutch Protestant farmers nevertheless own adjacent farms in blocks of land throughout the northern part of Perth county. The map in figure 21 shows the land owned by Dutch farmers in the vicinity of the town of Listowel, which straddles the boundary between Wallace and Elma townships.

I observed similar clusters of Protestant Dutch farms in the Clinton area (especially in the townships of Hullett and Goderich), where several families own farms on both sides of Highway 8 and its nearby concession roads, and in the triangle of land between Norwich, Burgessville, and Salford (in Oxford County). However, the more typical pattern for the rest of Southern Ontario is smaller clusters of Dutch farms scattered throughout a whole township. For example, I found small pockets of Dutch farmers from different backgrounds, similar to those earlier described for Teeswater, along the 5th line near Thornton (Lake Simcoe region) and on both sides of Highway 6 between the towns of Mount Forest and Arthur (Wellington County), where there were one or two Dutch dairy or chicken farms in each block.

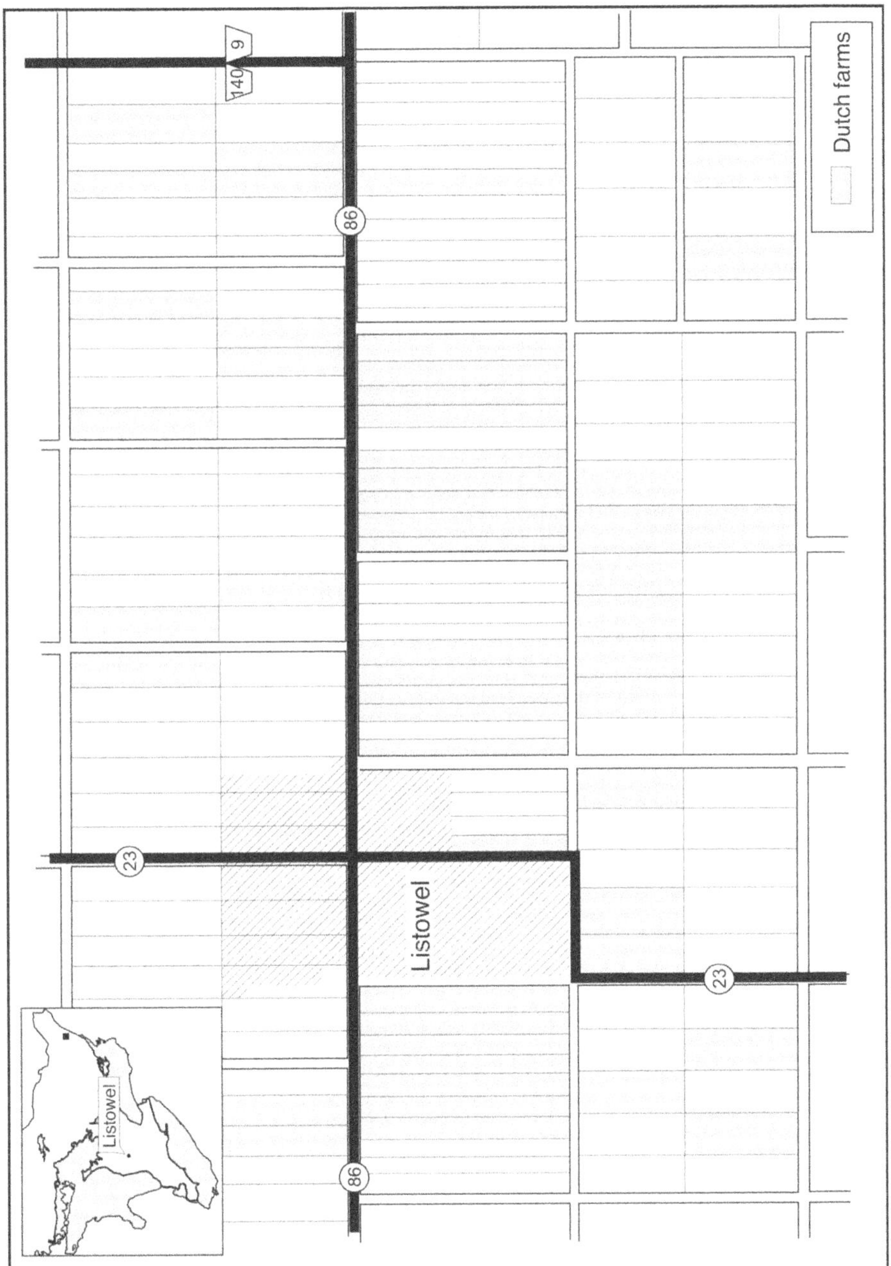

Figure 21. Dutch-Canadian Farms in Vicinity of Listowel (circa 1996).

Peoples' Perceptions

The concentration of farms owned by Dutch-Canadians in all of the places mentioned is consistent with the perception of people of both Dutch and non-Dutch descent. However, even in places with a strong presence of Dutch farmers, farmers themselves might not be aware of the presence of other Dutch-Canadian farmers in their own township, if they belong to different religious denominations. For example, in one township, an elderly Dutch immigrant farmer, now retired, accurately identified the location of all farms owned by Dutch-Canadians on a map indicating lots in his township. Most of these farmers belonged to two Reformed denominations. However, he did not recognize a cluster of Dutch Roman Catholic farmers in another section of that same township. There is also a tendency to overestimate the number of farms owned by Dutch-Canadians. When I asked how many farmers in a particular region were Dutch, common answers were: "most of the farms," "practically every farm around here is owned by a Dutchie," or "this township now has more Dutch farmers than non-Dutch farmers."[19] Some farmers gave me specific percentages, such as 90, 50, or 30. However, upon more careful investigation, it often turned out that the actual proportion of Dutch farmers was lower. Some people later qualified their original estimates by explaining: "well maybe not 30 percent, but they certainly have 30 percent of the influence,"[20] or "the Dutch tend to have the bigger farms." When we look at the actual acreage of farmland in Dutch hands, as opposed to the percentage of Dutch farmers, the original estimates are close to actual percentages.

Figure 22 shows a section of approximately 35,000 acres of land (showing eight concession lines) in Downie township (Perth county), with some clustering of Dutch farms. The percentage of the population of Dutch descent in this township is around 10 percent, which is not unusually high. In comparison, of the approximately three hundred 100-acres lots shown on this map, fifty-one are owned by Dutch farmers, representing about 16 percent of all farms. Thus, the percentage of farmland in Dutch hands in this part of the township is higher than the percentage of the population of Dutch descent. However, the actual number of Dutch farm families is much smaller than fifty-one since just one extended family, legally constituted as a corporation, owns 1,500 acres of land, spread out over the area (numbered from 1 to 15). The impression of such a

[19] This statement referred to Ashfield (in Huron).

[20] Interview with a Dutch-Canadian farmer in the Mitchell area (Perth county), 12 October, 1992.

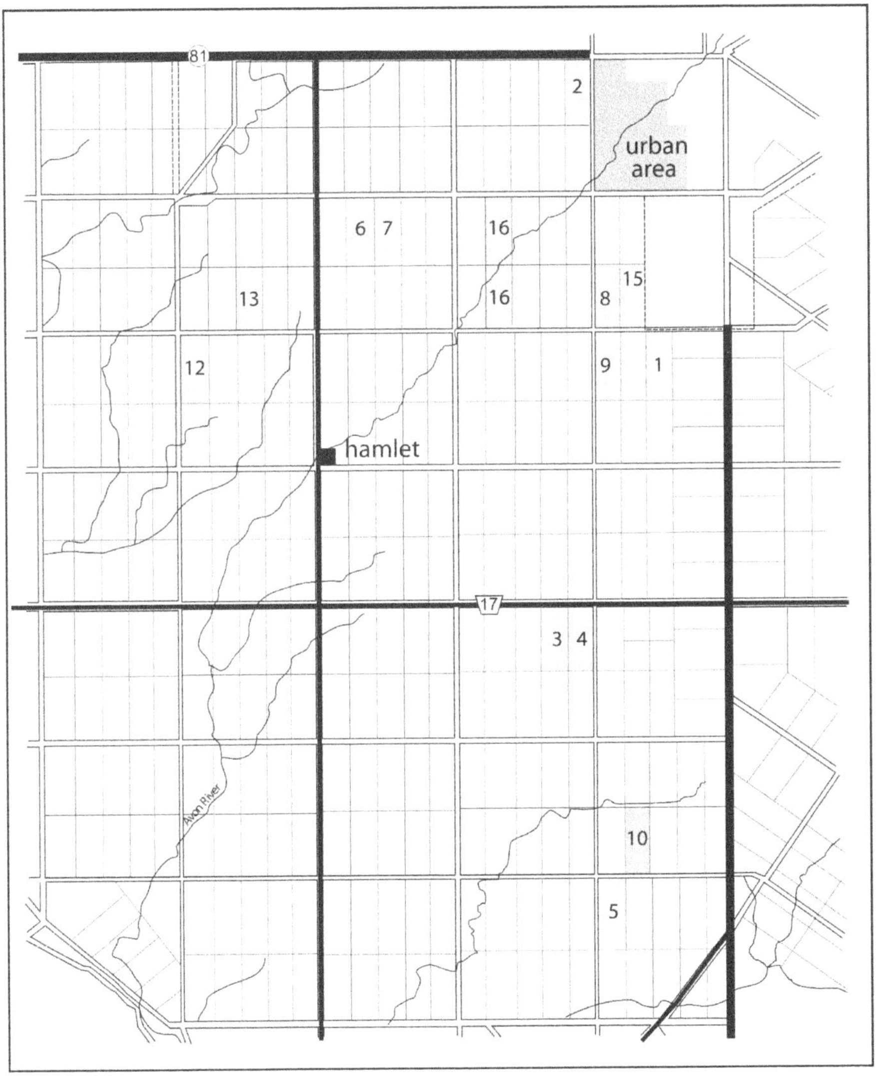

Figure 22. Dutch-Canadian Farms in Downie.

a region as being "completely dominated by Dutch farmers," is likely to be shared by many people precisely in places where the names of Dutch farm families are prominently displayed on big, modern barns. It is also hard not to notice that the names on mailboxes in front of farms with older, smaller barns, dilapidated shacks, and messy barnyards full of rusty farm equipment, are definitely not Dutch or Frisian. Such stark contrasts are not found in other townships with big, modern farm operations run by members of Canada's founding groups (English, Scottish, Irish and French).

Patterns of Change

While the maps and statistics of farm ownership presented so far provides a convenient snapshot of a definite point in time, the number of farms owned by postwar immigrants has changed over time. A typical scenario, repeated all over Southern Ontario, is one where only a single Dutch farm family moves in and fixes up a dilapidated farm. Within a decade, they buy two or three nearby farms, and then several of their children set up their own farm operations, and a small Dutch-Canadian cluster is created. One example is the area near Ariss, just west of Guelph. Over three or four generations, with inter-ethnic marriages, it will no longer be possible to distinguish between Dutch and non-Dutch farms in such areas. However, in some cases, two or three Dutch farm families who intermarry could expand enough to create a permanent cluster. Even if some of those offspring form their own family farm corporations and continue purchasing additional farms, they could end up dominating an entire township.

Another pattern of development, as illustrated by areas with high concentrations of Dutch immigrant farmers from the start, is one characterized by an increasingly intertwined network of Dutch-Canadian farm families with a common background resulting from ethnic intermarriage. In such areas, the proportion of Dutch-Canadian farmers might increase over time as non-Dutch farm families start to leave. Such areas, like the region around Drayton or the areas surrounding Warwick, Arkona, Watford, and Park Hill, most closely resemble other ethnic enclaves where English-speaking "Canadian" farmers are still predominantly of Catholic German, or Protestant Irish, descent (see chapter 2) and aware of their ancestry. On my way to an interview in Park Hill, I stopped at a fast-food restaurant located at the side of Highway 79. When I mentioned the presence of Dutch farmers in the region, a young woman whom I would not have identified as Dutch-Canadian said, "We are all Dutch around here."

Another scenario is that of areas with a high proportion of farmers of Dutch descent prior to 1970 or 1980, where they have since disappeared. An example is the area surrounding the village of Hickson (Oxford County), where seventy Dutch families once owned farms, mostly on the 13th line just east of the village.[21] Most retired or sold their farms because of a lack of successors. Many had simply not expanded, making it difficult for their children to get involved in the family enterprise. In contrast, the one Dutch immigrant farmer left on the line did buy an additional six farms. That farmer was able to accommodate several sons, who now run the larger operation. A handful of the Dutch families remaining on that line are part-time farmers. This area, which represents a small part of the township of East-Zorra/Tavistock, illustrates how the ethnic composition of a rural area can change in under two decades. Most of the Dutch farmers who originally displaced non-Dutch farmers eventually sold their land back to non-Dutch farmers or to non-Dutch newcomers, including a recent immigrant from England. However, the increase in number of farms owned by people with no Dutch connections was still greater than the six farms bought by the only postwar immigrant who expanded. Even the purchase of several additional farms by new immigrants from the Netherlands did not restore the level of Dutch ownership in the Hickson area. Only if these new Dutch immigrants start expanding, could this area again become as "Dutch" as in the seventies. Yet, there is still a significant Dutch presence in other parts of the township, in the vicinity of Innerkip and Drumbo (further south), and all along the 8th line, to the west of Hickson. That Dutch presence, in turn, cannot match the even larger proportion of land owned by both second-generation Dutch-Canadian farmers and Dutch newcomers in the region south of Woodstock, in the southern half of the county.

Finally, we can look at the location of farms owned by both established, postwar Dutch-Canadians and more recent farm immigrants (both Dutch and Swiss) in an even larger region. The adjacent townships of Grey and Elma, located in the northeastern part of the region of London, belong to the counties of Huron and Perth respectively. Both townships fall into the intermediate, and thus more typical, category of townships where the percentage of people of Dutch descent is between 5 and 10 percent. The map shown in figure 23 provides further evidence that the percentage of land in Dutch hands (almost a quarter for Elma) is much greater than the proportion of Dutch-Canadians in the population of

[21] Interview with Martin Verkuyl, at his farm near Hickson, 24 November, 2000.

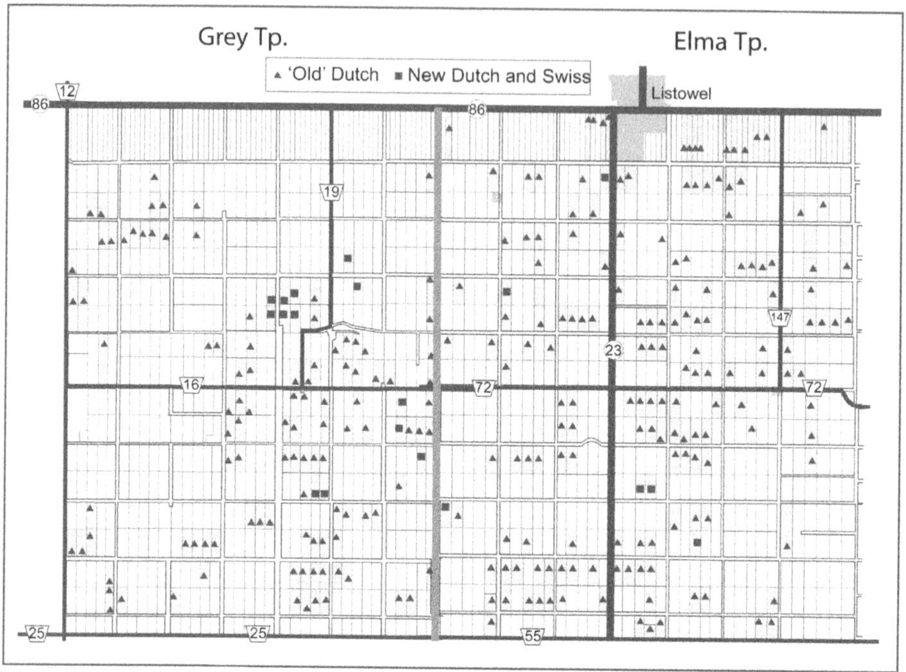

	Grey Tp.	Elma Tp.	Total
Percent population of Dutch decent	9.11	5.78	7.40
Percent of land owned by old Dutch	17.46	24.96	21.24
Percent of land owned by new Dutch and Swiss (combined)	2.06	0.94	1.49

Figure 23. Dutch Ownership of Farmland in Grey and Elma Townships.

this area. Moreover, the amount of land owned by established Dutch-Canadians ("old Dutch") vastly outnumbers that of all new, more recent, immigrant farmers (see chapter 15), even if we combine the Swiss and the Dutch. By the time this book is in print, more Dutch newcomers will have bought additional farms from both "old" Dutch and non-Dutch farmers.

Whether we look at the figures of the population census highlighting the numbers and percentages of Dutch-Canadians in rural census units, or maps of townships or sections of townships showing where Dutch farms are located, it is apparent that the Dutch farm immigrants who came to Ontario, and their descendants, are both spread out throughout the province and found in specific regions. There are more rural Dutch-Canadians in the southern, as opposed to the northern parts of the province. Within the more fertile arable sections of the former region, they are both thinly dispersed (in most rural areas) and concentrated (in several townships and villages). Dutch farmers are not everywhere, but it is easy to imagine how some people could assume they are, especially those who attend meetings of farm organizations or marketing boards where the Dutch presence is particularly conspicuous.

XIII. Crop Specialization and Agribusiness

While it is useful to chart the locations of Dutch immigrant farmers and their descendants, further light is shed on the Dutch presence in Ontario agriculture by examining their involvement on marketing boards and producers' organizations. Here too, the picture of their presence in farming and horticulture can be fine-tuned, by narrowing in on various branches of agriculture and agribusiness, which constitute sub-fields of the agri-food industry. It is in these sub-fields that we can best track how and, to what extent, Dutch-Canadian farmers have made names for themselves. Dutch farm immigrants entered many branches of agricultural production and agribusiness such as processing, distributing and marketing, where their visibility and impact varied. Not only were Dutch producers more likely to specialize in specific commodities, they were not equally represented in all organizations. Even when Dutch-Canadian farmers and processors were involved in these organizations, they did so much earlier in some than in others. In each specialization, the presence of farm families of Dutch background and the accomplishment of members of those families, contributed to the emergence of positive group stereotype, to a greater or lesser extent.

COMMODITY SPECIALIZATIONS

It is difficult to determine the exact proportion of farmers in various commodities who are Dutch. Membership lists do not break down by nationality, and smaller organizations do not keep records more than two or three years. Nevertheless, it is possible to extrapolate from surnames, in combination with interviews with people acquainted with the history of various organizations. I also asked farmers with both Dutch and non-Dutch backgrounds to provide rough estimates and then compared the figures to those obtained with written records.

The Greenhouse Floral Industry

The specializations with the highest Dutch representation are flowers growers and shrub nurseries. Four people I interviewed told me that at least 80 percent of the 200 greenhouse flower growers in the Niagara

Peninsula, who together are responsible for more than a third of Ontario's floral production, are of Dutch origin. This predominance of Dutch flower growers is not immediately apparent when examining the names of many of the larger family-owned companies that both grow and distribute flowers in the Niagara region and other parts of the province: Bayview Flowers, Westbrook Floral Ltd., Lakeshore Inc., Northend Gardens, Rose Flora Ltd., Valleybrook Gardens, Epic Plan Company, and many others. However, their owners have Dutch names and Dutch immigrants founded them all, as noted on the company web sites that contain historical information. For example, the history of Bayview Flowers goes back to 1948, when Clarence and Metta Van Staalduinen started growing gladiolas in Brantford for the cut flower market in Toronto. This business became a major producer and distributor of cut flowers, with greenhouses in Brantford (The Lake Simcoe region), Simcoe (Lake Erie region), and Jordan Station (Niagara Peninsula) where their headquarters are located. A fleet of trucks ships potted plants and cut flowers to other parts of Ontario and several locations in the U.S.[1]

The overrepresentation of the Dutch in greenhouse flower production on the Niagara Peninsula is clearly evident in a recent book, *Floral Passion*, celebrating the flower industry in the Niagara region. This coffeetable book, with lots of illustrations, includes short descriptions of almost one hundred families or individuals who run greenhouse operations dedicated to floral crops, potted foliage, cut flowers, bedding plants, and nursery stock. The author tried to be as representative as possible by including both high-profile producers (with over fifty employees), as well as smaller, strictly family-based, businesses. *Flower Passion* does not focus strictly on the Dutch, although it can not ignore their presence or their country of origin: one of the twenty-three short chapters is devoted to the Netherlands, the "garden of the world" and "the county of origin of many of Niagara's greenhouse growers."[2] What is more revealing is the preponderance of Dutch examples. Of the ninety-plus cases presented, I counted one farmer descended from a prewar Swiss immigrant farmer, plus one Polish, one Italian, three English, and two Ukrainian postwar immigrant farm families. There were likewise five cases of Canadian-born farmers (non-Dutch-Canadian), some of whom were descendants of Mennonite farmers who came to the region in the nineteenth century. Together, these farmers without any Dutch connections constituted less than 15 percent of

[1] See short description of their business in Van der Mey et al., *Floral Passion* (St. Catharines: Vanwell Publishing, 2004), 86-87; and their website www.bayviewflowers.com.

[2] *Floral Passion*, 6, 254.

the flower growers mentioned in the book. The rest consisted of Dutch immigrant farmers or their immediate descendants. Such figures are consistent with the various estimates I had already been given by numerous people well before that book was written. *Flower Passion* also provides sufficient information to determine the relative proportions of the different cohorts and generations of these Dutch immigrant farmers. Of the Dutch greenhouse growers described, ten originally came in the forties, forty-two in the fifties, seven in the sixties, eight in the seventies, one in the nineties, and finally, two after the year 2000. These figures are fairly consistent with the proportion of Dutch farm immigrants who settled in other parts of Ontario during these periods. Another trend to emerge is the large number of greenhouse growers who originated in the Westland region of the Netherlands, a region already well known internationally for its greenhouse industry. Thus, while many immigrant farmers in other parts of Ontario learned new forms of farming, or encountered new breeds of animals, those who specialized (and almost took over), the floral industry were able to directly transfer the experience and technology of their home region.

Other Specializations

The Dutch are also overrepresented in landscaping enterprises on the outskirts of bigger cities, as well as market gardening in places like the Holland and Grand Bend Marshes. Another specialization, introduced into Ontario directly from the Netherlands after the war, is milk-fed veal, also known as "white veal," almost completely dominated by Dutch farmers since its introduction into Ontario, as we shall see in chapter 14. Such overrepresentation does not apply to other agricultural commodities, although Dutch-Canadian farmers are certainly well represented in almost all branches of agriculture. For example, in terms of absolute numbers, most farmers of Dutch background specialize in the same three or four commodities typically associated with Ontario agriculture: chickens, hogs, dairy, and specialized cash crops. More than half of Ontario's 1,100 chicken producers are of Dutch origin,[3] and they are well represented in the hog and dairy industry. My impression is that Dutch-Canadians are underrepresented in cash cropping, although they pioneered the introduction of newer specialty cash crops, such as soybeans and edible beans, as a secondary specialization.

---

[3] Communication via e-mail with Chicken Farmers of Ontario (info@cfo.on.ca) 16 August, 1999.

A crop not generally associated with Dutch farmers is burley tobacco, once grown by Canadian descendants of Flemish immigrants who emigrated prior to the Second World War. Nevertheless, postwar immigrants from the Netherlands also started growing this crop in Elgin, Norfolk, and part of Middlesex counties, while Dutch-Canadian hog producers in Kent County today are as likely as their non-Dutch counterparts to cultivate black tobacco as a sideline. Many Dutch immigrants also specialized in apple orchards, once mainly the domain of farmers of English background, and were at the forefront of developing new varieties of dwarf trees and more careful pruning methods. Similarly, with the development of a rejuvenated wine-making industry, Dutch farmers were among the first to cultivate the new varieties of vinifera grapes imported from Europe in the mid-seventies. Indeed, there are few agricultural specializations, including Christmas tree farming, without at least one or two Dutch farmers. While not perceived to dominate theses sectors of farming, even a minor presence can reinforce the positive group stereotype of Dutch farmers as successful in all sectors of the world of agriculture. This image is further enhanced when one or two Dutch farm immigrants happen to achieve prominence in the processing of a product not normally associated with Dutch farmers. This happened in the case of edible beans, and to a lesser extent, apples, as we shall see later.

REPRESENTATION IN MARKETING BOARDS, PRODUCER ORGANIZATIONS, AND GENERAL FARMERS' ORGANIZATIONS

The relative import of the presence of farmers of Dutch background in different commodities is reflected in their participation in producers' organizations, particularly in local or regional committees and provincial boards. Although few Dutch immigrant farmers held such posts in the fifties and sixties, with the exception of eggs and chickens (see chapter 9),[4] an increasing number of Dutch-Canadians ran for, and were elected to, such posts starting in the late seventies. To better understand the variations over time and across different commodities, we can examine the number of Dutch and Frisian names that appear in the records of these organizations.

---

[4] Unfortunately, I was unable to verify this involvement through written records, except for the minutes of the 1966 annual meeting which shows C. Maaskant as a director for district 2 (see also chap. 8).

Dairy Farmers of Ontario

The most complete records, going back to 1966, are available for Dairy Ontario, previously known as the Milk Marketing Board. The year it was established, there were fifty-five local committees (reduced to fifty in 1997), each with anywhere from five to sixteen members, including a local executive consisting of a chair, vice-chair, and secretary-treasurer. At the provincial level today, a board of directors, with its own executive represents twelve districts. From its inception, established Canadian (non-Dutch) dairy farmers dominated the provincial board and most local committees. The only member of Dutch background on the provincial level, in 1940, was Francis Redelmeier (see chapter 9). By the mid-seventies, he was serving as vice-chairman, around the same time that another Dutchman joined the board. In 1981, Bill Schouten became a board director.

The growing influence, and presence of, Dutch-Canadian dairy farmers on the provincial board is reflected in membership figures of local committees. Dutch dairy farmers start to appear as members of most local committees from the time of the formation of the Milk Marketing Board, although they started to serve as local executive members this early only in Thunder Bay and Rainy River (see chapter 7). Nevertheless, in sixteen districts, a Dutch dairy farmer served as a local executive at least once between 1969 and 1979. Their participation was more substantial in other places: Dufferin, Dundas, Elgin, Grey, Haldimand, Kenora-Fort Francis, Kent, Lambton, Leeds, Middlesex, Niagara North (Lincoln), Niagara South (Welland), Prince Edward County, Victoria, and Wellington (which includes Drayton). In all of these districts, two or three Dutch dairy producers served in that capacity at one time or another, together with non-Dutch dairy farmers. By the eighties, Dutch farmers routinely served as local chair, vice-chair, or secretary in most counties and held more than 50 percent of local executive positions through the eighties and nineties in Middlesex, Carleton, Haldimand, Wellington, Prince Edward County, and, to a lesser extent, Frontenac (which includes Wolfe Island), Oxford and Perth. The names I recognized for the most part, were postwar immigrant, although the role of younger, first-generation Dutch-Canadian dairy farmers born in Canada become more prevalent in some regions. Overall, then, there is a strong presence of Dutch farmers, especially after 1980. This presence is also reflected in figures related to Dutch involvement in the Ontario branch of the Holstein Association, founded in 1981. Dutch farmers served as executive members in nearly all of the Holstein clubs in Eastern Ontario, and to a lesser extent, in the western regions.

The "Feather" Boards

Chapter 9 described the early involvement of Dutch immigrant farmers in the setting up of marketing boards for chickens, and turkeys. The only statistics I was able to obtain for district committees and the board of directors of the Ontario Chicken Producers' Marketing Board was for 1994-1995, at which time, anywhere from a low of zero out of 2 to a high of four out of five members of the district committees were of Dutch background. The higher levels of involvement (over 40 percent) were in district 2 (Huron and Perth), district 4 (Haldimand-Norfolk), district 5 (Niagara), district 6 (Brant and Hamilton-Wentworth), and district 9 (most of Eastern Ontario). Dutch chicken producers representing these districts also served on the provincial board, where they comprised well over half of the directors. These figures indicate a continuity of Dutch participation in this marketing board consistent with their early involvement.

Available statistics regarding the provincial executive of the Ontario Turkey Producers' Marketing Board, going back to 1966, can provide a better picture of the changing Dutch presence in this segment of the "feather" industry. For the period between 1966 and 1977, I was only able to obtain information about directors (as opposed to chairs or vice-chairs), when only one Dutch producer served in such a capacity. However, beginning in 1978, and continuing until 1997, there was always at least one Dutch director on the four or five member board, and often as many as three or four (i.e. the majority), particularly in the nineties. Dutch turkey producers constituted either half or a majority of the four executive positions throughout these two decades. This trend of increasing participation over time on the executive of such boards is consistent with what happened in other agricultural organizations.

Other Producer Organization and Marketing Boards

The Ontario Apple Marketing Commission, known prior to 1968 as the Apple Marketing Board, had a membership of 802 producers in the late nineties. An examination of a list of directors and committee representatives for nine provincial districts from 1974 to 1998, indicates a low level of involvement by Dutch producers, ranging from a low of 6 to a high of 15 percent as committee members, and a low of 8 to a high of 16 percent as directors,[5] with participation increasing throughout the

[5] This was a high estimate, including what might have been Flemish names rather than strictly Dutch.

nineties. While unable to determine the exact proportion of producers of Dutch background (since I was unable to locate membership lists), my overall impression was that the Dutch did not represent even close to a sizable minority of the apple growers in the province. They have nonetheless made a name for themselves, since several prominent producers of Dutch descent are also packers or dealers in Whitby, Bowmanville, Lambeth (near London), Watford, and Port Stanley. I immediately recognized their names, which had come up in earlier interviews, in the records I consulted.

The Ontario White Bean Producers' Marketing Board, which represents 2,000 white bean producers, has had Dutch board members since the late sixties, although a Dutch producer served as its top executive member only in 1987 and 1988. The level of Dutch participation in the Soybean Marketing Board, for which I have figures dating back to 1965, is lower than that of white beans, but with a similar pattern of increasing involvement over time. Membership on local committees was rare in the sixties and early seventies, but from then on, at least one Dutch name appears on the lists of most district committees every couple of years. Moreover, I encountered at least one Dutch farmer (also actively involved in the Farmers' Union) who not only served on the committee for Lambton throughout the early seventies, but was also a provincial board member from 1977 to 1984. Another Dutch immigrant farmer who was a committee member throughout the eighties was its chair from 1993 to 1995. However, I was only able to find five other names of Dutch soybean producers who served in that capacity and two other board members (directors), both in the eighties. I was not able to generalize about Dutch involvement on the Wheat Producers' Marketing Board, which represented about 17,000 active producers in the nineties, due to insufficient data. Some figures, available for committee members and their alternates for 1987, 1988, and 1990 indicate a Dutch participation ranging from 5 to 10 percent.

The Ontario Fruit and Vegetable Growers' Association (OFVGA) is somewhat better represented by farmers of Dutch descent. In 1999, this umbrella commodity organization, representing fruit and vegetable growers, greenhouse vegetable growers, and other horticultural industry members,[6] had 8,000 members, who in turn, represented the branches and commissions of affiliated organizations. A list of posts dating from 1945 (with 1963, most of 1968, 1969 and other years prior to 1980, missing)

[6] Ontario Fruit and Vegetable Growers Association (FVGA) Homepage [Online]. www.ofvga.org/index.html (3 October 1999).

includes 108 names that are Dutch or Frisian.[7] The proportion of members of Dutch background does not seem to be greater than that of farmers not of Dutch background, although in some sections of the organization, such as the vegetable growers' association of the Holland Marsh and surrounding areas, they are definitely overrepresented. This can be explained by the structure of the OFVGA; if member organizations consist mainly of Dutch-Canadians, this will be replicated in the larger organization. In terms of the approximately one hundred-member board of directors (figures are only available from 1969 to 1972), the proportion of Dutch directors representing member bodies increased from 11 to 14 percent. However, a higher proportion (25 percent) of the eight board members directly appointed by marketing boards or commissions for 1971 and 1972, was Dutch. An examination of member organizations gives a more accurate picture of the Dutch presence in these various commodities. In the case of the Ontario Vegetable Growers' Marketing Board (which started off as the Ontario Tomato Growers' Marketing Board), the figures from 1989 to 1999 indicate that the percentage of Dutch on the board of directors at the provincial level varied from 11 to 33 percent. The Dutch were also well represented on the provincial executive (as chair, vice-chair, and secretary-manager); the proportion of Dutch on the executive was usually higher than for membership only, although in some years (e.g. 1995 to 96) there were no Dutch farmers on the executive. At the district level, the Dutch presence on local executives was consistently strong (between 33 and 50 percent) in district 3 (Huron, Perth, Bruce, Lambton, Oxford, and Middlesex), a region that has a higher concentration of people of Dutch background (see chapter 12).

In the Grape Growers' Marketing Board, for which statistics were available going back to the seventies, long-term trends become more evident. In the late seventies, only one of six districts, all within the Niagara Peninsula, had a Dutch representative, although that person also served on the provincial board. This means less than 2 percent (one out of sixty) representation. Because that Dutch grape grower also served on the nine member provincial board, his presence at that level counts as 11.11 percent. However, there were no Dutch members at any level of this negotiating board throughout the eighties and early nineties. Then, starting in 1993, Dutch grape growers were again elected to committees in six districts, for totals ranging from one to three per year, giving district percentages ranging from 7.14 to 15.8. In 1997/98, the Dutch

[7] A few names might be German or Mennonite.

presence at the provincial board level also increases to two members out of eleven (18.8 percent), indicating the increasing involvement of Dutch-Canadian farmers even in a specialization not associated with the Dutch. What these figures do not show is that, starting in 1978, at least five Dutch-Canadian grape producers were awarded the title of "Grape King,"[8] and that two Dutch grape producers set up their own wineries.

The more recent presence of Dutch producers in both the Potato Growers' Marketing Board (Processing) and the Fresh Potato Growers' Marketing Board shows a similar trend. These two closely interconnected boards were formed in 1976 and 1979 respectively. Although many Dutch potato producers (who must have holdings of at least five acres to qualify) were members from the beginning, their involvement on the board was minimal until the late eighties, when they began holding executive positions. Since then two or three out of about nine positions on the board were usually filled by Dutch farmers. The last organization for which detailed figures are available, going back to 1948 (although only for provincial directors), is Ontario Pork, known at various times as the Hog Producers' Association and Ontario Pork Producers' Marketing Board. The first Dutch director (out of fifteen board members) appears in 1975, joined by a second in 1976, dropping back to one between 1979 and 1986. Dutch board members do not appear again until 1993 (one out of fifteen) but their numbers then increase, to two in 1994, three in 1995 and four out of fifteen in 1997. My interview data indicate that the level of involvement of Dutch hog producers on committees at the local level was much higher. I was only able to obtain a list of county directors for the period of 1971 to 72, but it contained over fifty Dutch names, including several county presidents, vice-presidents, and secretaries.

Another indicator of the level and nature of Dutch activity in Ontario agriculture is their involvement as elected representatives in general farmers' associations. Chapter 8 showed how one segment of the Dutch immigrant farm community (mainly Calvinist) set up farmers' associations based on religious principles (the Christian Farmers Federation of Ontario or CFFO) which eventually accepted non-Dutch, and non-Calvinist, members. Yet, even by 2002, the vast majority of directors of twenty-three district Christian Farmers Associations (CFA) and the executive of their provincial umbrella organization (CFFO) were still of Dutch background,

---

[8] John Marynissen was the first to receive this award in 1978 (personal communication); later recipients include Peter Buis (in 1987) and a Huisman, Vandelaar, and Bijl. See also article by John Schofield, inside the front cover of the mimeographed bulletin of "Club the Netherlands" (October, 1987).

mostly with connections to various Reformed churches. However, the majority of Dutch farm immigrants and their descendants, including those affiliated with such churches, belong to the larger Ontario Federation of Agriculture.

The Ontario Federation of Agriculture (OFA)

It is my impression that the participation of Dutch farmers in this federation has also increased over time. However, to my knowledge, there has not been any Dutch involvement in the top positions at the provincial level, compared to past leadership roles in the Ontario Farmers' Union (see chapter 9) and their domination of the Christian Farmers Federation. Indirect evidence, based on conversations with farmers of Dutch background, is that those without church affiliation, as well as members of the Catholic Church and various Protestant denominations without Dutch connections, are more likely to belong to the OFA than their Reformed (Dutch Calvinist) counterparts. Unfortunately I was unable to locate historical records of past office holders at the district level, to provide written support for these assertions.[9]

The only way to determine the recent involvement of the Dutch in district leadership positions of the OFA, is to examine the 1997 phone directory and the 2002 website of the OFA. Both contain information on regional directors and county executives for forty-nine county or regional federations. In 1997, there was a Dutch presence on the executive in thirteen areas: Bruce, Durham, Essex, Haldimand, Halton, Hastings, Middlesex, Niagara North, Norfolk, Ottawa-Carleton, Perth, Prince Edward County and Stormont. In 2002, there were no longer any Dutch members (judging by surnames) for Essex, Norfolk, and Ottawa-Carleton, although the number of county or district federations with Dutch executive members had risen to twenty-two. I suspect that Dutch participation in districts has fluctuated over time, but the figures for these two time periods are not sufficient to detect long-term trends. However, if we focus on just two regions, we can see that the participation of farmers of Dutch background in the OFA is consistent with the greater numerical importance and previous political involvement of Dutch farmers in those regions; there was a higher proportion of Dutch directors or Dutch members of ancillary committees (e.g. regional directors of commodity and rural organizations) in both Middlesex and

---

[9] An exhaustive search of the library at Queen's University, the official repository of the historical records of the OFA, failed to yield any lists of post-holders at the local, regional or provincial level.

Thunder Bay. Both of these areas saw an early influx of Dutch immigrant farmers, and their subsequent displacement of non-Dutch farmers.

DUTCH-CANADIAN INVOLVEMENT IN AGRIBUSINESS

The vast majority of Dutch immigrant farmers and their offspring who stayed in farming did not expand beyond a full-time family-farm operation, with occasional sidelines. Nevertheless, a sizable minority developed large-scale, highly capitalized operations, but without other agribusiness components (processing, distribution, or marketing). Eventually, some established quite substantial farm-related businesses, thus giving postwar Dutch immigrants an even higher profile.

Landscaping and greenhouses

The number of small nurseries, sod farms, and landscaping businesses owned by postwar Dutch immigrants in Ontario are too numerous to mention. We have already seen one example of the preponderance of greenhouse flower operations on the Niagara Peninsula. Some of these small businesses, usually with a farm component, grew into substantial agribusinesses geared to the wholesale market or for export. For example, Bill Vermeer, who started a greenhouse operation with a partner in the fifties (see chapter 5), had three greenhouse operations in the eighties, including two in Beamsville. He bought out his Dutch partner in the seventies, but formed two new partnerships to establish additional greenhouse operations in the United States.[10] Over the course of three decades, he went from building his own greenhouses to cut down on expenses, to manufacturing greenhouses and greenhouse equipment in his own factory in Grimsby for export all over the world. Although Bill had been schooled entirely in the Netherlands, his greenhouse operation was developed solely within Canada, independent of Dutch agribusiness and Dutch investment. Initially, he bought most of his equipment from the Netherlands, supplied by several immigrants based in the Toronto and Hamilton area. But Bill did not return to his old home in De Lier, in the Westland region, until fourteen years after coming to Canada. Yet, he has always read Dutch horticultural magazines and sent his oldest son, his only child who learned written Dutch, to a Dutch horticultural school for three years. Like other greenhouse flower producers in the Niagara region who came from the Westland, Bill has kept up with the latest developments in horticultural research in the Netherlands.

[10] Interview (FS) with Bill Vermeer, in his home in Grimsby, 28 November, 1989.

However, the influence has not all been in a Europe-North America direction. Bill's inventions, such as the use of curved glass, ventilators, and forced air, were later adopted in Holland.[11]

Bill Vermeer had to operate in an increasingly competitive international business operating in a global market. When it was no longer possible to compete against countries specializing in fresh flowers, Bill switched from growing carnations, bulbs and chrysanthemums to potted plants. Roses were the last to go. Starting around 1975, he started importing flowers for his own retail operations. As did so many other farmers in Canada, he also became increasingly dependent on new sources of labour. For the first ten years, all of his workers were of Dutch descent. By 1990, the majority of his largely year-round labor force was from Vietnam and Central America. At that time, only his top employees were of Dutch background, including members of his own family.

The Vermeer case illustrates the value of prior Dutch experience and learning, plus ongoing contact with the Netherlands, in developing a successful agribusiness in Canada. Being young and adaptable – he was twenty when he came to Canada – is also a factor. However, not all young Dutch immigrants who set up agribusinesses in Ontario had previous training in the Netherlands. A good example is the Dol brothers of Cookstown,[12] who developed a small sod operation into a large company that seeded, and/or sodded, Ontario's highways. Initially, they were strictly a sod operation, mainly supplying landscapers, but they started specializing in sod for highways, and hydro seeding, which involves spraying a combination of seed, paper mulch, and fertilizer. In the early seventies, the Dol brothers signed a contract with the Ontario government, and their business expanded. Their company, which peaked around 1990, was divided into various divisions (laying sod, bulldozing and dousing).

Processing

We have seen examples of Dutch immigrant farmers who bought feed mills or elevators, such as Bill Van Wely (chapter 5) or Harry

[11] Interview (MF) with Bill Vermeer, in Grimsby, 30 March, 1994. Bill Vermeer is not the only success story of a farm immigrant who expanded from agricultural production into the manufacturing of agricultural equipment. Jerry Brouwer, who settled in Keswick, is another well-known Dutch sod producer, who invented a sod harvester. He even set up a plant in Ohio in 1994, specializing in the manufacture of turf equipment.

[12] Joe Dol, one of several brothers who ran this business, was mentioned in interviews with other Dutch-Canadian immigrants, with both farmers and non-farmers, who knew the Dol family. I found out about their humble beginnings in an interview with a Dutch immigrant couple who knew the family when they were living in Aurora in the fifties.

Vanderzanden (chapter 6). For a much larger agribusiness that operates globally, we need to turn our attention to the Great Canadian Bean Company established in 1978 by Peter Twynstra, the same person who was leader of the Ontario Farmers' Union in the sixties (see chapter 9). An early member of the White Bean Marketing Board (he served between 1968 and 1970), Peter later became one of the five biggest processors and exporters of white beans and other edible beans, buying from farmers all across Ontario. Almost every edible bean producer I came into contact with had heard of, or dealt, with Peter Twynstra, although farmers not of Dutch descent might be unaware that his name is Dutch (actually Frisian). Besides his involvement in marketing and processing, Peter and his wife, Dinie, specialized in producing pedigree edible bean seeds supplied to farmers.

The amount of land cultivated by their family farm corporation, Twilight Acres Farm Ltd., increased from several hundred to 2,500 acres, and more employees were hired. In 1995, they constructed a separate building for their office, and expanded further with the purchase of a new processing plant in London and a joint venture in Quebec.[13] In 1997, they had storage capacity for more than 800,000 bushels, and six satellite receiving stations, as part of an increasingly export-oriented business shipping beans to thirty-five countries. The Great Canadian Bean Company has become a recent player in the new global economy.

Another example of a gradual move from farming to processing, though on a smaller scale, comes from the wine industry. A number of Dutch immigrant farmers on the Niagara peninsula had already switched from more diversified farming, including hog raising or dairy farming, to growing grapes. Two of these growers, established wineries.

The first, Marynissen Estates, illustrates the transfer of skills originally acquired in the Netherlands. John Marynissen was born and raised on a medium-sized farm in the western part of Noord Brabant. He started working on his dad's farm after primary school and was away from home for most of the war, after joining the underground resistance movement. At the end of the war, at the age of twenty, he took accounting courses and obtained his *middelstand diploma.* He received further practical training working for a garage, where he learned to weld, after obtaining a driver's license. He also drove Italian and French buyers to see cattle available for export, while working for a cattle dealer. This exposure facilitated his adjustment to Canada when he emigrated in 1952, at the age of twenty-eight.[14] John started off working on a fruit farm on

[13] See Van der Mey and Mol, *The Dutch Touch*, 144.

[14] Interview with John Marynissen, at his farm in Niagara-on-the-Lake, 18 July, 2000.

the Niagara Peninsula. His wife also picked fruit and pruned. After three months, John earned more money welding in Hamilton and driving a truck for a fruit and vegetable distribution firm. Initially he was not planning to become a farmer, but his brother, who joined them a year after coming to Canada, did want to farm. They pooled their resources for a down payment for a 27-acre fruit farm with a house and a big barn, and raised hogs on contract. John continued to work out while his brother stayed home to look after the farm and did pruning for fruit farmers in the winter. They split up the partnership after a year, and John's brother then moved to Listowel to buy a dairy farm. John continued to combine working off the farm with raising hogs, together with wife. His off-farm jobs included assembly-line work at General Motors, and more welding. John Marynissen started working less in the city, starting in the late fifties, and became more enterprising as a farmer. The Marynissens made a switch to grape production in 1970. An open, fifteen-year contract with one of the new wineries, Château Gai, provided financial stability, and in 1978, John became the first Dutch grape producer to be chosen as "Grape King" for producing quality Chardonnay.

With ratification of the North American Free Trade Agreement (NAFTA) in 1990, John decided to set up a commercial winery, Marynissen Estates, by converting his old hog barn. Like other small wineries, Marynissen Estates found a market niche by producing specialty wines using quality grapes, while older, more established wineries were mixing different varieties of grapes and importing grape juice from the United States. They also opened a boutique to attract the many visitors who come to Niagara-on-the-Lake each summer,[15] and their winery got its first international exposure when a Dutch journalist featured their winery on Dutch TV. John has also won many awards for his wines, both before and after becoming a commercial producer. John's prior work experience, entrepreneurial skills, and broader exposure to people and places allowed him to make the right business decisions, culminating in his family-run business.

Another Dutch-Canadian immigrant farmer, who also started a winery (Stonechurch Vineyards) in 1990, made a similar transition from one form of farming to another, to remain viable. Lambert and Grace Hunse, who emigrated in 1950, originally had a fruit farm, and changed crops until adding the new varieties of grapes in 1972. Originally from Beilen, in Drenthe, they ended up producing an award-winning ice wine, with

---

[15] I obtained this information on 17 June, 2002, from a website that is no longer active.

their own grapes.[16] They imported a wine master from Germany.[17] Like the Marynissens, they worked hard together as a couple over several decades, and their children also joined the business. Their son, Rick Hunse, and his wife Fran took over the family business in 1995. Today their wines are available at 150 outlets of the Liquor Control Board of Ontario and they also ship to distributors as far away as Taiwan. Likewise, Sandra, their youngest daughter, is vice-president of the company, while his son Ron looks after the grape growing and supervises their three Mexican workers.

Collectively, Dutch immigrant farmers have had a greater impact on, or achieved a better reputation in, some sub-fields of agriculture, than others. Thus in Ontario Dutch processors are better known in the greenhouse industry than they are in wineries or other branches of agribusiness. Similarly, in the field of agriculture, Dutch greenhouse flower producers have a higher profile than their counterparts in cash cropping, or broiler operations, regardless of their respective numerical importance. Yet the high visibility of the greenhouse operators, combined with the reputation of Dutch dairy farmers, is sufficient to make most people convinced that all Dutch farmers are rich and successful. Even in sub-fields where they do not have a strong numerical presence, but where they have assumed leadership positions as processors or political representatives, the presence of even a handful of Dutch-Canadians has a multiplier effect in the creation of a positive group stereotype.

[16] VanderMey and Mol, *The Dutch Touch*, pp. 145-46.
[17] I obtained this information June 18, 2002, from a website that is no longer active.

XIV. The European-Dutch Corporate Connection

The Netherlands used to have an economic elite that derived its profits from overseas investment in commercial agriculture. Dutch colonialists set up plantations and carried out land reclamation in places like Java and Guyana. Dutch companies were later involved in surveying and land development in North America long after they had lost their possessions there. More recently, just prior to the Great Depression, a Dutch mortgage corporation, the Netherlands Investment Company, did business in the Canadian province of Manitoba and collaborated with North American partners in financing the reclamation of the Pit Polder in British Columbia, also in Canada, to earn dividends by selling land to Dutch farmers. After World War Two several thousand acres of the Pit Polder were again drained and developed under Jan Blom's direction.[1] Less well known but quite relevant to this study is the presence of Dutch capital in Ontario. The vast majority of rural Dutch immigrants who came to Ontario in the 1940s and 1950s worked for Canadian farmers. When they started buying their own farms, they had dealings with banks, farm machinery dealers, and feed mills owned by Canadians not of Dutch descent. Yet, some Dutch immigrant farmers did have contact with representatives of Dutch capital as a result of a renewed interest by Dutch investors in Canada after the Second War World. These Dutch-owned agribusinesses did not penetrate very far into the Canadian, much less the broader North American market until the mid-eighties, and then only as a supplier of greenhouse equipment to already well-established farmers of Dutch descent. By then, Dutch immigrants had already made their impact on Ontario agriculture. Not only were they starting to be known as successful farmers, capable of turning run-down farms into viable commercial operations, a sizable minority had expanded and developed agriculture related sidelines. In the early fifties, no one would have predicted that international agribusinesses based in the Netherlands could only survive in North America because of the expertise of postwar Dutch immigrant farmers.

---

[1] Interview (FS) with F. Jensma in Nepean, November 18, 1989. See also VanderMey, *To All Our Children*, p. 449.

THE HOLCAN COMPANY

I encountered the name Holcan during my first trip to the Hamilton area. When this name appeared again during research in Elgin county, I realized I had to learn more about this early Dutch-Canadian company, which had disappeared long before I initiated my study.

Holcan (Holland-Canada Rural Development and Investment Company), a joint business venture started in 1954,[2] was the brainchild of a group of men who met in Canada. One was a graduate of the agricultural university in Wageningen. He had prior experience with plantation investments in Java (Indonesia) and later worked for the HVA (*Handels Vereneging Amsterdam*). Another partner was an agricultural engineer who worked under Abe Tuinman at the Dutch embassy. They obtained financial backing from the Mercantile Bank, an offshoot of the *Nederlands-Indisch Handelbank*, based in the former Dutch East Indies.[3] Jan Heersink, a post-war immigrant became the key player. A farmer's son, Jan grew up in Hummelo (near Steenderen), in the Achterhoek region. In 1945, he served as the youngest mayor ever in the Netherlands. Nine years later he and his family moved to Canada with several other families from Steenderen.[4] He had already made a trip to Canada on behalf of the Dutch Reformed Church (*Hervormd*) and was recruited as director of immigration for the Reformed Church of America upon arrival. An investment advisor and agricultural engineering consultant, Heersink was appointed honorary Dutch vice-consul in Burlington and developed an extensive network of connections with Dutch professionals and businessmen in Toronto.[5] He and the two other Holcan founders recruited a fourth man then acting as advisor for the Christian Farmers Federation of Ontario (see chapter 8). While the four men were supposed to become equal directors, Heersink took control and within a decade he was in charge. In 1965, he hired a woman trained as a secretary in the Netherlands, to be his receptionist and do paperwork for the consular office and several businesses, including Holcan.[6]

Holcan was involved in a combination of farm development, land speculation, and cattle import. It bought farms, which were in turn sold

[2] The story of this company was pieced together from five interviews with people directly involved as employees, business partners or customers. There were several other passing references. I was unable to access documentation.

[3] Additional capital was provided by the AMRO (Amsterdam-Rotterdam) Bank and Dutch government sources.

[4] For a short biography, see Van der Mey, *To All Our Children*, 282-83.

[5] In Toronto, he was one of the co-founders of the Luncheon Club, for middle and upper class Dutch immigrants. See Schryer, *The Netherlandic Presence*, 187.

[6] Telephone conversation (FS) with Ann Klein, April 10, 1997.

to immigrants. Holcan held the mortgages and provided credit, seeds, and farm equipment. One of the original partners, who became a full-time employee, visited numerous farms scattered between Hamilton and Windsor. He quit in 1957, disagreeing with the way the company treated the farmers,[7] but also to develop a new career. Nevertheless, Holcan continued to operate in different areas of Southwestern Ontario. They exported cattle, including stud bulls, and imported stockers (cattle destined to be butchered). A woman whose father did business with Heersink remembers how the crates for shipping cattle held in quarantine were stored upstairs in their barn.[8] Holcan got the twenty percent return expected during the first five years of operation.[9] However, business floundered after that. According to those who had dealings with him, Heersink was friendly, outgoing, and energetic, but he did not have much experience in the world of agriculture. One of the former agricultural attachés in Ottawa told me Holcan experienced financial difficulties because they did not anticipate that their Dutch borrowers would be unable to make their payments in time.

The troubles experienced by the Holcan Company could have been due to a policy of extending long-term mortgages to immigrant farmers when most of them were struggling to get established. A company dependent on financing from international banks could not be as flexible as the retired farmers willing to help Dutch immigrants with private mortgages at low interest rates and flexible payments. Canadian farmers were more forgiving if an immigrant farmer was unable to make payments on time (see chapter 5). Several businesses run by Holcan, such as a feed mill in Burford (in Brant County), folded. Holcan had to sell this mill and a farm used to hold heifers. The failure of such ventures demonstrates the problematic nature of concerns run by people not familiar with North America agribusiness, as shown by the attempt of an even bigger, completely Dutch-owned corporation to enter the North American market.

PROVIMI AND VICON

The Provimi name dates back to the 1930's, when a Dutch entrepreneur, Harry Bonda, set up a company specializing in producing pig and chicken feed by recycling by-products of dairy and fishery industries.

---

[7] I was told by someone in Toronto whose father volunteered for one of the immigration societies, of many farmers coming to their house with horrific experiences with Holcan.

[8] Interview with a farm woman in Oxford County, March 1995.

[9] This was mentioned by one of the partners in this venture.

The Bonda family owned a holding company (the Bonda Group) in the Netherlands which included a textile factory and a pearl essence[10] factory. The company was well established in the Netherlands when they began expanding into other European countries, the Americas, and elsewhere in the world. Their holdings in Nova Scotia included a fish meal factory, where they made a feed rich in proteins, vitamins, and minerals (hence the name Provimi) for distribution to other locations across Canada. They also brought in trawlers from Holland to catch fish for the plant. Harry Bonda owned a farm in Nova Scotia with a mansion built by Americans, where he and other members of his family would spend several weeks each year. The farm enterprise, which employed twenty-five people, had hogs, pulpwood, and greenhouses with strawberries and rosebushes. In 1959, it became apparent that the Bonda farm was operating at a loss, and required another manager.

Harry sent his brother, Jan, to Canada to look for a manager. He had heard about a young man who had moved to Ontario in 1947. Jan knew the young man's father, a cash-crop farmer in the Haarlemmermeer Polder. So he flew to Canada to look up Martin Verkuyl, who was operating a dairy operation in Utica, near Port Perry. Martin was surprised when Jan Bonda appeared on his doorstep, and accepted an invitation to take a trip, together with his wife, to decide if he wanted to become the manager of their Nova Scotia farm operation. The offer of a job with a salary greater than the gross income of his farm was too good to turn down. His first decision as manager was to add a dairy herd, and during the first four years their losses declined to the point where they were breaking even. But the long-term prospects of the Bonda Company were not promising, and Martin wanted to move back to Ontario.[11] In 1963, he asked for a transfer to Woodstock, Ontario, where he helped to set up a Provimi outlet, together with Mr. P. Bartelse, Bonda's business associate in charge of feed operations. Martin's job was to persuade farmers, most of whom lived in the vicinity of Burgessville and Norwich,[12] in Oxford County (London region), to buy their feed from Provimi. Initially Provimi distributed fishmeal trucked in from Nova Scotia, until the company established a plant in Ontario to remain competitive. By that time, Martin was no longer working for Provimi. A farmer at heart, he bought a farm near Hickson in 1965. In 1971, Provimi was sold to Central Soya, a company that was taken over by

---

[10] Extracted from the skin of herring and other fish, pearl essence is used in the making of cosmetics, paints and other products.

[11] Interview (FS) with Martin Verkuyl, on his farm near Hickson, 23 November, 2000.

[12] Interview (FS) with Jerry Bartelse, in his office in Cambridge, 6 August, 1998.

an even bigger international corporation in 1987. Their animal nutrition division, which employed 6,629 people in 2004, still operated under the name Provimi.[13] However, while the name Provimi survived in North America, the Bonda enterprises, including the farms and fish plant, did not. Prior to hiring Martin Verkuyl, Bonda managers were recruited from the Netherlands. Those who were good, and capable of adapting to North American forms of farming and doing business, left to set up their own farms. Others returned home. The Bondas pulled out of Canada, although their feed division survived and has since expanded into Belgium and France. However, one other Dutch agribusiness managed to capture part of the Canadian market for a somewhat longer period.

Vicon is a Dutch-based multi-national corporation that manufactures farm equipment. Their products are found in farm retail outlets throughout North America. The owners and founders, three brothers by the name of Vissers, started off as independent blacksmiths. By 1960, they were a sizable company and expanding into Australia, Japan, and India. Their specialties were hay-rakes, potato planters, fertilizer spreaders, and power harrows. They had not yet, however, tackled the North American market. One of the brothers made a deal with Massey-Ferguson, a Canadian farm machinery company, to use Vicon's hay-rakes, but with the Massey-Ferguson label and colour. Since the rakes proved popular, Vicon started selling the same rakes with a different color under its own label. After first entering the North American market as a wholesaler in the early sixties, the Vicon Company ran into difficulties. Despite previous experience in other parts of the world, they were unfamiliar with the consignment system, where the manufacturer is paid only for equipment that is sold. Soon there were thousands of unpaid for Vicon rakes in dealerships scattered all over North America. They fired their Dutch salesman, and looked instead, for a Dutch-Canadian familiar with North American farming techniques and with prior experience in sales. They also wanted someone trustworthy. Coincidentally, one of the Vissers brothers was married to the sister of Martin Verkuyl, who had earlier worked for Provimi. Several of Vicon's senior management staff knew Martin, so they suggested offering him a job.

Martin Verkuyl was just as surprised at receiving a phone call in the middle of the winter in 1966, as he had been when someone showed up seven years earlier. He was just setting up his farm operation in Hickson. He and his wife were ready to put down roots and Martin

[13] The international company has various websites under the name "provimi".

was becoming involved in the Christian Farmers Federation of Ontario (see chapter 8). Yet, with four teenagers at home, and the increasing costs of farming, his budget was somewhat stretched.[14] The job offer was attractive, but the idea of traveling all over North America was unappealing. To sweeten the deal Vicon let Martin take time off each year to plant and harvest his crops. He sold his sows and paid others to do custom work on the farm when he could not be there. Martin was good at his new job, but traveling took its toll and a recession in the late sixties resulted in cutbacks in the equipment industry. He wanted to get back to farming full-time and his oldest son, now eighteen years old, was ready to farm. So in 1970, he quit his job. They started growing more corn, edible beans, and wheat, and returned to hog production.

The entry of Vicon into the North American market illustrates the close contacts and interdependency between Dutch-based agribusiness and postwar farm immigrants. Young and middle-aged Dutch farm immigrants, most of whom ended up in Ontario, had the right combination of Dutch technical training and North American experience to serve as managers and representatives for Dutch companies wanting to do business in Canada and the United States. However such agribusinesses with strong Dutch connections had a competitive advantage only if they could offer expertise, services, or products not yet available in North America. This is the case for the milk-fed veal industry.

THE MILK-FED VEAL INDUSTRY

Prior to the Second World War, Canadian consumers used the term "veal" to describe anything from a newborn to a 1,200-pound bull calf, which was grain-fed (also known as "red" veal). In restaurants it was called "baby beef". Red veal was a sideline for beef farmers, who might continue to raise calves for regular beef, since dealers often paid poorly for young cattle. The bull calves from dairy cows were considered almost worthless. The superior texture and taste of the meat of such calves – with proper care and feeding – was virtually unknown. This milk-fed veal business (once known as "white" veal) started in the Netherlands in the fifties when Dutch entrepreneurs saw an opportunity to use surplus milk powder shipped from North America as a milk replacement to feed calves. In Ontario, only farmers already exposed to the veal industry in Europe were interested in this type of farming. In the late sixties, a group of Dutch farmers obtained the special feed for their white veal operations from the Provimi outlet in Woodstock. Their best customers

[14] Interview (FS) with Martin Verkuyl, at his farm near Hickson, 23 November, 2000.

were other European immigrants, particularly Italians, Portuguese, and Jews, groups that prefer this type of meat. However, the technical challenges posed by white-veal production have made this one of the most vertically integrated agricultural specializations in Ontario and Quebec, presently dominated by a company headquartered in Cambridge. The career trajectory of its founder, Jerry Bartelse, can provide further insights into the links among Dutch investment, Dutch expertise, and immigrant farmers.

We came across the name Pieter Bartelse, in connection with the establishment of a Provimi feed outlet in Ontario. His son Jerry (whose father was the associate of Harry Bonda mentioned earlier) first came to Canada in 1966, at the age of seventeen. That summer he worked with one of the Provimi salesmen at its Woodstock operation, and befriended several Dutch farm families in the area, some with sons around his own age. In 1969, at the age of twenty, he returned to Canada, this time on his own, as an immigrant, ending up in Woodstock. Up until that time, Provimi had provided feed for chickens and pigs, but Jerry's job was to establish a veal program whose customers included the same farmers he had met earlier. Two years later, when Central Soya bought out Provimi, Jerry returned to Holland, where he enrolled in a school with an applied program in animal nutrition (*praktijkschool*).[15] Around the same time, an increasing number of immigrant farmers were getting involved in the newly established white veal industry. Little did they know that the young Dutchman who worked for Provimi would end up controlling most of that industry in Ontario.

The First White Veal Shippers: Trials and Tribulations

Several Dutch immigrant farmers started white veal operations in the area between Woodstock and Brantford: Jerry and Harry Veldjesgraaf, Lloyd Moesker, and Kees Verbeek. They had been in Canada for more than a decade, mostly doing off-farm work or renting farms before buying their own. Lloyd, who had bought and sold several farms before moving to Burford,[16] had prior experience in dairy farming and raising pigs, while Jerry Veldjesgraaf managed a turkey farm before working for himself, raising pigs.[17] These men, who came to Canada around the

[15] Interview (FS) with Jerry Bartelse, in his office in Cambridge, 6 August, 1998.

[16] Interview (FS) with Lloyd and Verdina Moesker, at their home near Eastwood (near Woodstock), 8 October, 1998.

[17] Interview (FS) with Al and Alice Veldjesgraaf, on their farm near Burgessville, 10 March, 1995.

same time, were from the Veluwe region in the Netherlands. They all remodeled their barns to raise white veal around the same time that Provimi came on the scene. Provimi sold them the feed. But, if a calf became sick because of the feed, they found that no one from the University of Guelph could help them, and even recommendations from Provimi, based on Dutch experience, did not always work. They had to experiment to find the right proportions for the milk supplement they used to feed their calves. Wives, who spent hours nursing the delicate calves, played an important role, and later ran their own white veal operations. By 1973, a lot more people were growing calves for white veal and Lloyd and Jerry started supplying other farmers with calves. They took turns driving a truck to Montreal, to buy the many bull calves shipped there from all over that province. When the men were on the road, their wives took care of their farms with the help of their children. Jerry Veldjesgraaf, who had a farm in Norwich, was also busy in organizational efforts. In order to co-ordinate their sales to the packers buying the end product, the veal producers formed an association and Jerry became their president when they had around fifty producers between Norwich to Belleville.[18] However, the producers were not able to form a united front against the packers. The white veal business, while potentially lucrative, continued to be a risky and competitive operation.

Lloyd and Jerry both experienced difficulties. Jerry met and did some business with Jan Heersink of the Holcan Company (see above), who also started raising veal and acting as an advisor to the White Veal Association. They became good friends, but had a falling out over a business deal.[19] Lloyd then developed a close relationship with the new feed company that took over from Provimi in 1971. Central Soya wanted to experiment with mixing soya feed and milk powder for white veal operations, so they rented Lloyd Moesker's barn and paid him on commission to continue working there. However, with so much activity in the barn, the calves were not able to develop properly, so Lloyd started working on his own again. His partnership with Jerry Veldjesgraaf dissolved, and for a while he sold feed with another group. Simultaneously, Jerry Bartelse, who returned to Canada, was busy setting up his own feed company.

---

[18] One person gave me an estimate of 150 and someone else said there were thirty-five white veal producers. Even factoring in the high rate of attrition between 1969 and 1980, I used a more conservative number. I was unable, however, to find any written records to substantiate my guess.

[19] Interview (FS) with Al Veldjesgraaf, at his farm near Burgessville, 10 March, 1995.

Delft Blue/Grober

The reader last heard that Jerry Bartelse had gone back to the Netherlands to go back to school. That was in 1971. In 1974 he married a teacher, and four days later they boarded a boat destined for New York, from where they would then drive to Canada. Originally Jerry was supposed to have helped set up a veal operation in Costa Rica for his family' business, but that fell through.[20] During the boat trip to Canada, he met Aat Groenenveld, one of their business partners, who had been stationed in Watertown, Wisconsin. Aat's father had earlier established a Promivi outlet in the States that continued as a separate operation after its Canadian counterpart had been sold to Central Soya. He and another partner and Dutch milk technician, Guy Tober, had incorporated under the name Grober (a combination of letters from their surnames). Aat persuaded Jerry to use his connections in Canada to set up a division of Grober in Canada.

When they landed in New York, Jerry and his wife bought a car, crossed the border to Canada at Windsor and started looking for a place to locate. By nightfall they had reached Kitchener. The next day, after taking a wrong turn on the way to Toronto, they drove through an area with townhouses for rent, and decided to live there. They have been in the Kitchener-Waterloo-Cambridge area ever since. Jerry Bartelse started a feed plant producing veal feeds and milk replacers for other animal species in Cambridge under the name Grober, and re-established his contacts with Dutch veal farmers. That is when he also met Jerry Veldjesgraaf for the first time. A year later, in 1975, Jerry Bartelse established his own company, Delft Blue, and bought the feed mill and farm in Burford previously owned by Holcan. Initially, Jerry dealt strictly in feed, but in 1979 Delft Blue became a company specializing in calf production and meat processing. Grober specialized in providing animal nutrition.[21] He now ran two, interconnected companies, which both expanded, and then established Delft Blue Farms to produce more of his own veal. The success of his companies, which today supply veal to supermarkets all over North America, can be attributed to his ability to integrate a packing facility and feeding program, and maintain control over farm production. By 2005, his three companies, known as the Grober/Delft Blue group, added an additional production barn in Elba, New York, called Provitello. Their Delf Blue Inc. now has a US office

---

[20] Interview with Jerry Bartelse (FS), at his office in Cambridge, August 6, 1998.

[21] Gijs Tober, who later moved to Pennsylvania, still owns 50% of the company and runs its U.S. division. See Van der Mey and Mol, *The Dutch Touch*, 60.

in Green Bay, Wisconsin.[22] Jerry is a prime example of the owner of a small (by global standards) agribusiness that is adapting to, and surviving in, an increasingly competitive market. However, he has also displaced the many smaller veal producers and processors that used to operate in Ontario.

Jerry starting growing and packing milk-fed veal when the major meat packers (for whom white veal had been a sideline and loss leader), discontinued that part of their business due to declining demand. The white veal market became depressed in the seventies and eighties when many butchers catering to ethnic constituencies disappeared, and smaller and medium-sized supermarkets that sold white veal were absorbed into bigger chains. Smaller veal producers had difficulty staying in business because they lacked the necessary capital to expand and upgrade at a time when new forms of feeding and other technical innovations were making production more competitive. Unlike red veal, white veal requires substantial capital layout for specialized facilities and veterinary services. Pressures from the humane society resulted in new regulations requiring the building of larger stalls. Consequently, most dairy farmers who had had white-veal operations dropped that sideline, and producers specializing in white veal sold out or went bankrupt. That is where Jerry stepped in.

The veal producers we encountered earlier had gone bankrupt when Jerry Bartelse offered to help get them back into business, by working on contract for him, or becoming farm managers. He also set up his own barns, and by 2004, Jerry had built state-of-the-art facilities, with open pens, automatic feeding equipment and temperature control. But many of his calves were still bought from dairy producers, and eighty percent of the calves at his facilities in Cambridge come from Quebec, which has a strong, government-supported, dairy sector, whose dairy farmers have displaced Dutch immigrants as the main source of calves. At the other end of the production cycle (from farm gate to the consumer), Delft Blue's three meat processing plants now ship most of their veal as a counter-ready product, compared to a decade ago when their meat production was in carcass form.[23]

In 1980, there were thirty white veal producers in Ontario, but by the turn of the century, there was only a small number left: a Portuguese producer in Paris, and a few Dutch and Italian farmers. While an independent white veal operation can still succeed in a family farm setting, it requires at least 400 calves. However, even these producers buy their

---

[22] See http://www.provitello.com/ (2005).

[23] See http://www.delftblueveal.com/company.htm (2005), and linked web sites.

feed from, and sometimes ship their finished calves to, Jerry's outfit in Cambridge. They get technical advice from a nutritionist who is an employee of Grober. However, the Delft-Blue/Grober group owned by Jerry has not lost its Dutch connections. Jerry still deals with a Dutch company that produces vitamins for animal nutrition, and his company headquarters in Cambridge has Dutch and well as English-language farm magazines on a rack in the main waiting area of their office. And what could be more Dutch than the name of two of his companies (Delft Blue)?

Overall, postwar Dutch immigrants have been more successful in adapting than European investors who entered the North American market. The few Dutch agribusinesses that operated in Canada prior to 1980 never became a dominant force in Ontario. In contrast, a handful of young immigrant farm couples, or the offspring of older farm immigrants, with no contact whatsoever with Dutch agribusiness, became trendsetters and leaders, operating at different levels, often with international connections. We have also seen how the relocation, and subsequent transformation, of a Dutch agribusiness (Bonda/Provimi) into a Canadian one, was predicated on the migration of one of its family members. Jerry Bartelse, whose father worked for the Bonda Corporation, used his father's European business experience and modified it in a North American setting.

Not until the end of the twentieth century, did an already thriving Dutch-Canadian greenhouse industry in Southwestern Ontario, particularly in the Niagara region, facilitate the re-entry of a large-scale European-Dutch agribusiness into the North American market. Priva Computers Inc., a trans-national, Dutch-based, manufacturing company specializing in high tech equipment for the increasingly more automated greenhouse industry, first established an office in British Columbia in the late eighties. In March of 2001, they build an office and manufacturing complex in Vineland Station, on the Niagara Peninsula, to serve both the Canadian and American markets.[24] As a corporation based in the Netherlands, it is likely to reinforce the already established reputation of the Dutch in Ontario. Yet, unlike its predecessor in the fifties and sixties, its ability to consolidate its position in the broader world market by doing business in Canada is closely linked to the existence of a more mature, though now much smaller, contingent of both new Dutch farm immigrants and the descendants of postwar Dutch immigrant farmers in

[24] *Floral Passion*, p. 241.

Southern Ontario. European-Dutch corporate connections demonstrate how global forces shape and alter local social realities, yet they are in turn the product of what happens in communities and regions in various parts of the world. The presence of Dutch-based international agribusiness corporations in the Canadian province of Ontario also illustrate the important role of networks of personal connections that link old friends and relatives on both sides of the Atlantic.

XV. The Newcomers

By 1988, Dutch immigrants who came to Canada as adults just after the war were living in retirement homes or had passed away. Most younger farmers, especially those born in Ontario, were more "Canadian" then "Dutch", and new farm immigrants were a rare occurrence; the number of Dutch dairy farmers arriving between 1980 and 1987 was three per year.[1] This changed in 1988, when farm families from the Netherlands again started appearing in Ontario. The majority bought dairy farms, but they also include greenhouses producers and other types of farmers. Between 1988 and 1999, about 700 Dutch farmers came to Ontario, and at the time of writing, they were still arriving in full force.[2] Their presence both reinforced and modified the positive stereotype of the successful Dutch farmer, especially in the eyes of those not of Dutch background.

REASONS FOR RENEWED MIGRATION AND THE LOGISTICS OF MOVING

Larger-scale immigration from Europe dwindled to insignificant proportions by the early seventies. With increasing prosperity, most countries in Western Europe saw an influx of foreigners rather than an exodus of nationals. Workers from Northern Africa, the Caribbean, and the Near East came to work in the factories and farms of Germany, France, Britain, and the Netherlands. Living standards surpassed those of North America. Dutch farming, while still intensive, became highly mechanized. It no longer made sense for people to leave. Canada, which welcomed people from all over the world, likewise experienced a shift in immigration patterns, and its cities became more diverse. In 1978, the government introduced a new category of business immigrants, including the self-employed. Those willing to invest money were given preference,

[1] See Art Wolleswinkel, "Dutch Dairy farmers in Ontario: Reasons for an evaluation of emigration", (Research report, Department of Agricultural Economics, University of Guelph, 1999), 9.

[2] The Wolleswinkel report mentions between 400 and 600 dairy farmers alone. My calculation for all types of farmers is based on estimates provided by two real estate agents who sell farms to new immigrants.

resulting in the arrival of many entrepreneurs and investors, particularly from Hong Kong. The new policy was also designed for people willing to set up farm enterprises. However, it would take another decade before Canada once again became an attractive alternative for Dutch farmers.

While Canada's changed immigration policy made it easier for prospective Dutch farm immigrants to relocate, they only started returning in the late eighties, when they began to feel the impact of new policies in the European Community affecting the disposal of liquid manure, wetland preservation, and farm expansion. Hog producers, already paying for manure disposal liked the Canadian policies allowing them not only to produce more hogs, but also get paid for spreading manure on neighbors' fields.[3] The increasing price of Dutch farmland provided another incentive for moving to Canada. Another incentive to emigrate was the introduction of a milk quota in the Netherlands in 1984.[4] As the value of quota increased, dairy producers, with limited room for expansion, started looking seriously at moving to Canada, where they could start a larger operation with the proceeds from the sale of quota. Urban expansion resulting in the expropriation of farmland also put money in the pockets of Dutch farmers interested in relocating.

FARM REAL ESTATE AGENTS

The logistics of moving to Canada and finding a farm were quite different in the nineties than in the fifties. Information regarding farms for sale and support services was provided by several Dutch-Canadian real estate agents, many of whom are themselves farmers or former farmers. Such agents, with established links to their Dutch counterparts, regularly fly to Europe to put on slide shows and distribute promotional brochures. They arrange tours to different regions in Canada. The typical new farm immigrant makes several such trips, flying back and forth to Ontario and Alberta, before making up his mind. Once decided, the real estate agent provides the support services, once delivered by fieldmen and immigration societies, including arrangements for the unloading of the container, transfer of quotas, and opening a Canadian bank account. Initially, few salesmen in Ontario catered to Dutch farm immigrants. Prior to 1988, business was very slow, but as Dutch farmers started to come to

[3] Karl Froschauer and Jan Rath, "The Unusual Case of Dutch Immigrant Farmers in Canada: A Recent History of Self-employed Dutch Business Immigrants" (Paper presented at Fourth European Social Science History Conference, March 2, 2002 in The Hague, The Netherlands), 8.

[4] See Art Wolleswinkel, "Dutch Dairy Farmers", 22.

Canada again, more Dutch-Canadian real estate agents familiar with the world of agriculture sold farms to new immigrants. By the end of the nineties, the emergence of more real estate companies and agents increased competition, although agents tend to be first-generation Dutch-Canadians capable of communicating in both Dutch and English. Direct farm experience is essential.

Vanderzwaag-Den Hollander

The first real estate broker serving Dutch farmers was a man in Hamilton not of farm background. Ben Vanderzwaag started the farm component of his business in 1969, when few farmers were coming from the Netherlands. Some Swiss farmers were then thinking of relocating to Canada. From the beginning, Ben relied on several Dutch-Canadian salesmen who were, or had themselves been, farmers: Hank den Hollander, a chicken farmer in Kirkton; and Andy Veldman, who owns a dairy farm in Embro, Oxford County. Andy came to Canada in 1954, as a bachelor, with a *landbouwdiploma* and started a hog operation in 1958.[5] However, when he started working for Ben's realty company, business was slow. Andy Veldman left to joint Remax Realty. Hank left the real estate business altogether to work for Shurgain, selling feed. During the eighties Hank started selling farms again, this time as a full partner in the completely Dutch-Canadian owned Vanderzwaag-Den Hollander Realty. When business picked up in the mid-eighties, this realty took on more salesmen, including Harry Blijleven, who became manager. Harry came to Canada in 1948, at age eleven, with his parents and the rest of their family of five. In 1961, he took over their Canadian home farm, a dairy operation near Caledonia (Lake Erie region). Twenty years later he was running a farrow-to-finish hog operation and cash cropping. Harry sold his land to a new immigrant from Germany, and then studied for his real estate license, and joined Vanderzwaag realty. Since he was unsure of this new business, he continued cash cropping on land leased back from the German landowner.[6] Gradually the real estate business picked up.

By 1990, Andy Veldman had brought around one hundred new Dutch farmers to Ontario. Throughout most of the nineties, Veldman (for Remax) and Blijleven handled the bulk of the farm sales to new immigrants. By the turn of the century, Vanderzwaag-Den Hollander, the only sizable Dutch-Canadian realty disappeared from the scene when it

[5] Interview (FS) with Andy Veldman, at his farm near Embro, 7 July, 1990.

[6] Interview (FS) with Harry Blijleven, at his home in Woodstock, 19 April, 2002.

was taken over by Royal LePage. However, Blijleven stayed on as their chief salesman. In an increasingly competitive environment, these real estate salesmen offered assistance with introductions to farm associations, applying for a social insurance number, and car or equipment purchases. At the same time, the entry of new independent, Dutch-Canadian brokers was putting pressure on realty companies. Marius Verbeek and Gerry Breemen, who has a beef farm and pigs, were still selling farms to newcomers from the Netherlands in 2002. They both started their careers in real estate with Vanderzwaag.

ADAPTATION TO LIFE IN CANADA

The new immigrants are quite different from their postwar counterparts. They are well educated and have sufficient capital to buy a farm and all necessary equipment prior to arrival. Many speak excellent English and know about Canadian regulations, soil conditions, and heat units. They only buy land in areas already ideal for agriculture, and not likely to experience rapid urban expansion, such as the Oxford County (especially the Woodstock area), and Perth, both in the region of London. Like their predecessors, they not only look for suitable areas, but also want access to good schools for their children and, in most cases, the church of their choice. Those who are Calvinist have a wide choice of churches run by denominations with Dutch connections. Only members of the Dutch equivalent of the Netherlands Reformed Congregations have more restricted options, since only the area around Norwich has churches of that denomination. Like earlier immigrants, liberal Protestants can be accommodated in any mainstream North American Protestant congregation, and Roman Catholics generally find a parish to their liking. The Dutch-Canadian real estate agents know the best locations for farm families seeking a particular denomination. However, those real estate agents who are themselves Catholic or Calvinist are more likely to have the right connections, and be able to arrange introductions, for clients of their own religious backgrounds.

We can get a good picture of recent farm immigrants by looking at a study of twenty-four Dutch dairy farmers conducted by Art Wolleswinkel, a Dutch student who did a four-month internship at the University of Guelph. Twelve of these farmers lived in the area of Woodstock (Oxford), six in Listowel (Perth), and the rest of other parts of Southwestern Ontario.[7] They came from every province of the Netherlands, except

[7] Wolleswinkel, "Dutch Dairy Farmers", 19.

Noord Brabant,[8] and were all married. The ages of the adults ranged from twenty to fifty-eight, with an average age of 37.3 for men and 34.5 for women. They all had children at the time of the study (1999), and a third of the families had a son planning to take over the farm in the future. The average level of education of the farmers was typical for farmers in the Netherlands today, with 63 percent having the middle level (*middelbare*) agricultural school and 8 percent with university-level training. In terms of farming qualities, all but five had had their own farm operation in the Netherlands, and of those who did, 42 percent were sole proprietors, 26 percent junior partners and 32 percent renters.[9] In Ontario, they all owned their own farms, and the average milk quotas and areas of farming land were double that of Dutch farms. These attributes influenced how well newcomers adapted, as well as their level of satisfaction. According the Wolleswinkel report, farmers who came without children had the most negative experiences concerning social life in general (compared with what their expectations were), while those with children between the ages of ten and seventeen at the time of immigration found that their children had difficulty integrating into Canadian society.[10] This second finding is similar to the experience of earlier immigrants, whose teenage children also found it difficult to leave their friends. In contrast, the first finding is more indicative of the different experiences of two cohorts of farm immigrants. The young couples that came to Canada in the past two decades grew up in an era of prosperity, enjoying a high standard of living and an active social life. Without school age children, they had less opportunity to come into contact with parents and teachers in Canada. Consequently, they felt more socially isolated even though they had greater economic security. In contrast, young postwar immigrant couples were more likely to interact on a daily basis with their sponsors, as well as with the large number of other recent Dutch immigrants they met.

As in the fifties, not all the newcomers are the same. Although many newcomers buy big farms and immediately erect new barns and start expanding, others start with more modest operations. As in the sixties and seventies, some of the newcomers are former agricultural students not able to start their own farm operations without working off the farm for several years. Indeed, the variation in size of operation was considerable, as shown in the section of the latest (1996) national census that

---

[8] The absence of Noord Brabant suggests that Dutch Catholic dairy farmers were underrepresented in this non-random survey.

[9] Wolleswinkel, "Dutch Dairy Farmers", 20-21.

[10] Wolleswinkel, "Dutch Dairy Farmers", 43-44.

focused on agriculture. That census showed an almost dichotomous distribution of wealthy farmers versus those with smaller, often part-time operations.[11] Indeed, the polarization between big and small farmers was greater than any other immigrant group in farming.

Regardless of size of farm, or whether these newcomers came in the late seventies or early nineties, they did not always find it easy to adapt. Indeed, in some cases, a husband or wife might have little English. One man I interviewed thought he would never have the opportunity to learn English. Not only did he spend most of his time in the barn, he soon discovered that several of his neighbors were of Dutch descent. The man who picked up his milk and the contractor who remodeled his barn wanted to try their now rusty Dutch. Luckily, his wife was already fluent in English. Even those who already spoke excellent English had to adapt to a North American style of living, a different climate and way of farming. Like earlier immigrants, the newcomers continue to speak Dutch at home and with friends. Indeed, their presence in some locations has led to a temporary revival of some Dutch-Canadian clubs, as in the case of one in Mitchell (an offshoot of the Stratford club), which was nearly defunct. New farm immigrants in the Woodstock area helped to resurrect a Dutch-language folk theatre (toneelclub) in 1985.[12] People scattered over a wider area also assemble at big annual parties organized by one of the real estate agents.

Unlike postwar immigrants, new farm immigrants frequently interact with both relatives and professionals in the Netherlands. Some even prefer to consult Dutch veterinarians for difficult cases. However, while still attached to their homeland, they want to adjust to, and even make changes to, Canadian society. One newcomer in the London region threw a big party for all of his neighbors, "old" Dutch and non-Dutch alike. When these neighbors noticed that the new Dutch family always sat down for coffee around eleven and talked together, he explained that this gave them a chance to discuss their troubles and exchange information. He sometimes invited his neighbors to join them and even invited a man who claimed he rarely had time to socialize. While initially

---

[11] The census did not include many farmers who came to Canada in the fifties, since most of them would have already retired or died. Lonmo, "Farming is going Dutch", 297-301.

[12] This theatre club was first set up in 1973, but did not last for more than a few years. I attended one of their rehearsals at the end of November, 1989 and they were still putting on plays in 1996. All the players were members of farm families, who lived near Innerkip, Thamesford, Kintore and Woodstock. See also Frans Schryer, *The Netherlandic Presence in Ontario* (Waterloo: Wilfrid Laurier University Press, 1998), 173.

reluctant, this non-Dutch neighbor warmed to the idea and one day commented, "I wish we had that custom here." Several neighbors met at those gatherings, even though they had been living in the same area for many years. However, while learning to get along with their neighbors (not always a smooth process), the newcomers also had to learn how to do business in Canada.

The Dutch study group

Recent regulations regarding new immigrants required a written opinion on the ability to establish a successful farm operation. So real estate agents approached Jack Rodenburg, an extension worker in the Woodstock office of OMAF (Ontario Ministry of Agriculture and Food). Jack, who had come to Canada with his parents in 1960, when he was ten years old, could still speak Dutch. He and his wife also ran a part-time sow operation.[13] When he started working with several new farm immigrants in the mid-eighties, Jack provided information packages as part of a consultative process for new immigrants started by OMAF. This was provided on a one-to-one basis, and included oral explanations in Dutch about the Milk Marketing Board, rules and regulations for farmers, and how to register cattle. The sudden increase in the number of farm immigrants necessitated meetings for each new group as they arrived. Jack contacted people of Dutch background in various organizations, including the Western Ontario Breeders, to also give presentations. In the winter of 1990, a group of newly arrived Dutch farmers in Oxford County (London area) organized their own study group and took over the task of providing the orientations.

The study group set up by Dutch dairy farmers follows a long tradition of study groups in the Netherlands. Andy Veldman, one of the real estate agents, first suggested the idea to form their own organization.[14] Initially, there was not much interest and most early members were Dutch Catholic farmers (as is Andy). They decided they wanted a more inclusive group and also sent out letters of invitation to all new immigrant farmers from Europe. Jack Rodenburg, with whom many had already had contact, was asked to be their technical advisor, and Jack Van Ham became the first chairman. Five years later, the group had fifty members, almost all dairy farmers, whose farms are located within approximately a thirty-mile radius of Woodstock (from Tilsonburg to

---

[13] Interview (FS) with Jack Rodenburg, at the Ontario Ministry of Agriculture and Food (OMAF) office in Woodstock, Wednesday, 14 June, 1994.

[14] Interview (FS) with Jack van Ham, at his farm near Culloden, 7 June, 1995.

Stratford). As their membership grew, they became more diverse. By the mid-nineties, 40 percent of their members were Catholics, another 40 percent were affiliated with a Reformed denomination and the rest had no specific religious affiliation.[15] Since each category was represented on the executive, they allowed each outgoing executive to choose their successors and, since Catholics tended to pick Catholics, and Protestants, Protestants, their group retained a semblance of the Dutch pillar system (see chapter 1). However, they did not draft a formal constitution, and attempts to establish similar study groups in other counties did not materialize. The first year, there was a meeting in the Listowel-Drayton area for new farmers arriving in the northern parts of the counties of Wellington and Perth (mainly Waterloo/Guelph region), but not enough newcomers were then coming to that part of Ontario to warrant a separate group. Farmers from both counties continued to attend the Oxford county study group.

The meetings of the study group were conducted in English, since most of their speakers did not speak Dutch. Yet, during coffee breaks after the formal part of the meetings, most of the farmers switched to Dutch. It is therefore not surprising that their attempts to include non-Dutch farm immigrants failed. A few Swiss farmers showed up, but were uncomfortable when they realized how many Dutch farmers were involved. Nevertheless, they continued to hold meetings, inviting a variety of speakers, including researchers from the University of Guelph promoting low-till farming, and representatives of the milk industry. Dutch newcomers needed to know the technical terms and wanted to share their experiences with other newly arrived Dutch farmers. They also wanted to see changes to the system of supply management in Ontario. While their group was officially concerned only with practical issues, and did not want to be involved in political controversies, some members were quite outspoken. At one point, they wanted to call their group the Investigative Research Committee of the Milk Marketing Board, which did not sit well with that Board's executive. The opinions of newcomers from the study group were not well received by older, more established dairy farmers, including a Dutch-Canadian member of Dairy Ontario. Only some of the bigger milk producers, including a few postwar Dutch immigrant farmers, were sympathetic to their ideas. As the study group started to retool as a lobby group, their technical advisor suggested that they present themselves as representatives of progressive dairy farmers rather

[15] Interview (FS) with Jack van Ham, at their farm near Culloden (near Tillsonburg), 7 June, 1995.

than as Dutch farmers. They once again invited more non-Dutch farmers to their meetings and considered sending an observer to Ottawa when the bigger producers presented a brief to the government regarding the application of biotechnology. As of 1996, their group, which has no official status, had not made any formal declarations or written letters. These new Dutch dairy farmers are certain to influence both their industry and policy in the future.

IMPACT AND MUTUAL PERCEPTIONS

The increasing demand for farmland as a result of the arrival of Dutch farm immigrants in the nineties accelerated the already steady increase in land prices. Since farmland is relatively cheap in Ontario, compared to Europe, real estate agents could ask higher prices for farms. And, since so many of the new Dutch farmers are dairy producers, the price for quotas went up, in turn driving up the price of poultry quota; Canadian farmers who sell dairy farms to the newcomers often convert part of the profit to poultry quotas, to minimize the taxes owable. However, while selling land and milk quotas to newcomers may reap short-term benefits, the increase in farm prices is generally not welcomed, because of the difficulties for offspring buying their own land. The resentment directed towards recent Dutch farm immigrants is especially noticeable in counties like Perth and Huron.

In a study of the attitudes of non-Dutch farmers, as well as Dutch-Canadian farmers conducted in 2001, Michael Johnston[16] showed how a combination of resentment, admiration, and anxiety took the form of ethnic jokes directed against new Dutch immigrants. As in the fifties, resentment and hostility often revolved around changes in the religious composition of a community, such as when Dutch Catholics settle in predominantly Protestant areas. When Dutch Reformed newcomers do not join mainstream Protestant churches, they are seen as "keeping to themselves". Or Canadian farmers accuse newly arrived farmers of contributing to the breakdown of community bonds when enrolling their children in Christian schools set up by postwar Dutch immigrants. In Oxford County, there was a rumor that the new Dutch farmers wanted to set up their own Milk Marketing Board! Such inter- and intra-group tensions of the last decade are further complicated by the fact that many contemporary "Canadian" farmers are themselves of recent Dutch descent.

[16] Michael Ray Johnston, "In it for the Milk': The Attitudes of Canadian Farmers Towards Recent Dutch Immigrants" (MA thesis, Department of Sociology and Anthropology, University of Guelph, 2002).

The same ambivalence and diametrically opposed viewpoints among new and old farmers attending different churches are also found within rural parishes dominated by Dutch-Canadian farmers, or within predominantly rural Reformed congregations. While there is considerable social interaction between members of both cohorts of farmer immigrants in such faith communities, contrasting styles of farm practices, differences in levels of education and socio-economic differences, create internal divisions and misunderstandings. At the same time, older, more established, Dutch-Canadian farmers, even if uneasy about the newcomers, may themselves be aware of resentment emanating from non-Dutch neighbors whom they have known for over forty years. One such Dutch-Canadian farmer, who came to Canada at the age of thirteen recounted: "The other day I was talking to a man I have known since we came to Canada. I told him – excuse my language – 'you guys call me a f... Dutchman, but those new farmers from Holland call me a f... Canadian. So what am I?'

Some non-Dutch farmers, especially those not on the cutting edge of agriculture, may resent and criticize the newcomers. However, others see them in a more positive, even highly favorable light. I have heard laudatory comments from well-established farmers not of Dutch background, government officials, and researchers at the University of Guelph. But, they believe that recent Dutch farm immigrants are all rich, successful farmers, even though this is not the case as shown in the 1995 census.

It is safe to say that new farm immigrants from the Netherlands have not gone unnoticed. Initially, their arrival triggered a flurry of activity as local construction companies, often owned by the descendants of Dutch postwar immigrants, scrambled to keep up with the demand for new barns and renovations. Neighbors and the curious had a chance to see the new farm set-ups via open houses. Word soon spread that the newcomers were paying much higher prices for land than the norm. Dutch "millionaire" farmers became a topic of conversation for both established (postwar) Dutch-Canadian farmers and members of farm families not of Dutch descent. However, the way the newcomers are perceived by Dutch-Canadians is not the same way as how they are perceived by farmers not of Dutch background. Nor does the picture painted by either group have any bearing on how these newcomers see themselves, much less what the newly arrived Dutch farmers think of their new Canadian neighbors. An examination of such contradictory perceptions can provide insights into the changing dynamics of inter-group relations in rural Ontario, as well as a better understanding of the similarities and differences among different cohorts of farm immigrants.

Contrasting Images

When I first started hearing comments about the new Dutch farm immigrants, I was surprised at the level of consensus among farmers of Dutch background themselves. The same story came up time and time again.[17] Everyone with whom I spoke invariably emphasized how much money these farmers had. "They come with suitcases full of money," was a common expression. These newcomers were not considered "real immigrants," since they made so many trips back and forth to the Netherlands. It was believed that since the newcomers did not have to work for Canadian farm sponsors, the way post-war immigrants had done, they would not learn about soil conditions, so they would often make mistakes, and would soon go bankrupt. Established Dutch-Canadian farmers thought that the newcomers were "too opinionated", "too Dutch" and did not integrate well into Canadian society. Most people with these views had only heard about the new immigrants, or had seen or met a new immigrant farmer on only one or two occasions. Some Dutch-Canadians who sold their farms to newly arrived farm immigrants from the Netherlands, were flabbergasted to see the new owners demolish what the vendor considered to be a perfectly good barn. As I began including the topic of recent Dutch farm immigrants in my interviews, I started hearing more stories about how, "there are exceptions," and that there are newcomers who went out of their way "to fit in." However, overall, a prejudicial attitude and resentment was apparent even among postwar Dutch immigrants related to the newcomers through family ties.

The way postwar Dutch immigrants view themselves, as successful farmers and good Canadians, is in sharp contrast to the way they are seen by recent farm immigrants. These newcomers usually preface their remarks with a statement of how much they admire the hard work of the postwar Dutch immigrants, and how much they have accomplished. They recognize that their predecessors "had a hard time." However, their overall impression is that farmers who came in the fifties are, "backwards," "conservative," and "old-fashioned."[18] One farmer who came to Canada in the early nineties though that the earlier cohort of immigrant farmers had only gone so far, but then "stood still," and how their offspring who have inherited the farms, "'want to take it easy and have

---

[17] In my fieldnotes files, I have eighty-one excerpts dealing with the topic of newcomers.

[18] These opinions were also expressed in a Radio Nederlands radio documentary, "Less Cows than Kids: Stories of Dutch Settlers in Canada Throughout the 1900s", and "Dutch Farmers in Canada," The Netherlands: Wereldomroep, a compact disk, 2001.

no education." Newcomers see established Dutch-Canadian farmers as no different from Canadian farmers not of Dutch background.

In making sense of these diametrically opposed viewpoints, one should remember that the farmers who came to Canada after 1988 came from a very different type of Dutch society than those from a few decades earlier. Given the different dispositions of these two groups of Dutch farmers, it is not surprising that their respective members feel they have little in common. Nevertheless, as they come into greater contact with each other, their mutual impressions become more complex. Postwar immigrant farmers who have themselves adopted the same farm technologies, and who are in a better position to compete with the newcomers, have a somewhat different perspective. One farmer who came to Canada as a young man in the late forties, and who now owns one of the biggest farms in his township, told me he admired the newcomers and wished them well. He fully recognized that his new Dutch neighbors know what they are doing, and that they are well educated. However, he was also conscious that they looked down on him and considered him "backwards".

Despite such frictions between recent Dutch farm immigrants and those who came in the period following the Second World War, there are many parallels between these encounters and the earlier contact between postwar immigrants and the "old timers" (prewar immigrants)'. In the late forties, immigrants who had arrived in the twenties, argued that they were "real pioneers," having survived the Great Depression. They looked upon the postwar immigrants as "spoiled," because the Dutch government subsidized part of their passage, and because they arrived with so many belongings. The "old timers" saw that the postwar immigrants were better educated, since so many of them had attended the *lagere landbouwschool*, not accessible to most rural students growing up around the time of the First World War. Like the newcomers, postwar immigrant farmers displaced non-Dutch farmers, leading to resentment.

The feelings of resentment today against new immigrant farmers among all but the most prosperous and established Canadian farmers are similar to those once felt in the sixties by many Canadian farmers of Scottish and English descent (see chapter 6). Indeed, one farmer I interviewed, who had been in Canada for over forty years, thought that, compared to the newcomers, Dutch farmers of his generation faced even greater hostility and resentment from their neighbors. He insisted that the "rich newcomers" are either ignored or readily accepted by their non-Dutch neighbors. His argument is that people are more jealous when they are surpassed by newcomers who start out with nothing. Regardless of whether or not such views have any validity, it is true that Dutch immigrants faced resentment in both cases. In each case, they were blamed for the concentration of farm ownership and the decline of local communities. Just as non-Dutch farmers were once appalled when their Dutch neighbors started cutting down fencerows to stretch their land, today they accuse Dutch immigrant farmers of destroying Canadian rural heritage by demolishing old farm buildings. Their non-Dutch neighbors saw both postwar and recent immigrants as aggressive and overly ambitious. Yet, they were at the same time admired and respected since they conformed to the positive stereotype of the successful Dutch farmer.

Recent Dutch immigrants are not the only ones with more intensive dairy operations or building new Dutch-style barns with slotted floors. A prominent Dutch-Canadian farmer, who played a leadership role in the Milk Marketing Board in the eighties, established the same type of intensive dairy operation as the newcomers. The arrival of new farm immigrants from the Netherlands happens to coincide with technical and structural advances in agriculture happening throughout Europe and North America, as part of a process of increasing global integration. Yet, the impression in many parts of rural Ontario, especially in those regions where non-Dutch farmers or older Dutch-Canadians did not take the lead in introducing new methods and building new barns, people have the impression that Canada's agriculture is indeed becoming "Dutch" (see introduction).

Conclusion

In 1946, no one could have predicted the consequences of bringing Dutch farm workers to Canada. The Netherlands-Canada Settlement Scheme was designed as a short-term measure to address labor shortages, yet it set in motion a series of events resulting in an influx of Dutch farm families. Between 1950 and 1980, farmers of Dutch background not only became numerically significant, they left an enduring imprint on the social and physical landscape of rural Canada. In some places, particularly Southwestern Ontario, the Dutch were exceptionally prominent. This postwar cohort was responsible for the consolidation of a positive image of "the Dutch farmer" in the popular imagination.

Despite the existence of a positive stereotype, there is no typical Dutch farmer. Dutch farm immigrants came from distinct regions in the Netherlands, contrasting socio-economic backgrounds, and had different levels of education. Most had little social contact with people of Dutch background attending churches other than their own. Furthermore, they did not all become well-known, commercial, full-time farmers. Many Dutch immigrant farmers only ran part-time operations or never expanded beyond a 100-acre operation. Nor were they all involved in agricultural politics or community service. There was considerable variation in the contribution of spouses in running farm operations. Nevertheless, despite differences in income, type of operation, and size of farm, on the whole the Dutch have done well. Not only did a surprisingly large number of people who started as farmhands in the fifties buy their owns farms – as much as a third of them – but a sizable minority of Dutch immigrant farmers and their offspring became prominent farmers in the province. These Dutch farm immigrants, who arrived in Ontario during the decade following the Second World War, came at the right time, when land prices were low and many Canadians farmers were no longer interested in farming. However, the resulting overrepresentation of the Dutch in agriculture is not just the result of a favorable set of circumstances. Their overall success, as a group, indicates that the Dutch had a competitive advantage over all but a handful of already highly capitalized Canadian farmers not of Dutch background. This book has shown why this was the case.

Emigrants willing to leave their friends and relatives behind to search for a better life are bound to be more motivated and energetic than the average person in the countries of origin, or destination. The fact that they had fewer ties to neighbors and fewer distractions than long-established Canadian farmers meant that they were able to dedicate more time to work and family. The Dutch tradition of children handing over their earnings to their parents prior to marriage meant that immigrant families pooled resources and saved money more quickly than their non-Dutch neighbors. Postwar Dutch farm immigrants brought with them a wealth of practical farm experience. A difficult period of wartime shortages taught post-war immigrants how to be thrifty; they could make do with just the basic essentials. Yet, even the poorest of Dutch farm immigrants had more resources than immigrants from other European countries. The Dutch arrived with most of their household belongings. Many smuggled in money or were later able to gain access to capital left in the Netherlands. The practical and theoretical education available in the Netherlands also ensured that Dutch farmers had another competitive advantage; many Dutch farmhands in the fifties knew more about the chemical composition of fertilizers, how to read blueprints or basic accounting than their employers! These skills not only enabled them to run their own farms, it allowed prospective farmers to get better-paying jobs and to save for a down payment on a farm.

The history of postwar Dutch farm immigrants in Ontario is unique, but their overall success is not atypical. Dutch farmers left their mark on the countryside of the United States as well, and agriculturalists of Dutch background are trendsetters in other parts of the world, including Australia and Eastern Europe (see introduction). The very recent arrival of Dutch farm immigrants to Canada, as a result of a federal policy designed to attract foreign entrepreneurs, is another example of this broader trend. Today, Dutch farmers are again purchasing land in Ontario, and other parts of Canada, replacing those farmers no longer able to stay abreast of the latest technological changes. The broader question then becomes, why this is so? What do Dutch immigrants farmers who emigrated at different times to different parts of the world have in common? Why do they continue to have a competitive advantage in so many different settings? The answer lies in the agrarian structure and culture of the Netherlands.

DUTCH FARMING: A GLOBAL COMPETITIVE ADVANTAGE

The rural and agricultural system of the Netherlands is the main reason why Dutch farm emigrants have been able to successfully adapt to, and occupy a leading position in, commercial agriculture. The Netherlands has a long-standing tradition of trade, regional interdependency, and commercial farming. The sandy regions, the reclaimed peat moors, and the river flood plains have always been interconnected. Such interconnections were strengthened throughout the twentieth century, when farmers and farmhands, even in less commercially-oriented Dutch rural regions were exposed to more advanced forms of agriculture. Many farmers from the sandy regions, who later moved abroad, had worked in the newly reclaimed clay soils of the Wieringermeer polder. Similarly, with the building of both a dike and a road connecting the provinces of Friesland and Noord Holland, people who grew up on dairy farms on permanent grassland in Friesland came into direct contact with farmers working in the tulip fields and so learned the techniques of intensive horticulture that tulip farming requires. While naturally more familiar with the farming systems of their home regions, young immigrant farmers thus experienced working with different types of soil, crops, and methods of husbandry. For such immigrants, the experience of having used new techniques was just as relevant when learning how to farm in a new country, as knowing how to milk a cow or run a greenhouse operation. Prior experience thus enabled many Dutch farm immigrants to quickly adapt to new soil and climatic conditions in Canada (see chapter 6).

Some aspects of agriculture in the Netherlands, especially horticulture and the dairy industry, have long been on the cutting edge. Dutch agriculture has historically been geared to export, so producers stayed competitive and innovative. The resulting ingrained habits and dispositions (the "habitus" of Dutch farmers, to use a term coined by Pierre Bourdieu), shaped their ability to improvise in new countries.[1] Another factor to consider is the mindset of farmers from a country that has confronted ongoing land shortages and water management challenges, which put a premium on being innovative. Dutch agriculture was characterized by pragmatic cooperation as well as individualism. The ability to co-operate when necessary, both between relatives and neighbors, combined with the desire of each person to excel, helped them to be flexible and to adapt to new environments. In the case of Ontario, the individualism emerges in the clear desire of young Dutch-Canadians to become economically independent as soon as possible, resulting in most of them establishing their own, separate farm enterprises. The ability to co-operate when necessary is illustrated by the many cases studies presented in this book showing how friends, siblings, parents and children formed partnerships that frequently lasted for decades. Members of large, extended

[1] The transposition, as well as modification, of a Dutch habitus into a Canadian context by postwar immigrants in Ontario (not just farmers) is a main theme in an earlier study. See Schryer, *The Netherlandic Presence in Ontario*, 68, 88, 179, 278, 317.

families, each running their own farm, continued to share farm equipment and provide specialized services to each other. A more recent example is the Dutch study group established in the nineties in Oxford County (see chapter 15).

It is difficult to separate economic factors from culture. Dutch rural immigrants, even those who belong to strict religious sects, share a cosmopolitan outlook. The historical experiences of the Netherlands, with its history of colonial expansion and trade, earlier wars with first Spain, then France and Great Britain, then a German occupation, fostered an awareness of the rest of the world. From the birth of the Dutch nation-state in the sixteenth century, its ruling classes and economic elite have been cosmopolitan and outward-looking, attitudes that influenced the rest of the population. More recently, the Dutch took the lead in the formation of a common market, culminating in the present European Community. Another aspect of Dutch society and culture that shaped the mentality and attitudes of Dutch farm emigrants is the Dutch system of social stratification (*standsverschil*). Dutch society used to have a system of status distinctions related to class and occupation more pronounced than that in England. This system of social stratification was still strong after the Second World War and continues in an attenuated fashion today. While they might have disliked absentee landlords with aristocratic pretensions, Dutch farmers continued to make invidious comparisons among themselves. In the Netherlands, farmers in turn had looked down on factory workers. This preoccupation with social ranking, combined with their competitive nature, gives the impression that Dutch rural immigrants are "go-getters" when they go abroad. I heard this expression on several occasions when Canadian farmers were discussing recent Dutch farm immigrants.

In the case of the postwar farm immigrants, those who had owned land or had had rental agreements in the Netherlands were survivors of several decades of increasing concentration of land ownership and cycles of alternating boom and bust. The offspring of small farmers, in particular, were very aware that they were not likely to inherit the family farm or have the opportunity to rent land. Many were resigned to learning a city trade (see chapter 1). Yet such children often still aspired to become farmers and were sometimes more anxious to emigrate than their parents. Even middle-aged or older farmers with relatively secure access to land back home had to struggle to expand in the Netherlands in order to survive or to give their children a chance to farm. Indeed, their strong desire to help their children find a place in agriculture was a prime motivating factor to start anew in another country. This motive still holds for the "rich" Dutch farmers coming to Canada today.

A strong drive to become or remain a farmer, and the desire to expand, are themes that surfaced repeatedly in discussions with or about Dutch immigrant farmers in Ontario. Non-Dutch farmers in the fifties perceived Dutch immigrant farmers as being "very hard working, but aggressive." In the nineties, Canadian farmers, including second-generation Dutch-Canadians, used these same expressions to describe the more recent cohort of farm immigrants from the Netherlands. Recent Dutch immigrants, like their postwar counterparts, in turn, regard most Canadian farmers as complacent and believe that Canadian farmers "like to take it easy." The postwar immigrants could not understand why their sponsors or neighbours would be content to operate a small farm operation: why did Canadian farmers, who seemed to have enough money, not immediately buy more cows or plant more crops? Those who started off working as hired hands in progressive, large-scale, Canadian farm operations admired and respected their employers. But they also wanted to learn from them, emulate them, and ultimately surpass them.

The high proportion of Dutch farm immigrants in Ontario who started their own farm operations in the fifties and sixties is all the more remarkable given the range of alternative opportunities. Many newcomers quickly discovered that they could get a better job in town, working on the assembly line or in the construction industry. Indeed, many people who had grown up in the Dutch countryside no longer wanted to become farmers in Canada once they were drawing a steady paycheck. They discovered that urban workers had a better standard of living than most farmers and that there was no shame in being a factory worker in Canada. Yet many of those earning good money in the city nevertheless quit their jobs as soon as they had saved up enough to buy a farm or to work full-time on their own land. They had "farming in their blood," is another expression that comes up repeatedly. However their attitude to farming was not a matter of genes but of deeply ingrained cultural values going back at least several hundred years.

Many of the Dutch immigrants who became full-time farmers came to Canada with a strong identity as professional, full-time, farmers, and passed that identity on to their children. Dutch fathers who attended a *landbouwschool* still proudly display their diplomas in their new homes in Canada. According to one man who was four years old when he came to Canada, and who grew up in a close-knit Dutch-Canadian farming community in South-western Ontario, his parents' generation thought it was a "terrible thing" if someone could not be a full-time farmer. This attitude persists among the second-generation Dutch-Canadian farmers. One man I interviewed was particularly sensitive to the stigma attached to off-farm work because his own family was the exception. The primary occupation of his father, who lived in the midst of Dutch immigrant farmers, was that of bricklayer, and the son was acutely aware of their family's lower status in the community, even though they made good money. The social pressure on anyone who grew up in a rural Dutch-Canadian household to become a farmer was so great that even those more interested in other occupations felt compelled to enter agriculture. For example, one man developed a viable commercial farm even though he worked full time as a schoolteacher.

Despite such cultural values and aspirations, the likelihood of actually accomplishing the goal of becoming a full-time, as well as a prominent, commercial farmer depended in large part on socio-economic background. Dutch immigrants who became large-scale, full-time farmers in Ontario had, for the most part, been full-time farmers or the children of farmers (as opposed to full-time farmhands) in the Netherlands, although there were notable exceptions. There are many cases of Dutch immigrants who had been part-time farmers, or who came from very small farms in the Netherlands, becoming full-time commercial farmers in Ontario, but they were less likely to rise to the top. I suspect that a larger proportion of those Dutch farmers might have failed were it not for the very favorable set of economic circumstances in Canada during the sixties and seventies. Nevertheless, Dutch immigrant farmers were more likely to remain in farming and survive the debt crisis of the early eighties than their non-Dutch counterparts, as shown by the statistics showing their overrepresentation in the farm population.

Immigrant farmers applied their Dutch education, and to a lesser extent their Dutch experience, to Canadian farming. However, they did not automatically replicate Dutch methods of farming. Rather, they adopted many, but not all, Canadian farming techniques. Their Dutch training made them receptive to new developments in farming promoted by extension officers and representatives of feed companies. Consequently, the Dutch played a major role in introducing new crops, and working the soil in new ways. While Dutch immigrant farmers were not solely responsible for such innovations as planting new varieties of corn, edible beans, or alfalfa, rural Ontario would not look the same today without them.

Between the fifties and the eighties, differences between Dutch and non-Dutch farmers were attenuated, as non-Dutch farmers modified their methods of farming (with or without the example of their Dutch neighbors) in order to stay abreast of the newest developments in agricultural research and in response to market pressures. When asked to discuss differences between Canadian and Dutch-Canadian methods of farming, older immigrants thought that the stark contrasts they observed in the fifties – for example, how the Canadians mowed their hay much later in the season, and did not use fertilizers – no longer apply today. They pointed out that any lingering differences between Dutch-Canadian farmers and their non-Dutch neighbors are minor compared to the contrasts between both of them and the recent newcomers arriving from the Netherlands. Yet a minority of progressive Canadian farmers, including those with no Dutch connections, are today building the same kind of European-style barns and switching to more intensive forms of dairy production as these newcomers. The contrasts still found today in many Canadian townships presently witnessing an influx of Dutch farmers, will diminish over time as the offspring of these newcomers become Canadianized and non-Dutch farmers become more, "like the Dutch." With an emerging global culture, Canadian farmers and the farmers who remained in the Netherlands will also start to resemble each other more, despite the language difference.

Cultural influences go both ways. In Ontario, the Dutch directly or indirectly influenced Canadian farmers in some technical aspects of farming. However, Canadian farmers not of Dutch background also had an impact on their Dutch neighbors. Dutch immigrant farmers, when asked what they learned from Canadian farmers, admit that they learned to relax more in Canada. This sentiment was echoed in conversations with recent immigrants, who mentioned that in Canada they no longer worried as much about what their neighbors thought of them, while also wondering why those same neighbors were not trying to expand their operations. The changing attitudes and work habits of Dutch-Canadian farmers are even more pronounced beyond the first generation. The younger children of Dutch immigrant farmers grew up in a society that was radically different from the homeland of their parents. They played with the children of non-Dutch neighbors, and assimilated some of their values and lifestyle. Even the offspring of farmers who belonged to Reformed communities (and who attended their own schools) became more "Canadianized" over time, especially when they started studying in agricultural colleges or took courses related to farming in institutions without Dutch connections. Yet second- and third-generation Dutch-Canadian farmers did not completely lose or renounce their Dutch heritage, even if they no longer spoke Dutch. One prominent Dutch-Canadian farmer and businessman entirely educated in Canada, frequently refers to the values and work habits he learned from his father, who was an immigrant farmer. In a lecture to agricultural economics students at the University of Guelph, he talked about the influence of his Dutch heritage on his ability to take risks and make quick decisions.[2]

[2] I was present at this talk and later visited his farm office to conduct an interview (which was done anonymously).

THE REPUTATION OF DUTCH IMMIGRANTS FARMERS

The survival of a group identity and pride in being "Dutch" among farmers of Dutch background is attributable to the positive image associated with Dutch farming in Canada and abroad. One retired prewar Dutch farmer married to someone not of Dutch background (as are his three children), smiled when he told one of my research assistants that his grandson's future wife is of Dutch background. He emphasized that the family name of his partly Scottish offspring is Dutch and was pleased to see "how the Dutch blood is coming back into the family," and how proud he was to be Dutch.

The image of the Dutch as good farmers has become deeply entrenched in Canadian society, as reflected in the statements of prominent politicians. The first comment that Pierre Elliott Trudeau, the late Canadian prime minister, said when meeting the editor of a Dutch-language ethnic newspaper in the eighties, was, "Ah yes, the Dutch farmers have made such an important contribution to Canada." The Dutch-Canadian editor, who had no farm connections, resented the implication that all Dutch immigrants were farmers.[3] Yet Dutch-Canadians not of farm background cannot escape the fact that the Dutch are better known for farming than almost any other field of endeavor.

While the Dutch historically have been associated with successful, commercial agriculture, this positive stereotype did not always prevail in Ontario. Around the time of the First World War, the Dutch were a relatively obscure component of Canada's ethnic mosaic, although more educated Canadians would have heard of the Netherlands. In Ontario, Dutch farm immigrants first came to more prominent attention in the forties, with the development of the Holland Marsh just north of Toronto. At the time of the new millennium, two immigration cohorts later, it is difficult to ignore their presence. Successive cohorts of Dutch agrarian immigrants, the "Old Timers," the postwar generation, and the "Newcomers," have each left their own impressions and made somewhat different contributions to Canadian agriculture. Likewise, various groups of Dutch farm immigrants in other parts of the world have left their stamp, resulting in the creation or resurrection of old stereotypes. However, with the dramatic increase in the speed of communication over the last few decades, differing opinions on the reputation of a broad-based group (between countries or regions within the same country) may become a thing of the past. The positive stereotype of the successful Dutch farmer is becoming truly global.

---

[3] Personal communication from the former editor of this newspaper, July 15, 1994.

The prior reputation of the Dutch as good farmers is likely to have played a role in the decision of government officials to allow rural immigrants from the Netherlands to move to Canada. However, most of the Canadian farmers who sponsored Dutch families were not as convinced as the government officials that these newcomers would be able to farm on their own. In the five years following the end of the Second World War, the vast majority of these Canadian farmers saw their Dutch employees as yet another source of farm labor. They often did not realize or appreciate the fact that these Dutch farm immigrants wanted to farm for themselves. Some Canadian farmers were convinced their Dutch neighbors would fail. Nor were bankers initially willing to provide them with credit. The Dutch-Canadian farmers who came to Ontario had to earn the respect and recognition of their neighbors since most Canadian farmers in Ontario, outside of the Holland Marsh region, had not previously heard of the Dutch. Many associated the Dutch with Pennsylvania "Dutch" (which is really German). The subsequent concentration of groups of Dutch farmers in some parts of Southern Ontario and their involvement in farmers' organizations brought them to the attention of non-Dutch farmers throughout the province. Three decades later, most people in Ontario have at least heard of them. Today some people even believe that all Dutch are farmers and that all rich farmers are Dutch. I have often heard the comments, "the Dutch are found in every part of Ontario," and, "Dutch farmers don't stick together in one place like other groups."

A close examination of observable trends, as shown by statistical data, makes it clear that the Dutch are not equally represented in all parts of rural Ontario, nor do Dutch farmers have the same reputation in each branch of agriculture or agribusiness. Nevertheless, a combination of overrepresentation in the rural population as a whole, having a high profile in certain regions or agricultural sectors, and the involvement of Dutch farmers in various producers' organizations, have all contributed to the formation of a predominantly positive stereotype in the province of Ontario. It would be difficult to determine the extent to which this group stereotype developed as a result of an already existing reputation of Dutch farmers in English- and French-speaking countries. Should the postwar farm immigrants receive all of the credit for forging a group reputation, as they carved out a niche for themselves in the world of agriculture? Or was the subsequent overrepresentation of Dutch farmers itself a product of the reputation of their country of origin? These questions have no simple answers since the relationship between group reputation and the demographics and economic position of a group is one of reciprocal causation.

It is impossible to determine to what extent the actions of various types of Dutch immigrant families were responsible for the emergence or reinforcement of a positive stereotype of Dutch farmers in Ontario. This positive stereotype was as much the outcome of the presence of a few high-profile Dutch-Canadian farmers, as it was the result of a concentration of large numbers of Dutch-Canadian farmers in several parts of Southwestern Ontario. Nor can we discount the high number of Dutch who managed to establish their own farms within a decade of arrival in almost every county, and the even greater number of Dutch immigrants in small towns who ran part-time farm operations at one time or another. The group reputation of Dutch farmers in Ontario is the cumulative effect of all of these factors, combined with the prior reputation of their country of origin.

The role of stereotypes, even when positive, demands a look at the nature of the relationship between group reputation, the ultimate fulfilment of that expectation, and the objective life chances of their individual members. Objective factors, such as the education received by Dutch farmers, their ability to recruit family labor and the high level of economic and social development of the Netherlands (despite the Second World War) explain why Dutch farm immigrants were able to not to only insert themselves into Canadian agriculture, but to excel as a group. However, the reputation of their country of origin, and the example set by earlier farm immigrants, allowed more Dutch farm immigrants to establish themselves in farming than would otherwise have been the case. For example, starting in the mid-fifties, many bank managers, as well as private individuals, having heard positive reports about Dutch farmers, began extending more credit to Dutch farmers.

Using the theoretical language coined by the French sociologist Pierre Bourdieu, Dutch-Canadian farmers have a great deal of "symbolic capital", thanks to the reputation of their group. This symbolic capital is in turn derived from their prior cultural and social capital within the field of agricultural production. Their cultural capital consisted of a combination of formal agricultural training and practical knowledge related to commercial farming. Their social capital is embodied in the extensive network of kinship and friendship ties, especially, but not exclusively, with other farmers and horticulturalists of Dutch background. A minority of Dutch immigrant farmers accumulated a considerable amount of economic capital. Yet, the positive group stereotype that is the counterpart of a high level of symbolic capital, distorts and misrepresents social reality insofar as it creates a false impression of homogeneity (i.e., that all Dutch farmers are all rich and successful). Statistically, it is true that Dutch farmers, on average, have done well in Ontario. The ques-

tion then becomes, what is the connection between the differential success of Dutch farmers (on average) and the positive stereotype? A group reputation is not primarily responsible for how well members of that group do as individuals. However, any group member's potential is likely to be in part influenced by the stereotypes associated with their group. This holds true even for those on the bottom rungs of the socio-economic ladder, although not always as expected.

There is a negative side to a positive group reputation. The perception that a group has done well leaves all members of that group open to envy and resentment. Moreover, the image of the successful Dutch farmer can be detrimental to those unable to live up to that reputation. The desire and expectation to succeed explains why some Dutch immigrants with minimal resources, and few objective chances of success, nevertheless went into farming. Such farmers exerted great pressure on their children to contribute financially and physically, often at the expense of their schooling. If the parents went bankrupt, or if these children failed to carry on the family tradition of farming, they inevitably ended up in dead-end jobs. A positive group stereotype could thus result in some individuals of that group being less likely to succeed, in that it is not easy to live up to a reputation. I have encountered offspring of Dutch immigrant farmers who were deeply resentful of their parents' decision to move to Canada, or ashamed that they or their parents did not accomplish more in life. As Robert Murphy's notion of negativity, outlined in his *Dialectics of Social Life*, suggests, "the ardent pursuit of an ideal may lead to its opposite."[4]

My study dealt primarily with the postwar Dutch farm immigrants who remained in, or eventually returned to, agriculture after their one-year contracts with Canadian farm sponsors were over. By highlighting the accomplishments of those Dutch immigrants who became farmers in Canada, I am open to the criticism that my research neglected those wanting to become farmers but who failed to do so. Indeed, this book does not provide detailed case studies of postwar Dutch immigrant farmers who went bankrupt or who had to sell out to prevent bankruptcy. It does not dwell on recent farm immigrants who did not survive for more than one or two years. I am fully aware that the very act of writing this book, which focuses on those Dutch immigrants who became full-time farmers, may help perpetuate the positive stereotype I wish to examine in a critical manner. However, cases of farm failure, and the

[4] Murphy, Robert. *Dialectics of Social Life: Alarms and Excursions in Anthropological Theory* (New York: Columbia University Press, 1971), chap 3.

fact that many rural Dutch immigrants did not end up in farming, did not prevent the positive stereotype of Dutch farmers from becoming well established in Ontario in the last quarter of the twentieth century. I doubt if an academic book dwelling on the hardships and failures of postwar Dutch farm immigrants today, would diminish the already existing group reputation of Dutch-Canadian farmers.

A group stereotype, like any social image or label, simplifies a more complex social reality. Moreover, the connections between group stereotypes and the distribution of attributes of members of any group are indeterminate, with no clear causal direction. In the case of a positive stereotype, the way a group is regarded and how its members view themselves, rarely correspond to the observable attributes of all but a small minority of group members. I have demonstrated that even the full-time commercial Dutch farmers by themselves are a diverse group; no one farmer or farm family of Dutch descent can be accurately portrayed in terms of the positive stereotype of the resourceful, upwardly mobile farmer. A group stereotype, whether positive or negative, is like a statistical average that may not correspond to any of the specific individuals whose attributes are measured to derive such an average. However, a positive group stereotype has a different dynamic than a negative stereotype. The fact that a group of people is seen as poor or oppressed will not by itself prevent a small minority of that group from becoming well off and powerful. Yet, such a negative stereotype (like a statistically low average income) usually does hold true for a majority of the members of that group. A positive stereotype, especially one that portrays an ethnic or national group as rich, successful, and powerful, distorts social reality to a greater extent. Large-scale farmers of Dutch descent constitute but a tiny minority. The majority of farmers of Dutch descent have average-sized operations or are part-time farmers. Yet distorted group stereotypes, like the very existence of groups as social constructs, are an integral part of society.

A group stereotype may undergo drastic transformations. In the case of Dutch immigrant farmers in Ontario, a positive image of the Dutch as entrepreneurial, commercial farmers emerged in the forties as a result of the presence of a small group of pre-war immigrants in the Holland Marsh. Yet, that group reputation had little resonance in most of rural Ontario in the period between 1945 and 1950, when a much larger number of Dutch immigrant farmworkers appeared in most regions. Almost overnight, the most common image of the Dutch became that of a foreign farmworker. In 1948, Dutch farm immigrants were but one of several successive waves of these workers. Dutch farmworkers and Dutch immigrants who set up part-time farm operations were primarily seen as immigrants and erroneously called, "DP's" ("displaced persons", which came to have negative connotations). Then, somewhere between 1960 and 1970, the earlier image of the "hard-working, good Dutch farmer" came to the forefront. By the 1990s, when I was conducting my research, the predominant stereotype of "Dutch" (Dutch-Canadian) had become that of "the successful and rich farmer," an image further reinforced by the presence of yet another, albeit smaller, wave of immigrant farmers from the Netherlands. The new positive stereotype has taken on a dynamic and transformative life of its own, not only in Canada but in a new global culture.

IMPLICATIONS FOR PROSPECTIVE IMMIGRANT FARMERS

This book is primarily addressed to an academic audience, keeping in mind a broader readership, including the educated lay person. I particularly have in mind people with some training in, or familiarity with, agriculture, including those thinking about moving to another part of the world in order to fulfil their dreams.

Agriculture is for the most part not a lucrative option, in Canada or elsewhere. The agricultural sector, as well as agribusiness in general, is highly competitive and not as likely as the service or industrial sectors to provide someone with the opportunity to earn a good income, much less enable other family members to do so. The majority of farmers and horticulturalists in Canada, even in the prosperous region of Southern Ontario, depend on a second source of income. There are many advantages to living in the countryside, such as fresh air and open spaces, especially for those who can commute to a nearby town or city. However, men or women who farm full-time today are a tiny minority, and family farms that can provide a full-time job for older children still living at home is even rarer. Prospective immigrants should keep this in mind.

What is remarkable about Dutch immigrant farmers is that so many of them were able to make a good living from commercial agriculture, despite the risks and hazards of so doing. Anyone thinking about following their example must take heed that they may not succeed. Only those Dutch immigrants with the right combination of formal training and previous experience in farming, as well as strong kinship ties, not to mention a strong desire to make farming into a full-time occupation (plus a bit of luck), were able to do so. Nowadays one also needs to have financial backing (economic capital). Only then will he or she be likely to succeed. This book has provided evidence for my claim that Dutch immigrant farmers have a competitive advantage in the field of agriculture in Canada. I strongly suspect that this statement also applies to other countries, although only a comparative study of farm immigrants of different ethnic backgrounds, and in different parts of the world, would enable scholars to substantiate such a generalization. Nevertheless it does not make any sense for someone who does not have the necessary social and cultural capital, to start farming in today's global economy. Being Dutch or the fact that one happens to have a Dutch farm background is not sufficient by itself regardless of the reputation of Dutch farmers as a group.

Bibliography

Akenson, Donald Harman. 1999. *The Irish in Ontario: A Study in Rural History*. 2nd ed. Kingston: McGill-Queen's University Press.

Allingham, Elizabeth C. 2002. "Different Voices of Postwar Immigrant Women in Rural Ontario: victims of agrarian ideology and 'invisible' contributors to agricultural production." A paper for a course at the University of Guelph.

Bagley, Christopher. 1973. *The Dutch Plural Society: a comparative study of race relations*. London: Oxford University Press.

Biggs, Everett. 1990, *The Challenge of Achievement: The Milk Marketing Board's first 25 years of operation – 1965-1990*. Mississauga: Milk Marketing Board.

Boissevain, Jeremy, and Jojada Verrips. 1989. *Dutch Dilemmas: Anthropologists Look at the Netherlands*. Assen: Van Gorkum.

Bourdieu, Pierre. 1984. *Distinction: A Social Critique of the Judgement of Taste*. Cambridge, Mass: Harvard University Press.

——. 1977 (reprinted 1985). *Outline of a Theory of Practice*. Cambridge: Cambridge University Press.

Bourdieu, Pierre and Loïc I.D. Wacquant. 1992. *An Invitation to Reflexive Sociology*. Chicago: University of Chicago Press.

Burnet, Jean R. and Howard Palmer. 1988. *"Coming Canadians": An Introduction to a History of Canada's Peoples*. Toronto: McClelland and Stewart.

Chapman, L.J. and D.F. Putnam. 1966 (reprinted 1967). *The Physiography of Southern Ontario*. Toronto. 2nd ed. University of Toronto Press.

De Jong, Klaas. *Cauliflower Crown: The Cauliflower King of North America*. (Arranged by Martha Knapp.) Saskatoon: Western Producer Books.

Drummond, Ian M. et al. 1987. *Progress without Planning: The Economic History of Ontario from Confederation to the Second World War*. Ontario Historical Studies Series. Toronto: University of Toronto Press.

Favell, Adrian. 2001. "Migration, mobility and globaloney: metaphors and rhetoric in the sociology of globalisation." *Global Networks* 1 (4): 389-98.

Froschauer, Karl and Jan Rath. 2002. "The Unusual Case of Dutch Immigrant Farmers in Canada: A Recent History of Self-employed Dutch Business Immigrants." Paper presented a the Fourth European Social Science History Conference, 2 March, The Hague, The Netherlands.

Ganzevoort, Herman. 1983. Sharks in Wooden Shoes. In *Dutch Immigration to North America*, ed. H. Ganzevoort and Mark Boekelman. Toronto: Multicultural History Society on Ontario.

——. 1988. *A Bittersweet Land: The Dutch Experience in Canada, 1890-1980*. Toronto: McClelland and Stewart.

Gentilcore, R.L. 1972. "Settlement", Chap. 2 in *Studies in Canadian Geography: Ontario*, ed. Louis Trotier. Toronto: University of Toronto Press.

Gerrits, Gerrit H. 1996. *They Farmed Well: The Dutch-Canadian Agricultural Community in Nova Scotia 1945-1995*. Kentville, N.S.: Vinland Press.

Glazebrook, G.P. de T. 1968. *Life in Ontario: A Social History*. Toronto: University of Toronto Press.

Goudsblom, Johan. 1967. *Dutch Society*. New York: Random House.

Graaskamp, Ger. circa 1980. *Dutch Canadian Mosaic*. Niagara Falls, self published.

Harris, R. Cole, ed. 1987. *Historical Atlas of Canada:(Volume 1: From the Beginning to 1800*. Toronto: University of Toronto Press.

Hofwijk, J.W. 1952. *Canada: reis met onbekende bestemming* (Utrecht: Uitgeverij de Lanteern).

Johnston, Michael. 2000. "In it for the Milk": The Attitudes of Canadian Farmers Towards Recent Dutch Immigrants. MA thesis, University of Guelph.

Jones, Robert Leslie. 1946. *History of Agriculture in Ontario, 1613-1880.* Toronto: University of Toronto Press.

Kearney, Mark, and Otte Rosenkrantz. 2000. *Roots and Branches: St. Willibrord Community Credit Union, the first 50 years.* London, Ont.: St. Willibrord Community Credit Union.

Keddie, Philip D., and Julius A. Mage. 1985. *Southern Ontario Atlas of Agriculture: Contemporary Patterns and Recent Changes. Occasional Papers in Geography* 7. University of Guelph: Department of Geography.

——. 1991. Agriculture in Ontario. Chap. 7 in *Ontario: Geographical Perspectives on Economy and Environment.* ed. B. Mitchell. University of Waterloo. Department of Geography Publications Series no. 34.

Kenyon, Ron. 1976. *To The Credit of the People.* Toronto: The Ontario Credit Union League Limited.

Kloosterman, Abraham. 1961. "Verslag van de praktijktijd, doorgebracht op het bedrijf van Earl Elliott, Elmdale, Ontario, Canada, gedurende de maanden juli en augustus", student report written for University of Wageningen.

Kloosterman, Sijmen Paulus. (circa 2000). *Memoires, opgetekend door Sijmen Paulus Kloosterman.* Unpublished manuscript.

Lambert, Audrey M. 1971. *The Making of the Dutch Landscape: An Historical Geography of the Netherlands.* London: Seminar Press.

Lijphart, Arend. 1968. *The Politics of Accommodation: Pluralism and Democracy in the Netherlands.* Berkeley: University of California Press.

Lonmo, Charlene. 1999. Farming is going Dutch. In C*anadian Agriculture at a Glance.* Ottawa: Statistics Canada: 297-301.

Lucas, Henry. [1955] 1989 *Netherlanders in America: Dutch Immigration to the United States and Canada, 1789-1950.* Grand Rapids, MI: Eerdmans.

Magee, Joan. 1987. *The Belgians in Ontario: A History.* Toronto: Dundurn Press.

Mahmood, Cynthia Keppley. 1989. *Frisian and Free: Study of an Ethnic Minority of the Netherlands.* Prospect Heights: Waveland Press.

McHugh, Kevin E. 2000. Inside, outside, upside down, backward, forward, round and round: A case for ethnographic studies in migration. In *Progress in Human Geography,* 24 (1): 71-89.

Moore, Sally Falk. 1975. Epilogue: uncertainties in situations, indeterminacies in culture. In *Symbol and Politics in Communal Ideology.* Eds. S.F Moore and B. Myerhoff (eds). Ithaca: Cornell University Press.

Murphy, Robert F. 1971. *The Dialectics of Social Life: Alarms and Excursions in Anthropological Theory.* New York: Basic Books.

Nishisato, Shizuhiko. 1994. *Elements of Dual Scaling: An Introduction to Practical Data Analysis.* Hillside, NJ: Lawrence Erlbaum Associates.

Nishisato, Shizuhiko & Ira Nishisato. 1994. *Dual Scaling in a Nutshell.* Toronto: MicroStats.

Petersen, William. 1955. *Planned Migration: The Social Determinants of the Dutch-Canadian Movement.* Berkeley: University of California Press.

Rae, Kenneth J. 1985. *The Prosperous Years: The Economic History of Ontario, 1939-1975.* Toronto: University of Toronto Press.

Reaman, G. Elmore. 1970. *A History of Agriculture in Ontario* (in 2 Volumes). Toronto: Saunders of Toronto Ltd.

Reeds, L.G. 1972. "The Environment", Chap. 1 in *Studies in Canadian Geography.* Toronto: University of Toronto Press.

Sas, Anthony. 1957. Dutch Immigration to and Settlement in Canada: 1945-1955. PhD diss., Department of Geography, Clark University (MA).

Schryer, Frans. 1998. *The Netherlandic Presence in Ontario: Pillars, Class and Dutch Ethnicity.* Waterloo: Wilfrid Laurier Press.

——. 2001. Multiple hierarchies and the duplex nature of groups. *The Journal of the Royal Anthropological Institute* 7 (4): 705-21.

Schama, Simon. 1987. *The Embarrassment of Riches: An Interpretation of Duch Culture in the Golden Age.* London: William Collins Sons and Co. Ltd.

Shetter, William Z. 1971. *The Pillars of Society: Six Centuries of Civilization in the Netherlands.* The Hague: Martinus Nijhoff.
Sinclair, Peter. 1987. "The North and Northwest: Forestry and Agriculture", Chap. 5 in *Progress without Planning: the Economic History of Ontario from Confederation to the Second World War.* Ian M. Drummond et al. Toronto: University of Toronto Press.
Steinberg, Stephen. 1989. *The Ethnic Myth: Race, Ethnicity and Class in America.* Boston: Beacon Press.
Swierenga, Robert P. 1983. *Dutch Emigrants to the United States, South Africa, South America, and Southeast Asia, 1835-1880: An Alphabetical Listing by Household Heads and Independent Persons.* Wilmington, DE: Scholarly Resources.
——. 2000. *Faith and Family: Dutch Immigration and Settlement in the United States, 1820-1892.* New York: Holmes & Meier.
Tuinman, Abe Sybe. 1952. *Enige aspecten van de hedendaagse migratie van Nederlanders naar Canada.* (S'Gravenshage: Staatsdrukkerij).
——. 1956. The Netherlands-Canadian Migration. *Tijdschrift voor Economische en Sociale Geografie* 47 (8): 181-89.
Van der Mey, Albert. 1983. *To All Our Children: The Story of Postwar Dutch Immigration to Canada.* Jordan Station: Paideia Press.
——. 1994. *And the Swamp Flourished: The Bittersweet Story of Holland Marsh.* Surrey, B.C.: Vanderheide Publishing.
Van der Mey, Albert and Marten A. Mol. 1997. *The Dutch Touch in Ontario.* Toronto: Marten A. Mol and Associates.
Van der Mey, Albert et al. 2004. *Floral Passion.* St. Catharines: Vanwell Publishing.
Vanderschot, Margaret, and Mary A. Smith. 1991. "Netherlands Research Project: A Study of Postwar Immigration from the Netherlands to the St. Marys Area". A History of Ontario's People project report for St. Marys Museum, St. Marys, Ontario.
Van Leeuwen, Jerry. (circa 1995). "The Story of Jerry." Unpublished manuscript.
Van Stekelenburg, H.A.V.M. 1996. *Hier is Alles Vooruitgang: Landverhuizing van Noord-Brabant naar Noord-Amerika, 1880–1940.* Tilburg: Stichting Zuidelijk Historisch Contact.
——. 2000. *De Grote Trek: Emigratie vanuit Noord-Brabant naar Noord-Amerika, 1947-1963.* Tilburg: Stichting Zuidelijk Historisch Contact.
Verduijn, Arie. 1981. *Sojourners: A Family Chronicle.* Burlington, self-published.
Wall, Ellen. 1992. Agribusiness and Hired Farm Labour in the Ontario Tomato Industry, PhD diss., Department of Sociology, McMaster University. Hamilton, Ontario.
Whebell, C.F.J. 1972. "The Politico-Territorial Structure", Chap. 6 in *Studies in Canadian Geography: Ontario,* ed. Louis Trotier. Toronto: University of Toronto Press.
Wijdeven, Ad. (circa 1983). *Oogsten op vreemde velden.* Zutphen: Uitgeverij Terra Zutphen.
Wildenbeest, Gerrit. 1989. Recent Farmers' Protest in a Dutch Municipality: The Legacy of the Past. In *Dutch Dilemmas,* ed. Jeremy Boissevain and Jojada Verrips. Assen: Van Gorcum: 70-88.
Wolleswinkel, Art. 1999. "Dutch Dairy Farmers in Ontario: Reasons for and evaluation of emigration". Research report, Department of Agricultural Economics, University of Guelph.

Appendix (Dual Scaling)

This appendix provides enough explanation to allow a reader not familiar with the statistical analysis known at dual scaling to make sense of chapter 10. I begin with a brief description of how patterns in the data can be displayed in a series of diagrams, and follow with an explanation of the meaning of these diagrams.

DUAL SCALING

The diagrams created by dual scaling are visual representations of data recorded in the form of a rectangular table of numbers, with rows for individuals and columns for attributes associated with these individuals. Each individual becomes a dot, and each attribute is shown as a triangle, all of which appear on a two-dimensional graphic display resembling a scatterplot.[1] These attributes can be viewed as the answers to a series of multiple-choice questions, such as, "What type of crop do you grow?" (Answers: corn, wheat, or alfalfa) or "How big is your income?" These questions, known as "items" in dual scaling, represent variables, each of which can be assigned different values (e.g. one for "low", two for "average", and three for "high" income). However, such numbers do not imply a ranking order. In dual scaling, the answers to any of the multiple-choice questions, i.e. variations within each variable or item, are treated as separate attributes. For example, there is no reason why a triangle representing "average income" should appear half way between those standing for "low" and "high" income.[2]

[1] The dual scaling program can be used to create different versions of the same diagram whereby the distances between triangles or dots can be stretched out to aid interpretation. Normally the dots are stretched out more (i.e., "normed" versus "adjusted" rows), making it easier to identify different clusters. Similarly, stretched-out triangles are called "normed columns". It is also possible to create separate diagrams featuring dots only or triangles only. In all cases, the numbers will remain the same, as will the relative positions of the dots or triangles. For a technical discussion of the normed and adjusted vectors of weighted scores that underlie such normed versus adjusted rows and columns in a diagram, see Shizuhiko Nishisato, *Elements of Dual Scaling* (Hillside, N.J.: Lawrence Erlbaum Associates, 1994), 125-126.

[2] In the technical language of statistics, such relationships are "non-linear."

Figure 12 shows such a diagram. Each dot corresponds to one of sixty farmers and the twelve triangles represent different attributes. Dots and triangles are distributed among four quadrants created by intersecting lines that cross at right angles. These lines do not always divide the diagram into four squares, as in figure 12, since either line, or axis, may shift up or down as well as from side to side, increasing the size of some quadrants, in which case the four rectangles will not be perfect squares. In any case, dots that lie close together in the same quadrant represent individuals who share attributes (or choose the same answers), while triangles lying close together correspond to attributes shared by individuals.[3] Triangles on opposite sides of the intersecting lines represent contrasting attributes. The connections between attributes and individuals, and vice versa, are calculated together, hence the name dual scaling.

In figure 12, which respresents an entirely hypothetical situation, sixteen young farmers born in Friesland, most of whom grow tomatoes, appear as a cluster of dots in the top left corner. The triangle for Friesland as place of origin appears in the same quadrant near that cluster. In contrast, twenty older farmers, originally from Limburg, most of whom grow corn and also raise pigs, appear in the bottom right corner. Nine older farmers from Drenthe who grow tomatoes are shown in the bottom left corner. The relative distances between these dots indicate the extent to which farmers share attributes. For example, since the dots representing Frisian tomato growers, as opposed to pig farmers from Limburg, are quite far apart, individuals belonging to these two groups have little in common. A third cluster of dots (tomato growers from Drenthe) represents yet another combination of attributes. The distance of that third cluster from each of the other two is less than that between the Frisian and Limburg farmers, because farmers from Drenthe share at least one trait with individuals from each of these other groups; that is, farmers from Drenthe are older, just like those from Limburg, but they are tomato producers like those from Friesland. You will notice that the triangle for tomatoes on the left is equidistant from the two clusters of tomato producers. In diagrams derived from actual research, dots would be more spread out among the four quadrants.

Another way to read figure 12 is to focus on the two inner lines (i.e., the intersecting axes) dividing the diagram into four quadrants. Imagine each dot and each triangle projected first onto the horizontal axis at a perpendicular (90° angle) and then onto the vertical axis. The two stippled lines show one individual (a dot) projected onto the horizontal axis

[3] Indeed, if two people were exactly alike in all respects (i.e. had the same attributes), they would show up as only one dot! This is rarely the case.

and one attribute (the triangle for "Drenthe") projected onto the vertical axis. Examining these axes separately is a way of focusing on the contrasts between the right and left, as opposed to the top versus bottom halves of the plane. These two sets of contrast are shown at the bottom of figure A. For the horizontal axis (A), all dots and triangles now lie on a single line, with a mid-point where the vertical axis intersects with the horizontal axis, which brings us back to a one-dimensional image. The dots of interest appear on opposite sides; in this case representing farmers with contrasting attributes (those who grow tomatoes as opposed to corn and pigs). The proximity of the triangles for corn and pigs indicates that farmers who grow corn are also likely to raise pigs. If triangles lie closer to each other, but on opposite sides (where the two axes intersect) this indicates an item is of less significance. Thus, the positions for "lo" and "hi" (referring to level of education) indicate that there is a slight tendency for farmers from both Friesland and Drenthe to have high levels of education, compared to the farmers from Limburg with low education. The projection of dots and triangles on the vertical axis, shown as horizontal line (B), with the top of the vertical axis on the left and the bottom on the right, reveals a different set of contrasts. Note that this second axis does not reveal a contrast based on crops. Instead, it highlights a contrast between young farmers from Friesland and old farmers from Limburg and Drenthe.[4]

Each axis, called a "solution" in the terminology used in dual scaling,[5] is assigned a percentage numerical value indicating its proportional contribution to the total interpretation of the data. The numbers shown on the right side of figure 12 show that the first axis accounts for 13.5 percent of the variance associated with contrasting clusters, and the second solution, 10.2 percent. The first solution always has the highest value. In other words, type of crop grown is somewhat more important than age in explaining differences and similarities among farmers. Additional pairs of axes displayed in successive diagrams (3rd and 4th, 5th and 6th etc.), reveal further patterns, in which the dots and triangles shift positions. Dots and triangles lying on or close to the center in one diagram, might appear close to the outer edges in another diagram. For instance, a diagram combining axes 3 and 4 (not shown) might contrast a new cluster on one side representing middle-aged farmers who have assumed public office and raise chickens, with another cluster on the other side

[4] To indicate a single axis, or "solution", dual scaling shows dots and triangles along a single diagonal line.

[5] In an almost identical statistical method called Multiple Correspondence Analysis, each solution is called a "dimension."

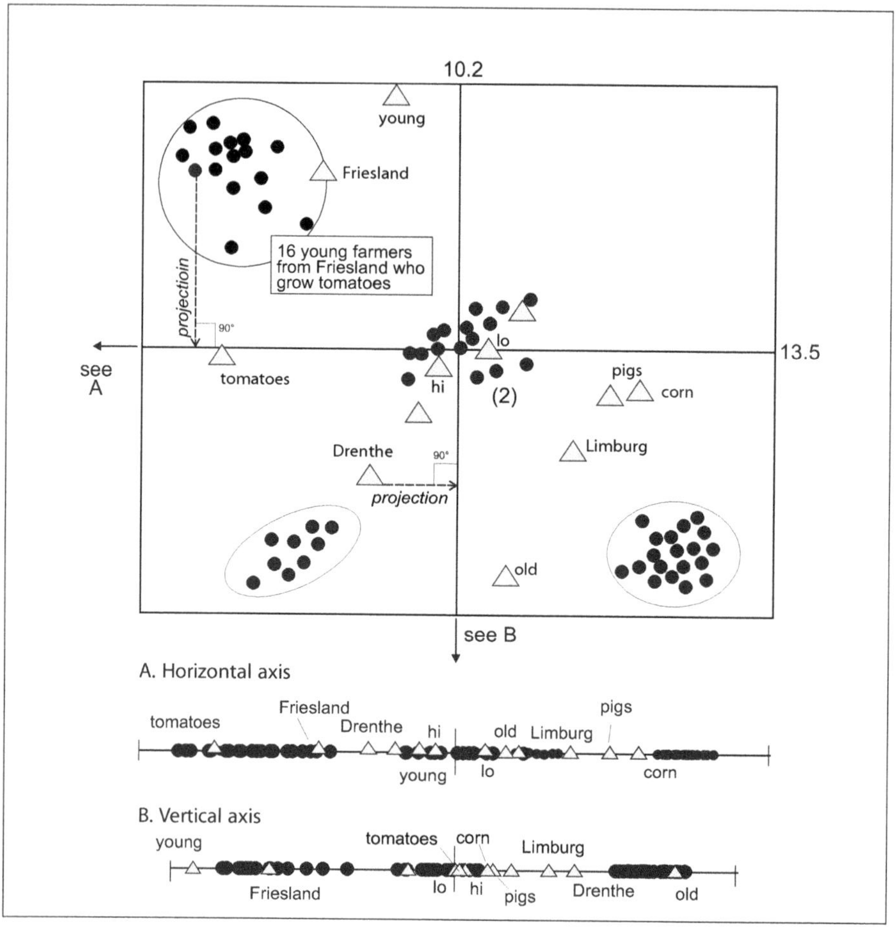

Figure 12. Hypothetical Dual Scaling Exercise.

for younger farmers with beef cattle but who are not politically involved. These additional axes, with diminishing yet still significant values, indicate secondary contrasts.[6] By adding up the contributions of all the axes, we obtain a numerical value for the total amount of information explained. It is important to note that with dual scaling, the researcher does not know ahead of time which variables (items) will show up, and in what order. The solutions, or sets of numbers are statistical "models"; however, the models are derived from or "follow" the data, not the inverse as in other forms of statistical analysis. With dual scaling, the researcher does not first develop a set of hypotheses and then "test" or confirm them. Nor does the researcher make assumptions about causation or direction of causation. This is why dual scaling is referred to as an exploratory or descriptive form of statistics.[7]

The diagrams created in dual scaling make it possible to visualize groups of individuals who share attributes, as well as the relative importance of variables such as age or type of education. However, the researcher must also examine the numerical output generated by the program. I have already referred to percentage of information accounted for, in each solution. This number, called the "delta partial,"[8] is the only one that appears in figure 12. The value of the partial deltas, shown in order of decreasing value, and corresponding to consecutive solutions, are summed to create increasingly larger "cumulative deltas" until we reach the total of variation explained. A set of closely related numbers (percentage homogeneity) is a better index of the merit of each individual solution because its values are higher and easier to interpret than the deltas, although they cannot indicate "total variance explained". A further number, with an upper limit of one, is the Reliability Coefficient Alpha. When that coefficient becomes negative, the computer run is aborted and no further solutions are generated. Another set of numbers indicates the relevance of each item (variable) to a solution. The higher the value of these item statistics, shown in a separate column as R(jt), the greater the relevance of the item in question. For example, for the first solution in figure 12, a higher R(jt) number for crop specialization and a lower one for place of origin would indicate that the former is more important than the latter for distinguishing among groups of

---

[6] Any clusters, and corresponding numbers, beyond the 4th solution must be analyzed with care, since they are more likely to represent spurious associations.

[7] Proper interpretation of the patterns revealed by this technique requires a good knowledge of the data.

[8] Based on Shizuhiko Nishishato and Ira Nishisato, *Dual Scaling in a Nutshell* (Toronto: MicroStats, 1994) 42-45.

farmers originating in the Netherlands. This number can be used to check the accuracy of the visual display. For example, in solution 1, if the R(jt) for the variable "type of crop" was .7201, the triangles representing different kinds of crops should be far apart, as well as on opposite ends of the axis. On the other hand, triangles representing the attributes of another item, for example education, with a R(jt) of .0902, would appear closer together.

To sum up, a careful inspection of a series of visual displays (and their corresponding sets of numbers) is a useful way or finding out regularities or patterns. Such a dual scaling exercise facilitates the interpretation of a large set of messy (complex) data.

Index

www.ingramcontent.com/pod-product-compliance
Ingram Content Group UK Ltd.
Pitfield, Milton Keynes, MK11 3LW, UK
UKHW020703041026
15590UKWH00001B/451

9 789004 151703